Lecture Notes in Computer Science 1580

Edited by G. Goos, J. Hartmanis and J. van Leeuwen

Springer
Berlin
Heidelberg
New York
Barcelona
Hong Kong
London
Milan
Paris
Singapore
Tokyo

Andrej Včkovski Kurt E. Brassel
Hans-Jörg Schek (Eds.)

Interoperating Geographic Information Systems

Second International Conference, INTEROP'99
Zurich, Switzerland, March 10-12, 1999
Proceedings

Volume Editors

Andrej Včkovski
Netcetera AG
Zypressenstr. 71, CH-8040 Zürich, Switzerland
E-mail: andrej.vckovski@netcetera.ch

Kurt E. Brassel
Spatial Data Handling
Department of Geography, University of Zürich
Winterthurerstr. 190, CH-8057 Zürich, Switzerland
E-mail: kbrass@geo.unizh.ch

Hans-Jörg Schek
Institute of Information Systems, ETH Zentrum
CH-8092 Zürich, Switzerland
E-mail: schek@inf.ethz.ch

Cataloging-in-Publication data applied for

Die Deutsche Bibliothek - CIP-Einheitsaufnahme

Interoperating geographic informations systems : second international workshop ; proceedings / INTEROP '99, Zurich, Switzerland, March 10 - 12, 1999. Andrej Vckovski ... (ed.). - Berlin ; Heidelberg ; New York ; Barcelona ; Hong Kong ; London ; Milan ; Paris ; Singapore ; Tokyo : Springer, 1999
(Lecture notes in computer science ; 1580)
ISBN 3-540-65725-8

CR Subject Classification (1998): H.2, H.4, J.2, E.1

ISSN 0302-9743
ISBN 3-540-65725-8 Springer-Verlag Berlin Heidelberg New York

Printed in Germany

Typesetting: Camera-ready by author
SPIN: 10703121 06/3142 – 5 4 3 2 1 0 Printed on acid-free paper

Preface

These proceedings collect the papers selected for the 2nd International Conference on Interoperating Geographic Information Systems held in Zürich, Switzerland, 10–12 March, 1999.

Interoperability has become an issue in many areas of information technology in the last decade. Computers are used everywhere, and there is an increasing need to share various types of resources such as data and services. This is especially true in the context of spatial information. Spatial data have been collected, digitized and stored in many different and differing repositories. Computer software has been developed to manage, analyse and visualize spatial information. Producing such data and software has become an important business opportunity. In everyday spatial information handling in many organisations and offices, however, interoperability is far from being a matter of fact. Incompatibilities in data formats, software products, spatial conceptions, quality standards, and models of the world continue to create as synchronicity among constituent parts of operating spatial systems. As a follow-up to the first International Conference on Interoperating Geographic Information Systems held 1997 in Santa Barbara, California, the Interop'99 tries to provide a scientific platform for researchers in this area.

The international program committee carefully selected 22 papers for presentation at the conference and publication in this volume. Additionally, this volume contains three invited contributions by Gio Wiederhold, Adrian Cuthbert and Günther Landgraf. Every paper was sent to three members of the program committee and other experts for review. The reviews resulted in a three-day single-track conference program that left some room for a few half-day tutorials on various topics regarding GIS interoperability.

Many people supported this event in various ways. We would like to express our thanks to the members of the program committee and the additional reviewers for their support in selecting the papers presented in this volume. Caroline Westort provided a lot of help handling all communication issues of Interop'99, as did Doris Wild with the organization of the conference. Hansrudi Noser supported many authors by translating their contributions into proper LaTeXcode.

We would also like to thank our sponsoring institutions for the various ways in which they provided support.

January 1999

Andrej Včkovski
Kurt E. Brassel
Hans-Jörg Schek

Organization

Chairs

Kurt Brassel, Dept. of Geography, Univ. of Zürich; General Co-Chair
Hans-Jörg Schek, Computer Science, ETH-Zürich; General Co-Chair
Andrej Včkovski, Netcetera, Zürich; Program Committee Chair

Advisory Committee

Andrew Frank, Tech. Univ. of Vienna
Michael Goodchild, Univ. of California, Santa Barbara (NCGIA)
Werner Kuhn, Univ. of Münster
David Schell, OpenGIS Consortium

Program Committee

Dave Abel
Gustavo Alonso
Kurt Buehler
Yaser Bishr
Klaus R. Dittrich
Max Egenhofer
Andrew Frank
Hiromichi Fukui
Michael Goodchild
Oliver Günther
Jiawei Han
John Herring
Cliff Kottman
Werner Kuhn
Robert Laurini
Jürg Nievergelt
Atsuyuki Okabe
Beng Chin Ooi
David Puller
Lukas Relly
Stefano Spaccapietra
Agnès Voisard
Andrej Včkovski
Xiaofang Zhou

Additional Reviewers

Martin Brändli
Felix Bucher
Ruxandra Domenig
Martin Huber
H.-Arno Jacobsen
Dirk Jonscher
Markus Kradolfer
Dean Kuo
Thomas Meyer
Corinne Plazanet
Dimitrios Tombros

Sponsoring Institutions

Inter-University Partnership for Earth Observation and Geoinformatics Zürich (IPEG)
OpenGIS Consortium (OGC)
Swiss National Science Foundation (SNF)
Swiss Organisation on Geo Information (SOGI)
C-PLAN

Environmental Systems Research Institute
GeoTask
Hewlett-Packard Company
Intergraph Corporation
Netcetera
SICAD Geomatics
Smallworld Systems

Table of Contents

Interoperability: Invited Contributions

Identification

Infrastructure

Implementation

Vectors and Graphics

Semantics

Heterogeneous Databases

Representation

Mediation to Deal with Heterogeneous Data Sources

Gio Wiederhold

Department of Computer Science
Stanford University
Stanford
gio@cs.stanford.edu

1 Introduction

The objective of *interoperation* is to increase the value of information when information from multiple sources is accessed, related, and combined. However, care is required to realize this benefit. One problem to be addressed in this context is that a simple integration over the ever-expanding number of resources available on-line leads to what customers perceive as *information overload.* In actuality, the customers experience *data overload*, making it nearly impossible for them to extract relevant points of information out of a huge haystack of data.

Information should support the making of decisions and actions. We distinguish *Interoperation of Information* from integration of data and databases, since we do not expect to combine the sources, but only selected results derived from them [27]. If much of the data obtained from the sources is materialized, then the integration of information overlaps with the topic of *data warehousing* [28]. In the interoperation paradigm we favor that the merging is performed as the need arises, relying on *articulation points* that have been found and defined earlier [13]. If the base sources are transient, a warehouse can provide a suitable persistent resource.

Interoperation requires knowledge and intelligence, but increases substantially the value to the consumer. For instance, domain knowledge which combines merchant ship data with trucking and railroad information permits a customer to analyze and plan multi- modal shipping. Interoperating over multiple, distinct information domains, as shipping, cost-of-money, and weather requires broader knowledge, but will further improve the value of the information. Consider here the manager who deals with delivery of goods, who must combine information about shipping, the cost of inventory that is delayed, and the effects of weather on possible delays. This knowledge is tied to the customer's task model, which provides an intersecting context over several source domains.

The required value-added service tasks, as selection of relevant and high-quality data, matching of source data, creating fused data objects, summarizing and abstracting, fall outside of the capabilities of the sources, and are costly to implement in individual applications. The provision of such services requires an architecture for computing systems that recognizes their intermediate functionality. In such an architecture mediating services create an opportunity for novel on-line business ventures, which will replace the traditional services provided by consultants, analysts, and publishers.

2 Architecture

We define the architecture of a software system to be the partitioning of a system into major pieces or modules. Modules will have independent software operation, and are likely located on distinct, but networked hardware as well. Criteria for partitioning are technical and social. The prime technical criterium is having a modest bandwidth requirement across the interfaces among the modules. The prime social criterium is having a well-defined domain for management, with local authority and responsibilities. Luckily, these two criteria often match. It is now obvious that building a single, integrated system for any substantial enterprise, encompassing all possible source domains and knowledge about them is an impossible task. Even abstract modeling of a single enterprise in sufficient detail has been frustrating. When such proposals were made in the past, the scope of information processing in an enterprise was poorly understood, and data-processing often focused on financial management. Modern enterprises use a mix of public market and service information in concert with their own data. Many have also delegated data-processing, together with profit-and-loss responsibilities, to smaller units within their organizations. An integrated system warehousing all the diverse sources would not be maintainable. Each single source, even if quite stable, will still change its structure every few years, as capabilities and environments change, companies merge, and new rate-structures develop. Integrating hundreds of such sources is futile.

Today, a popular architecture is represented by client-server systems (Figure 1). Simple *middleware* as CORBA and COM [24], provides communication among the two layers. However, these 2-layer systems do not scale well as the number of available services grows. While assembly of a new client is easy if all the required services exist, if any change is needed in an existing service to accommodate the new client, a major maintenance problem arises. First of all, all other clients have to be inspected to see if they use any of the services being updated, and those that do have to be updated when the service changes, in perfect synchrony. Scheduling the change-over to a data that is suitable for the affected clients induces delays. Those delays in turn cause that other updates needs arise, and will have to be inserted on that same day. The changeover becomes a major event, costly and risky.

Hence, dealing with many, say hundreds of data servers entails constant changes. A client-server architecture of that size will likely never be able to serve the customers. To make such large systems work, an architectural alternative is required. We will see that changes can be gradually accommodated in a mediated architecture, as a result of an improved assignment of functions.

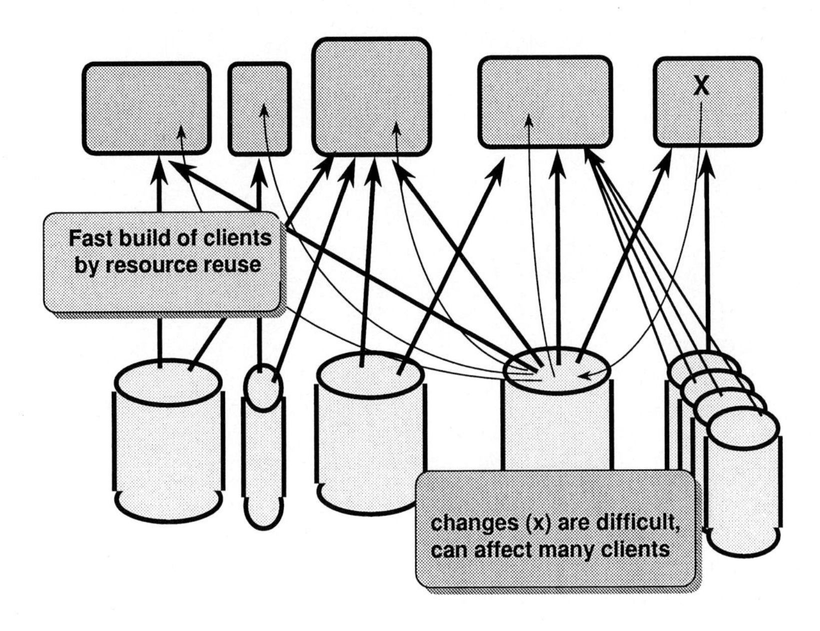

Fig. 1. A Client-Server Architecture

2.1 Mediator Architecture

The mediator architecture envisages a partitioning of resources and services in two dimensions, as shown in Figure 2 [42]:

1. horizontally into three layers: the client applications, the intermediate service modules, and the base servers.
2. vertically into many domains: for each domain, the number of supporting servers is best limited to 7 ± 2 [33]

The modules in the various layers will contribute data and information to each other, but they will not be strictly matched (i.e., not be *stovepiped*). The vertical partitioning in the mediating layer is based on having expertise in a service domain, and within that layer modules may call on each other. For instance, logistics expertise, as knowledge about merchant shippers, will be kept in a single mediating module, and a superior mediating module dealing with shared concepts about transportation will integrate ship, trucking, and railroad information. At the client layer several distinct domains, such as weather and cost of shipping, will be brought together. These domains do not have commensurate metrics, so that a service layer cannot provide reliable interoperation (Figure 3). The client layer and, in it, the logistics customer, has to weigh the combination and make the final decision to balance costs and risks. Similarly, a farmer

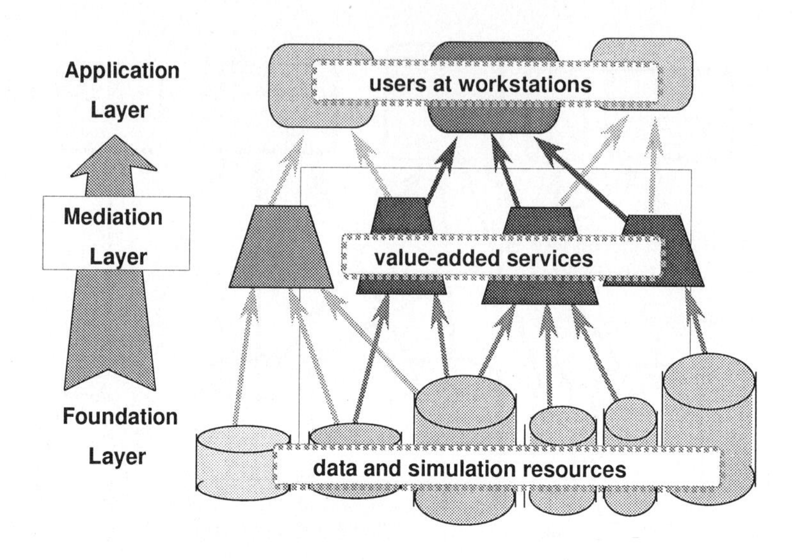

Fig. 2. A Mediated Architecture

may combine harvest and weather information. Moving the vagueness of combining information from dissimilar domains to the client layer reduces the overall complexity of the system.

2.2 Task assignment

In a 2-layer client-server architecture all functions had to be assigned either to the server or to the client modules. The current debates on thin versus fat clients and servers illustrate that the alternatives are not clear, even though some function assignments are obvious. With a third, intermediate layer, which mediates between the users and the sources, many functions, and particularly those that add value, and require maintenance to retain value, can be assigned there. We will review those assignments now.

Server. Selection of data is a function which is best performed at the server since one does not want to ship large amounts of unneeded data to the client or the mediator. The effectiveness of the SELECT statement of SQL is evidence of that assignment; not many languages can make do with one verb for most of their functionality. Making those data accessible may require a *wrapper* at or near the server, so that access can be performed using standard interfaces.

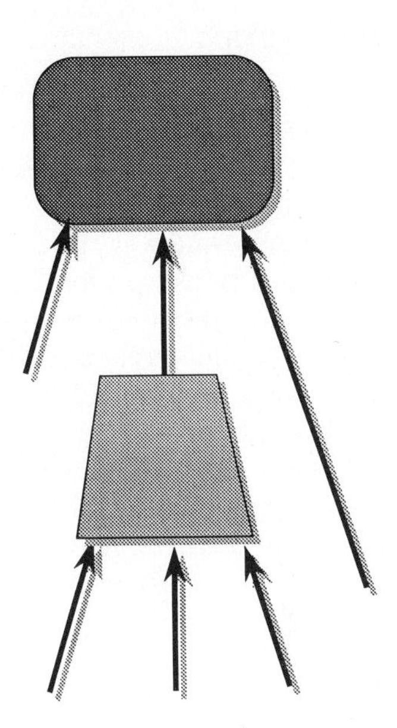
Application
• Informal, pragmatic
• User-control
Mediation
• Formal service
• Domain-Expert control

Fig. 3. Formal and Pragmatic Interoperation

Client. Interaction with the user is an obvious function for the clients. Local response must be rapid and reliable. Adaptation to the wide variety of local devices is best understood and maintained locally. For instance, moving from displays and keyboards to voice output and gesture input requires local feedback. Images and maps may have to be scaled to suit local displays. When maps are scaled, the labeling has to be adjusted [3].

Mediator. Not suitable for assignment to a server nor to a client are functions as the integration of data from multiple servers and the transformation of those data to information that is effective for the client program. Requiring that any server can interoperate with any other possible relevant server imposes requirements that are hard to establish and impossible to maintain. The resulting n^2 complexity is obvious. Similarly, requiring that servers can prepare views for any client is also onerous; in practice the load of adaptation would fall on the client. To resolve this issue of assignment for interoperation we define an intermediate layer, and establish modules in that layer, which will be referred to as mediators. The next section will deal with such modules, and focus on some of the roles that arise in geographic-based processing.

3 Mediators

Interoperation with the diversity of available sources requires a variety of functions. The mediator architecture has to accommodate multiple types of modules, and allow them to be combined as required. For instance, *facilitators* will search for likely resources and ways to access them [44]. To serve interoperation, related information that is relevant for the domain has to be selected and acquired from multiple sources. *Query processors* will reformulate an initial query to enhance the chance of obtaining relevant data [5, 12]. Text associated with images can be processed to yield additional keys [22]. Selection then obtains potentially useful data from the sources, and has to balance relevance with cost of moving the data to the mediator. After selection, further processing is needed for integration and making the results relevant to the client. In this exposition we will focus on issues that relate to spatial information and focus on two topics, integration and transformation. The references given can be used to explore other areas.

3.1 Integration

Selection from multiple sources will obtain data that is redundant, mismatched, and contains excessive detail. Web searches today demonstrate these weaknesses, they focus on breadth of selection and leave the extraction of useful information to the user.

Omitting redundancy. When information is obtained from a broad selection of sources, as on the web, redundancy is unavoidable. But since sources often represent data in their own formats, omitting overlaps has to be based on similarity assessment, rather than on exact matches [20]. When geographic regions overlap, the sources that are most relevant to the customer in terms of content and detail are best. Assessing the similarity of images requires new technologies, wavelets appear to be promising [11].

Quality of data is a complementary issue. A mediator may have rules as 'Source A is preferable over Source B', or 'more recent data are better', but sometimes differences of data values obtained cannot be resolved at the mediating level, because the metrics for judgement are absent. If the differences are significant, both results, identified with their sources can be reported to the client [1].

Matching. Integration of information requires matching of articulation points, the identifiers that are used to link entities from distinct sources. Matching of data from sources is based mainly on terms and measures. We now have to link complementary information, say text and maps. When sources use differing terminologies we need ontological tools to find matching points for their articulation [13].

While articulation of textual information is based on matching of abstract terms, when systems need to exchange actual goods and services, physical proximity is paramount. This means that for problems in logistics, in military planning, in service delivery, and in responding to natural disasters geographic markers are of prime importance.

Georeferencing. Unfortunately, the representation of geographic fiducial points varies greatly among sources and their representations. We commonly use names to denote geographic entities, but the naming differs among contexts. Even names of major entities, as countries, differ among respected resources. While the U.N. web pages refer to "The Gambia", most other sources call the country simply "Gambia". If we include temporal variations then the names of the components of the former USSR and Yugoslavia induce more complexity. Based on current sources we would not be able to find in which country the 1984 Winter Olympics were held [25]. When native representations use differing alphabets another level of complexity ensues.

The problems get worse at finer granularity. Names and extents of towns and roads change over time, making global information unreliable. For delivery of goods to a specific loading dock at a warehouse local knowledge becomes essential. Such local knowledge must be delegated to the lowest level in the system to allow responsive maintenance and flexibility. In modern delivery systems, as those used by the Federal Express delivery service, the driver makes the final judgement and records the location as well as the recipient.

Using latitude and longitude can provide a common underpinning. The wide availability of GPS has popularized this representation. Whiled commercial GPS is limited to about 100 m precision, the increasing capabilities of ground-based emitters (pseudolites), used in combination with space-based transmitters can conveniently increase the precision to a meter, allowing, for instance, the matching of trucks to loading gates [17]. The translations required to move from geographical named areas and points to areas described by vertices is now well understood, although remains sufficiently complex that mediators are required to offload clients from performing such transformations.

Matching interacts with selection, so that the integration process is not a simple pipeline. The initial data selection must balance breadth of retrieval with cost of access and transmission. After matching retrieval of further articulated data can ensue. To access ancillary geographic sources the names or spatial parameters used as keys must be used. When areas are to be located circumscribing boxes must be defined so that all possibly relevant material is included, and the result be filtered locally [19]. Again, many of these techniques are well understood, but require the right architectural setting to become available as services to a larger user population [16].

3.2 Transformation

Integration brings together information from autonomous sources, and that means also that data is represented at differing levels of detail. For instance, geographic results must be brought into the proper context for the application domain. Often detailed data must be aggregated to a higher level of granularity. For instance, to assess sales in a region, detailed data from all stores in the region must be aggregated. The aggregation may require multiple hierarchical levels, where postal codes and town names provide intermediate levels. Such a hierarchy can be modeled in the mediator, so that the client is relieved from that computation. The summarization will also reduce the volume data, relieving the network and the processors from high demands.

Summarization. The actual computation of quantitative summaries can again be allocated to the source, to the mediating layer, or to the client. Languages used for server access, such as SQL, provide some means for grouping and summarization, although expressing the criteria correctly is difficult for end-users. Warehouse and data-mining technology is addressing these issues today [2], but supporting a wide variety of aggregation models with materialized data is very costly. The mediator can use its model to drive the computation. However, server capabilities may be limited. Even when SQL is available, the lack of an operator to compute the variance, complementing the `AVERAGE` operator also motivates moving aggregating computations out of the server. While in 90% of the cases the average is a valid descriptor of a data set, not warning the end-user that the distribution is far from normal (bi-modal or having major outliers) is fraught with dangers in misinterpretation. Knowledge encoded in a mediator can provide warnings to the client, appropriate to the type of service being provided, that the data is not trustworthy.

While numeric processing for summarization is well understood, dealing with other data types is harder. We now have experimental abstractors that will summarize text for customers [29]. Such summarizations may also be cascaded if the documents can be placed into a hierarchical customer structure [39].

Aggregation may also be required prior to integrating data from diverse sources. Autonomous sources will often differ in detail, because of differing information requirements of their own clientele. For instance, cost data from local schools must be aggregated to county level before it can be compared with other county budget items. The knowledge to perform aggregation to enable matching is best maintained by a specialist within the school administration; at the county budgeting level it is easy to miss changes in the school system. Other transformations that are commonly needed before integration can be performed are to resolve temporal inconsistencies [18], or context differences, as seen in data about countries collected from different international agencies [25].

The complexity of these transformations is such that they are not appropriate for assignment to the client. Transformations performed on results of integration can, of course, not be assigned to servers.

Object-structuring. Anyone using the web today can attest to the complexity that linearly presented data imposes on the customer who seeks relevant information. Most clients are best served by structuring their information in object-oriented form. That means not only carrying forward the top-level summarization, but also the details that contribute to the summaries. Structural modeling tools can transform relational source data into diverse object-oriented formats, as needed by the client [6]. The base model can cover multiple sources.

Differing contexts require alternate hierarchies. In geography we distinguish political, social, topographical, and other hierarchies. While geographically-based hierarchies are common, other aggregations may be based on social criteria, as income or age of customers. Layering of geographic criteria and social criteria is also common.

Digital Libraries. Related research is being performed within the Digital Library Project, supported by NSF, DARPA, and NASA. For publications as journals and books mediating selection services were traditionally provided through reviewers and editors, while libraries, through their indexers, local storage capabilities, provided dissemination services to the clients. The technical challenge in automating the process is again dealing with the lack of common structure [23], heterogeneity of sources [35], and the redundancy [40] in the source data. For geographic libraries the base material is graphics and images, identified by related text [41]. There are many opportunities for innovative value-added services in this area [8].

3.3 Interfaces

For building and maintaining multi-layer systems, interface standards are crucial. When legacy files can be structured into tables, SQL will become the access language, as is being done by many extensions of relational system [9]. In addition to accepted standards for data, as SQL, ODL [10], and CORBA [36], a number of new interfaces have appeared. For instance, a transmission protocol for knowledge and data querying and manipulation being used in related research is KQML [30]. KQML provides for specification of the ontology being used in a transmission, to assure that the contents can be understood by communicating modules. Currently XML is gathering much momentum [14]. When data cannot be structured well, the XML format provides an alternative. Such semi-structured data have been the topic of much recent research [38]. XML structures can be defined for specific domains, using domain-specific type descriptions (DTD). Those DTDs will be developed by specialists, and will help in matching the meaning of the information being shared.

The alternative server-based technology, provided by pure Java, does allow uploading of functions to the client, but maintaining support for all user applications in the server or mediator is costly, as is shipping of all presentation alternatives for all client types. Furthermore, since we envision that pragmatic integration and processing will occur in the client, we must transmit information in a form suitable for further processing, and not just for display. As a language, however, Java is attractive, and we are likely to see Java programs in the client interoperating with Java-beans in the mediators.

Many new conventions are being considered for standardization, which will provide stability, and solidify market share. However, it is wise to wait before imposing any such standards on the community until adequate practical experience exists. It remains an open question how beneficial researcher involvement in the standards development process will be, but researchers will certainly be affected by the outcomes [31].

4 Status

Capabilities for data collection are increasing rapidly, advances in communications accelerate the flow, the situations that the clients must deal with are increasingly varied. Military intelligence systems were among the first users of this technology, even before solid research results were obtained and documented. *Fusion* of sensor data and images was already common. Geographic systems were integrated in several of these systems, but the interfaces to other data sources are still not very smooth.

Most operational mediating systems have been explicitly programmed. This means that the knowledge the mediators embody is in the form of computer codes. Moving to more formal descriptions is the objective of much current development. Building new systems can become more effective if there is reuse of technology and knowledge [34]. Use of rules makes the mediator easier to manage, important when the number of potential sources is large [37]. The leverage offered by modest, domain-specific knowledge bases should be substantial, but still has to be proven. In geography, such concepts have been proposed, but their use for interoperation has not yet been shown [7].

As software suppliers gain experience there will be spinoffs into pure commercial work [15]. An early example is the use of matchmaking mediators leading now to application in the Lockheed-sponsored venture for distribution of space satellite images [32]. A list of software suppliers was prepared for [45] and is maintained in related web pages (`http://www-db.stanford.edu/LIC/mediator.html`).

4.1 Effectiveness

Commercial dissemination of mediating modules will only occur if the information service paradigm proves to be effective. Interposition of a mediating layer into the client- server model incurs costs. A system's performance cost may be offset through reduction in transmitted data volume, as the information density increases.

Crucial benefit/cost ratios are in balancing service quality and system maintenance [43]. The bane of artificial intelligence technology has been the cost-versus- benefit of knowledge maintenance. Mediation provides a focus for such maintenance, in divorcing it from the operational pressures at the servers and the short-range needs at the clients, as shown in Figure 4. Reduced long-term maintenance costs may become the most powerful driver towards the use of mediating technologies, since software maintenance absorbs anywhere from 60% to 90% of computing budgets, and increases disproportionally with scale.

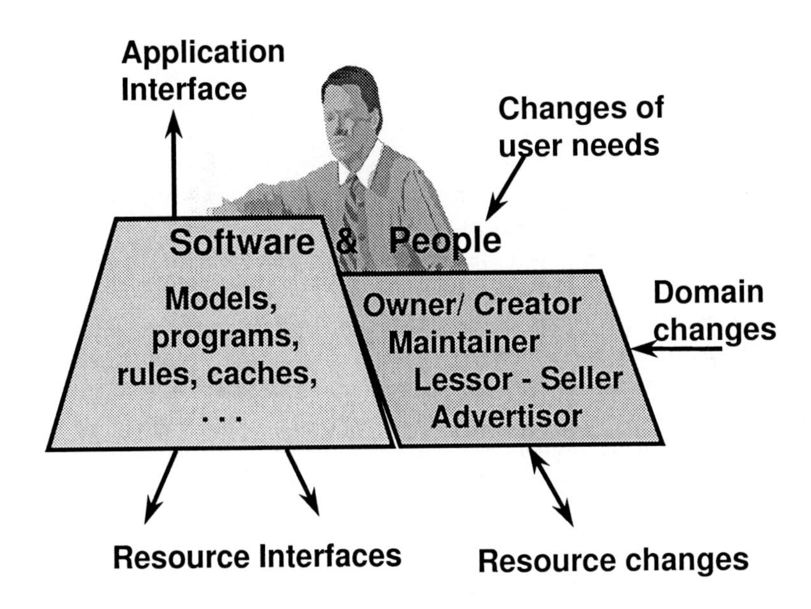

Fig. 4. Mediation assigns responsibility for maintenance

4.2 Privacy

Interoperation, while adding value, also adds risks. Combining information from multiple sources, aided by helpful agents that retrieve relevant information which was not directly requested, increases the risk of violation of individual and commercial privacy. Issues of privacy protection [26] and security must be addressed if broad access to valuable data is to become commonplace. A project on security mediation focuses on this issue [46]. A security mediator is a distinct module in an enterprise firewall, which complements traditional access protection with mechanisms to filter results before releasing them to the outside world. In a security mediator the owner is the security officer in charge of an organizationally defined domain [21].

5 Research

Having a *need* itself is not an adequate motivation for research investment; there also has to be a reasonable hope of moving towards solutions. In many areas, say in dealing with problems of strife and hunger, we are frustrated by complexity and a lack of leverage points. Providing information to agencies to effectively marshal and deploy their resources is a motivation for our research. Finding the right balance of the possible and the ideal is the major strategic issue in defining fundamental research. A tactical issue is finding the right time-point.

Research to solve problems that industry recognizes tends to be futile for academics. Industry will be able to devote sufficient resources to provide adequate, focused solutions. If academics can determine what solutions industry will adopt, then there are opportunities to go beyond. Going beyond can involve depth or breadth. Going in depth may mean dealing with likely omissions. In integration that might be providing for translation of terms that do not match, but not providing the triggers when domains change so that translations have to be updated. Going breadth in the same problem domain may mean devising rules that can work for multiple domains, rather than for some specific translation.

These tasks in information generation are complex and have to be adaptable to evolving needs of the customers, to changes in data resources, and to upgrades of system environments. The number of research issues needing solutions in the field is great.

5.1 Semantics

As the technical and syntactical problems of interoperation are being dealt with in industry, the semantic issues come to the forefront. Data resources, and especially databases, carry implicit or explicit domain definitions — no database customer expects a merchant shipping database to deal with interest rates. Similarly, a financial database is expected to ignore details about ships and a weather database is innocent of both. In all three domains the knowledge needed to adequately describe the data is manageable, but great leverage is provided by the many ground instances that knowledge-based rules can refer to.

5.2 Alternate Sources

While integration started out in dealing with well-structured databases, much current focus in on semi-structured data, and the textual contents of those data. Images, maps and graphs are brought in mostly through associated keys. Content analysis of these sources is making progress, and will become input to data integration. Video and speech are being analyzed as well, their volume makes integrated delivery to clients more problematical.

For planning and decision-making results from simulations also need to be integrated [4]. That will allow the clients not only to view the past, but also extrapolate timelines into the future [47]. Today this function is left wholly to the clients, and the tools they have, as spreadsheets, are not well integrated into their processing systems.

In the meantime, mediated systems are being built where alternatives are not feasible, for instance, where source data is hidden in legacy systems that cannot be converted, or where the planning cycle needed for data system integration is excessive.

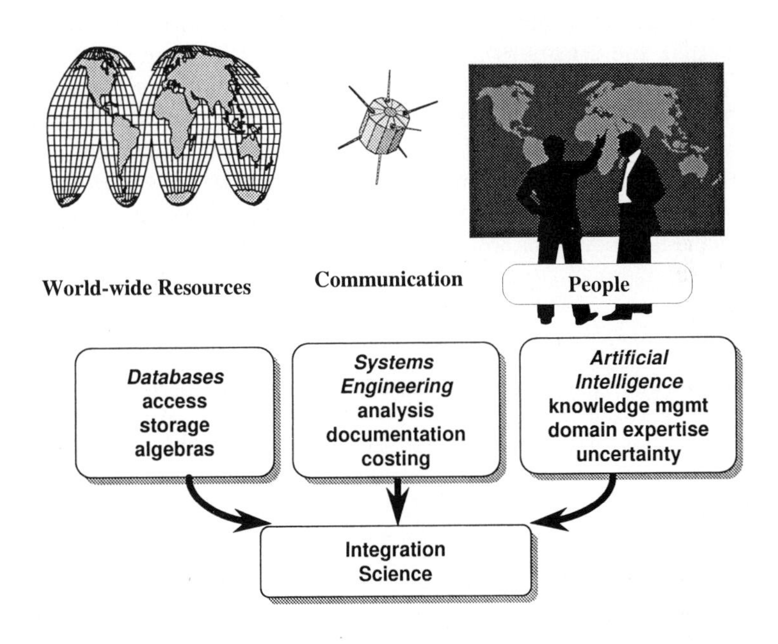

Fig. 5. Moving towards a New Science

6 Background

Starting in 1992 the Advanced Research Projects Agency (ARPA, now DARPA), the agency for joint research over all service branches of the U.S. Department of Defense, initiated a new research program in Intelligent Integration of Information (I3). Many results cited in this paper were initiated with ARPA support. Later research presented here was supported by NSF CISE and AFOSR. We thank the many participants and students who have helped in developing and realizing the concepts presented here.

7 Conclusion

Mediated systems are still in their infancy. We hope that ongoing development and deployment will fuel an effective research cycle. Having a clean architecture allows also a partitioning of research tasks, since the overall problem presented by information systems is greater than any single project can handle. Interoperation will require a variety of articulation points among sources and domain-specific knowledge.

The architecture we presented allows multiple application hierarchies to be overlaid, so that the structure forms a directed acyclic graph from client to resource, although the information flow is in the opposite direction. The complexity is still an order less than that implied by arbitrary networks, simplifying composition both in terms of research and operational management. The final vision is summarized in Figure 5, indicating the inputs we have in order to move towards a new science, focusing on integration of capabilities and competencies needed to make large systems work.

References

1. Shailesh Agarwal, A.M. Keller, K. Saraswat, and G. Wiederhold: "Flexible Relation: An Approach for Integrating Data from Multiple, Possibly Inconsistent Databases" ; *Proc. IEEE Data Engineering Conference*, Taipei, Taiwan, March 1995.
2. Rakesh Agrawal, Tomasz Imielinski, and Arun Swam: "Mining Association Rules Between Sets of Items in Large Databases"; *Proc. ACM SIGMOD Conference*, 1993, pp.207-216.
3. Horomi Aonumi, Hiroshi Imai, and Yahiko Kambayashi: "A Visual System of Placing Characters Appropriately in Multimedia Map Databases", in T.L. Kunii: *Visual Database Systems*, North-Holland 1989, pp.526-546.
4. Yigal Arens, Chin Chee, Chun-Nan Hsu, Hoh In, and Craig Knoblock: "Query Processing in an Information Mediator"; *Proc. ARPA/Rome Laboratory Knowledge-Based Planning and Scheduling Initiative Workshop*, Tucson AZ, Morgan Kaufmann, 1994.
5. Yigal Arens, Craig Knoblock, and Wei-Min Shen: "Query Reformulation for Dynamic Information Integration"; in Wiederhold (editor) *Intelligent Integration of Information*, Kluwer Pubs.,1996.
6. Thierry Barsalou, Niki Siambela, Arthur M. Keller, and Gio Wiederhold: "Updating Relational Databases Through Object-based Views";*ACM SIGMOD 91* Boulder CO, May 1991, pages 248-257.
7. Kate Beard, Terry. Smith, and Linda Hill: "Meta-information models for georeferenced digital library collections"; *Proc. 2nd IEEE Metadata Conference*, Sept.16-17, 1997. `http://computer.org/conferen/proceed/meta97/papers/kbeard/kbeard.html`.
8. Barbara P. Buttenfield and M.F. Goodchild: "The Alexandria Digital Library Project: Distributed Library Services for Spatially Referenced Data"; *Proceedings of GIS/LIS '96*, Denver, Colorado, November 1996, pp. 76-84.
9. R.G.G. Cattell: *Object-oriented Data Management: Object-Oriented and Extended Relational Systems*; Addison-Wesley, 1991.

10. R.Cattell (ed.): The Object Database Standard: ODMG (3 9version 1.1); Morgan Kaufman, May 1994
11. Edward Chang, Chen Li, and James Z. Wang: "Clustering for Searching Near-Replicas of Images on the World-Wide Web"; submitted for publication 1999, available from `http://www-db.stanford.edu`
12. W.W. Chu and Q. Chen: "A Structured Approach for Cooperative Query Answering"; *IEEE Transactions on Knowledge and Data Engineering*; Vol.6 No.5, October 1994.
13. C. Collet, M. Huhns, and W-M. Shen: "Resource Integration Using a Large Knowledge Base in CARNOT"; *IEEE Computer*, Vol.24 No.12, Dec.1991.
14. Dan Connolly (ed.): *XML: Principles, Tools, and Techniques*; O'Reilly, 1997.
15. Mike Debellis and Christine Haapala: "User-centric Software Engineering (USE)"; *IEEE Expert*, February, 1995, Vol.10 No. 1.
16. R. Dolin, D. Agrawal, A. El Abbadi, and L. Dillon: "Pharos: a Scalable Distributed Architecture for Locating Heterogeneous Information Sources"; in Golshani & Makki (eds.) *Proc. 6th Intern. Conf. on Information and Knowledge Management (CIKM'97)*, Las Vegas, NV, Nov. 1997, pp. 348-55. `http://pharos.alexandria.ucsb.edu/publications/cikm97.ps`.
17. Bryant D. Elrod and A.J. Van Dierendonck: "Pseudolites"; in Parkinson et al., (eds.) *Global Positioning System: Theory and Applications, Progress in Astronautics and Aeronautics*, Vol 2.., AIAA, 1996.
18. Cheng Goh, Stuart Madnick and Michael Siegel: "Context Interchange: Overcoming the Challenge of Large Scale Interoperable Database Systems in a Dynamic Environment"; *Proc. 3rd Inter. Conf. Information and Knowledge Management (CIKM'94)*, ACM Press, November 1994.
19. Volker Gaede and Oliver Gunther: "Multidimensional Access Methods"; *ACM Comp.Surveys*, Vol.30 no.2, June 1998, pp 170:231.
20. H. Garcia-Molina , L. Gravano , N. Shivakumar: "dSCAM: Finding Document Copies across Multiple Databases"; *Proc. 4th International Conf. on Parallel and Distributed Information Systems (PDIS'96)*, Miami Beach, Florida `http://www-db.stanford.edu/pub/papers/dscam.ps`
21. L. Gong and X. Qian: "Computational Issues in Secure Interoperation"; *IEEE Transactions on Software Engineering*, IEEE, January 1996.
22. E.J. Gugliemo and Neil C. Rowe: "Natural Language Retrieval of Images based on Descriptive Captions" *ACM Trans. on Information Systems*, Vol. 14 No.3, May 1996, pp.237-267.
23. J. Hammer, M. Breunig, H. Garcia-Molina, S. Nestorov, V. Vassalos, R. Yerneni: "Template-Based Wrappers in the TSIMMIS System"; *Proc. SIGMOD 26 Int. Conf. Management of Data*, Tucson, Arizona, May 1997, ACM.
24. Abdelsalem Helal and Ravi Badrachalam: "COM versus CORBA: Will Microsoft Come Out on Top"; *IEEE Computer*, Vol.28 No.10, Oct.95, pp.61-62.
25. Jan Jannink, Pichai Srinivasan, Danladi Verheijen, and Gio Wiederhold: "Encapsulation and Composition of Ontologies"; *Proc. AAAI Summer Conference*, Madison WI, AAAI, July 1998.
26. Vicky E. Jones, N. Ching, and M. Winslett: "Credentials for Privacy and Interoperation"; *Proc. New Security Paradigms '95 Workshop*, La Jolla, California, August 22-25, 1995.
27. Won Kim (ed): *Modern Database Systems: the Object Model, Interoperability and Beyond*; ACM press, Addison Wesley, 1995.
28. Ralph Kimball: *The Data Warehouse*; Wiley, 1996.

29. J. Kupiec, J. Pedersen, J. , and F.R. Chen: "A Trainable Document Summarizer"; in Fox, Ingwersen, Fidel: *Proc. ACM SIGIR Conference 18*, 1995; Seattle, WA, pp. 68-73.
30. Yannis Labrou and Tim Finin: "Semantics and Conversations for an Agent Communication Language"; in Huhns and Singh (eds) 'Readings in Agents', Morgan Kaufmann Publishers, Inc., 1997.
31. Martin C. Libicki: *Information Technology Standards*; Digital Press, 1995.
32. William Mark, Sherman Tyler, James McGuire, and Jon Schlossberg: "Comittment-Based Software Development"; *IEEE Transactions on Software Engineering*, October, 1992.
33. George Miller: "The Magical Number Seven +- Two"; *Psych.Review*, Vol.68, 1956, pp.81-97.
34. M.A. Musen: "Dimensions of Knowledge Sharing and Reuse"; *Computers and Biomedical Research*, Vol.25, pp. 435–467, 1992.
35. Shamkant B. Navathe and Michael J. Donahoo: "Towards Intelligent Integration of Heterogeneous Information Sources"; *Proc. 6th International Workshop on Database Re-engineering and Interoperability*; March 1995.
36. Object Management Group: The Common Object Request Broker: Architecture and Specification; OMG Document 91.12.1, OMG and X/Open, distributed by QED-Wiley, Wellesley MA, 1991.
37. Priya Panchapagesan, Joshua Hui, Gio Wiederhold, Stephan Erickson, Lynn Dean, and Antoinette Hempstead: "The INEEL Data Integration Mediation System"; to be presented at *AIDA'99*, Rochester, NY, June 1999.
38. Y. Papakonstantinou, H. Garcia-Molina and J. Widom: "Object Exchange Across Heterogeneous Information Sources"; *IEEE International Conference on Data Engineering*, pp. 251-260, Taipei, Taiwan, March 1995.
39. Wanda Pratt: "Dynamic Organization of Search Results Using a Taxonomic Domain Model"; *AAAI-97: Proc. 14th National Conference on Artificial Intelligence*, Providence, RI, AAAI, July 1997.
40. N. Shivakumar and H. Garcia-Molina: "Building a Scalable and Accurate Copy Detection Mechanism"; *Proc. 1st ACM Conference on Digital Libraries (DL'96)*; ACM, March 1996.
41. T.R. Smith and J. Frew: "The Alexandria Digital Library"; *Communications of the ACM*, Vol.38 No.4, pp.61-62.
42. Gio Wiederhold: "Mediators in the Architecture of Future Information Systems"; *IEEE Computer*, March 1992, pages 38-49.
43. Gio Wiederhold: "Modeling and System Maintenance"; in Papazoglou (ed.): *OOER'95: Object-Oriented and Entity Relationship Modelling*; Springer Lecture Notes in Computer Science, Vol. 1021, pages 1-20, December 1995.
44. Gio Wiederhold and Michael Genesereth: "The Conceptual Basis for Mediation Services"; *IEEE Expert*, Vol.12 No.5, Sep.-Oct. 1997, pages 38-47.
45. Gio Wiederhold: "Weaving Data into Information"; *Database Programming and Design*; Freeman pubs, Sept. 1998.
46. Wiederhold, Gio, Michel Bilello, and Chris Donahue: "Web Implementation of a Security Mediator for Medical Databases"; in Lin and Qian: *Database Security XI, Status and Prospects*, IFIP / Chapman & Hall, 1998, pp.60-72.
47. Gio Wiederhold, Rushan Jiang, and Hector Garcia-Molina: "An Interface Language for Projecting Alternatives in Decision-Making"; *Proc. 1998 AFCEA Database Colloquium*, AFCEA and SAIC, San Diego, Sep. 1998.

OpenGIS: Tales from a Small Market Town

Adrian Cuthbert

Laser-Scan Ltd.
Cambridge CB4 0FY, UK
adrian@lsl.co.uk
http://www.laser-scan.com

Abstract. The OpenGIS Consortium, OGC, has over one hundred and fifty members drawn from both the user and vendor communities and seeks to develop specifications for providing interoperability for geospatial data access and geoprocessing. This paper unashamedly adopts a biased perspective, that of a vendor active in the OGC. It attempts to explain the importance of the OGC in raising issues that go far beyond the writing of specifications.
Unlike many of the papers included here, the subject of this paper does not readily submit to a rigorous academic analysis. Issues are argued by example rather than by proof. To many vendors, the worth of the OGC lies in its recognition of key commercial realities. Since the OGC does concern itself with implementation and because it is trying to use the best of emerging technologies but is not tied to one particular platform, it faces many of the same problems that a vendor encounters. It is familiar with the compromise and pragmatism required to make progress. Consequently it provides one of the few forums where these issues are discussed.
This paper gives time to underlying issues that, although raised in the commercial world, impinge directly on technical developments. Many of these issues remain current and deserve a wider audience. They represent tales of the OpenGIS process, both past and present, told by a vendor located in a small market town.

1 Introduction

1.1 What Makes its so Difficult?

The OpenGIS Consortium, OGC[1], was founded in the belief that new and emerging technologies could fundamentally alter the way in which geospatial data and geoprocessing could be accessed. Technologies, such as component software, provide new ways of splitting up old problems. The OGC is not in existence to develop these new technologies, merely to harness the potential of those that emerge from the Information Technology mainstream. Component architectures include the promise of solutions built from best-of-breed components. However if such components are to work together they need to do more than share the same technology base, they need to agree on some core set of specifications.

Clearly the development of such specifications is far too important a task to be left to individual vendors. Ideally an open process should control it. The OGC provides such a process. Thus the OGC embodies a desire to use the latest technology and a desire to develop pragmatic specifications through consensus. It also includes the natural dynamic that exists as a result of combining the two.

The constitution of the OGC most closely resembles that of the Object Management Group, OMG[2]. The OMG is a consortium of many hundreds of organizations committed to providing a distributed computing environment. However that intent reveals a marked difference between the two. The OMG seeks to develop a technology, the OGC seeks to exploit technological developments. This leaves the OGC more susceptible to changes in the underlying technology base.

Indeed one of the major difficulties that the OGC has is that it cannot take refuge in a dogmatic adherence to its own developments. It does not yet possess an established base of users that can command industry to support it. This might be contrasted with the situation enjoyed by Java, where Sun Microsystems Inc.[3] can count on the world Java community[4] to make the development of new APIs highly desirable. The OGC must necessarily be responsive to the changes in the environment. It is in this regard that it faces many of the challenges faced by vendors, and therein lies its worth. As well as the problem of defining specifications it must constantly anticipate and analyze developments in the wider IT environment.

1.2 Specifications based on Interfaces

The OGC provides an open process for defining specifications designed to help promote interoperability between geospatial data and geoprocessing. These specifications allow the development of multi-tier solutions. The 'open' in OpenGIS refers to the fact that these tiers need not all be implemented by the same vendor/developer. When one is considering a single specification, it is common to distinguish between the two tiers involved as 'client' and 'server'. This terminology is used through out this paper, but should not be taken to indicate that the entire application conforms to a conventional client-server architecture.

The OGC was founded in August 1994, when the technological wave was component software. Technologies such as the OMG's Common Object Request Broker Architecture (CORBA)[5] and Microsoft's Component Object Model (COM)[6] promised platform and language neutral mechanisms for plugging components together. These represent an object-oriented generalization of an approach pioneered by SQL. The OGC refers to these as Distributed Computing Platforms, DCPs. These are the technologies against which specifications can be written.

Both COM and CORBA are broker based solutions that allow the definition of 'interfaces' using a language neutral Interface Definition Language. From this both stub and skeleton (or client and server) code can be generated for a variety of programming languages. The broker transparently allows different tiers of a solution to communicate, whether it be between different languages and/or different processes/machines. The OGC refers to these as 'interface' based solutions. With a good mapping between interface and programming language, this approach provides a near seamless joining of tiers. For example exceptions can be propagated from one tier to another. In addition these solutions provide a natural mechanism to handle state. A reference from one tier to an object in another tier provides the mechanism for one tier to retain state in the other.

However there are alternative approaches on which one might base a DCP. One might consider 'message' based solutions. In this approach communication between tiers is entirely via messages that conform to a structure that can be decoded. This means that the approach is not only language and platform neutral, it is also much less reliant on a common infrastructure shared between the tiers. About all that is required is an ability to send and receive messages. Nevertheless this approach does not provide the deep integration with languages provided by the interface approach, nor does it do a good job of retaining state between communicating tiers. Typically all the information required to define the request must be encoded in the out-going message.

1.3 Do Interface based Solutions have a Future?

In the current climate of technological change, there is an on-going responsibility to review fundamental assumptions. The OGC began with an assumption that there were a small number of DCPs that could be exploited. Since then two developments challenge that view: Java and the Internet.

- Java is increasingly being presented as a platform in its own right. The problem is not that Java would represent a language specific DCP, rather that it increases the number of DCPs to be supported. Without 'bridges' between different DCPs, supporting multiple DCPs becomes increasingly untenable.
- The characteristics of the Internet, most notably its latency, require changes in emphasis. Most obviously there is an on-going need to minimize the number of communications between client and server. Part of the solution is to be able to communicate more complex, structured data.

There are features found in Java that are not provided by either COM or CORBA. The most obvious examples are its ability for introspection, the ability to discover and dynamically load new class definitions and, as a consequence, the ability of an object to serialize and distribute itself. Those who wish to take advantage of these features would prefer to see Java regarded as a platform in its own right.

By contrast, advocates of a message-based approach are looking to reduce the dependence on platform specific capabilities. They say that the dual problems raised by Java and the Internet can be solved be together; the Internet is providing the tools to encode arbitrarily complex structured data (for example the eXtensible Markup Language, XML)[7] that can be used in messages between tiers that may be running on different platforms.

While it is true that CORBA provides a mechanism to describe complex structures, the manipulation of these in a programming language rapidly betrays its CORBA origins. Nor does the inclusion of a CORBA based broker as part of the Java 2 specification irredeemably tie Java and CORBA together. There is still much work required to make the interaction between Java and CORBA as seamless as many would like. For example, although COM has always provided for classes supporting multiple interfaces, CORBA will only provide support for this in CORBA 3.

2 Platforms

2.1 How much functionality should be included?

The OGC is committed to providing interoperability not only to access of geospatial data but also for the processing of geospatial data. Consider a simple scenario:

> *a server is accessible through an OpenGIS conformant interface. Data from the server has been displayed on the screen and two features, each with an area geometry attribute, have been identified interactively. The application is designed to highlight the area common to both features, if any. We might reasonably expect a system that maintains and can retrieve geometric data, a system that is able to perform queries with a spatial constraint, to handle all our geometric requirements. In this example the requirement is to generate the area representing the intersection of two other areas.*

The question that must be answered by those that write specifications is 'where to stop?'. From the client's perspective a large and all encompassing specification may be desirable, but it is of little worth unless it can be agreed on and delivered. This makes vendors a relatively conservative group. A minimum requirement might be stated as:

> *It should not be necessary for a client to develop functionality that already, albeit implicitly, exists within the server.*

For example if a server is capable of performing a spatial query that involves a test for whether geometries overlap, then it should be possible, given two geometries, to test explicitly whether they overlap. However, if the server cannot query on the area represented by the overlap, then one would not expect to be able to generate a new geometry representing the area of overlap. The purpose of this requirement is to avoid situations where a client must implement functionality implicitly available in the server. Not only does this reduce duplicated effort but also it minimizes the risk of incompatible implementations.

2.2 What about Simple Features?

At the time of writing (December 1998) the only specification adopted by the OGC was that to provide access to 'simple' features[8]. The 'simple' in the title is something of a misnomer and merely indicates recognition that the specification does not cover access to all aspects of a feature. The specification is provided in different 'flavors', one for each of three DCPs. Formally the specifications are titled:

The OpenGIS Simple Features Specification for

- COM
- CORBA
- SQL

The decision to develop specifications for a number of identified DCPs was taken at the time that proposals were requested. All the DCPs have the characteristics of being both language neutral and of being able to support different tiers on different machines.

SQL has demonstrated such strengths for many years. Furthermore the emergence of the object-relational model required the SQL specification to have two variants; for 'SQL92' and for 'SQL92 with Geometry Types'[8]. In the object-relational model, the columns in a table are no longer restricted to simple types, but may include more complex 'objects'. The inclusion of objects into relational databases allows them to offer unlimited functionality and ensures their position as a viable DCP.

The 'SQL92 with Geometry Types' variant acknowledges the emergence of vendor specific approaches to the object-relational model in advance of standards; for example Informix with its Datablade technology and Oracle with its Cartridge technology. SQL-3[9] seeks to support these innovations through its introduction of Abstract Data Types, ADTs. Currently the OGC is involved in a harmonization effort between future versions of 'The OpenGIS Simple Features Specification for SQL' and the 'multimedia' version of SQL-3, SQL-3/MM.

All the OpenGIS Simple Features Specifications (with the exception of the 'SQL92' only variant) provide the functionality required in our example scenario. The geometry interface has an 'intersection()' method (or the Geometry ADT has a function) that takes 'anotherGeometry' as a parameter and 'returns a geometry that represents the point set intersection of the source geometry with anotherGeometry'.

Despite all these similarities, there are profound differences between the three specifications. This is seen most clearly in considering how each of them deal with the concepts of 'feature', the basic unit of geospatial data, and 'geometry', a complex type of value describing a location in space. Features may have a number of attributes, zero or more of which may be geometric. The requirement that features must have precisely one geometric attribute had been rejected as being too restrictive. For example it would not be possible for a feature representing a real world entity such as a road to have a 'location' modeling its extent on the ground and a 'centerline' modeling its idealized route. Nor would it be possible to model aspatial features in the same way as more conventional features.

The three DCPs for which OpenGIS Simple Feature Specifications exist provide different approaches:

- SQL: the specification is based on SQL92 with provision for databases that support ADTs. Features are represented as rows in tables or views. Geometry is handled either as explicit coordinates in additional 'geometry' tables, as Binary Large OBjects (BLOBs) or as an ADT.
- COM: the specification is based around OLE/DB. This is Microsoft's strategic platform for all data storage, query and access. This provides COM interfaces to common relational elements. For example the result of a query can be returned a 'Rowset' COM object. Again rows are interpreted as features. However COM objects are used to represent geometries directly.
- CORBA: the specification provides an object-oriented model for geospatial features built from the ground up. Both features and geometries appear as objects accessible through CORBA interfaces.

OLE/DB represents the culmination of the process of integrating SQL and Microsoft technology by way of OBDC, DAO and the like. Nevertheless OLE/DB remains tied to a tabular model of data. It is closer in approach to the SQL specification than it is to the CORBA one. Rather than present a high-level object model of geospatial data, it represents a low-level model for data access.

Indeed no clear concept of a feature exists in the COM and SQL specifications. Rows in tables and views are interpreted to represent features, but they are not represented directly by objects or values in the underlying system. Thus features flit in and out of existence as views are created and deleted. Given that there are no geometric constraints on what constitutes a feature, in theory at least, a view with a single column of integers could be argued to represent a set of features.

Of course the decision to use interpretation on top of a tabular model was not taken lightly. The ultimate aim of the OGC is to see geospatial data more readily incorporated into everyday applications. The adoption of OLE/DB as the basis of the COM specification provides immediate integration with a wide range of Microsoft tools and products.

Given these differences, how does the OGC justify that these specifications represent different flavors of the same thing? A test of closeness between the various specifications might be expressed this way:

> *For a user familiar with the native capabilities of two DCPs and proficient at using an OpenGIS specification on one of those DCPs, how much more do they need to learn to be proficient at using the corresponding OpenGIS specification for the other DCP?*

Essentially this requires that, for those areas where the DCP provides no prior approach, the specifications for each DCP should be the same. However it is not inappropriate to use the native capabilities of a DCP directly. Thus for OpenGIS Simple Feature Specifications, the ability to store, query and manipulate geospatial data may differ from one DCP to another, because most of them are based on database technologies which provide precisely these capabilities. However there is no precedent for handling geometries in these DCPs. Consequently all flavors of a specification should provide semantically equivalent interfaces to geometries. This basically means that if the COM interface provides an 'intersection' method, then so should the CORBA interface.

3 The Internet

3.1 Where is the Internet?

The OGC was founded upon the recognition that the IT mainstream generated technologies that could fundamentally alter all branches of computing. The OGC was founded in August 1994 when 'component technology' was the major innovation. Since that time the world has been exposed to another paradigm-shifting technology: the Internet. From the perspective of end-users, this is far more profound a development than the ability of components that make the task of building software easier. How then, in all this discussion of DCPs, has there only been passing reference to the Internet?

Primarily because the Internet does not, in itself, constitute a DCP[10]. This is most clearly seen in the fact that other DCPs (for example COM and CORBA) include Internet technology within their remit. For example Microsoft provide Remote Data Services (RDS) to allow a visual COM component in a Web browser to maintain the fiction of a connection with COM components hosted on a remote server machine. This approach allows OLE/DB services to be 'remoted' to a Web client. At a more general level, CORBA brokers can communicate between themselves over the Internet using the Internet Inter-Orb Protocol, IIOP. Those that regard Java as a platform in its own right, are able to argue that it has the advantage of being developed at the same time the importance of the Internet was becoming apparent.

Consequently COM and CORBA specifications can, in principal, be applied across the Internet without change. This has been demonstrated to be true with the OpenGIS Simple Features Specifications:

- One demonstration used the COM specification and RDS to build a thick client running in a Web browser. Since the OpenGIS Simple Features Specification deals with data access rather than geoprocessing, it does not provide for rendering on the server. Consequently clients typically pull data across to the client and render it there. By Internet standards, the resulting client is 'thick' due to the quantity of data that it manages locally.
- Another demonstration made use of the CORBA specification to manipulate and interrogate features in a geospatial datastore without the need to pull all the feature data over to the client. Although this more closely represents the thin client model common to many Internet solutions, it required additional elements to allow the rendering to be carried out at the server.

These demonstrations show that defining specifications for DCPs that do not specifically include the Internet, does not preclude such solutions being deployed over the Internet. However they also demonstrate that the Internet cannot be made completely transparent. The resulting solutions do not necessarily conform to Internet based expectations and/or require extensions to existing interfaces.

The problem does not stem from the inability of the DCPs to utilize the Internet, but rather from the fact that OpenGIS Simple Features Specification is not complete enough to describe solutions architected for the Internet. Indeed this is only an exaggeration of what happens when the specification is applied on the desktop. Although one might expect desktop applications to naturally break into tiers based on data access and presentation, the use of the simple features specification mandates that rendering occurs in the presentation tier. Consequently all the data required for rendering needs to be passed through the interface. Obviously when both tiers are provided by the same vendor it is possible to add additional interfaces to allow the two to communicate, but this cannot be the case when one is attempting to make the two tiers interoperate.

GIS over the Internet is an area where new solutions are being experimented with, for which there is currently little consensus. This is in contrast to the area represented by the OpenGIS Simple Feature Specification. The problem of representing geospatial data in a database has been solved so many times before, there was little debate about the form of a solution. As the OGC tackles areas that are still under development, the task of reaching consensus becomes more difficult. GIS over the Internet is a prime example.

In response to this the OGC has, in conjunction with a number of sponsors, organized a Web mapping test bed that will allow interested parties to demonstrate the merits of their approach. The hope is that this will readily reveal basic similarities in approach that will themselves provide the basis of a common framework for GIS over the Internet. This use of a test bed is only possible as a result of the way in which the OGC has provided an environment in which traditional rivals can work together.

3.2 Where are the Maps?

When one thinks of geospatial data and the Internet, it is difficult to ignore one of the most compelling expectations; the ability to view geospatial data from anywhere in the world, of anywhere in the world based on data located anywhere in the world. The traditional way to view geospatial data has been as a map. A ‘map’ is something more than the data from which it is generated. It includes the choice of what data should be displayed and how it should be presented. Indeed one of the ways publishers of geospatial data add value is the manner in which they are able to present the data visually. Sometimes this means presenting the same data in different ways for different tasks.

In some areas, development of Internet related standards have been slower than many might have hoped[11, 12]. When the Internet began its explosive growth, it was already possible to construct pretty Web pages incorporating text and images. Now we also have the ability to transfer applets, video and even models of a virtual world using VRML. But we still do not have an agreed format in which to exchange vector based images.

Consequently virtually every GIS vendor has been obliged to develop their own mechanism to communicate 'maps' across the Internet. This duplication of effort appears even more absurd when one realizes that there are many domains, other than GIS, that require an ability to communicate vector-based images. As a reflection of this, the World Wide Web Consortium (W3C) has received a number of submissions in the area of 2D graphics[13], including:

- *Precision Graphics Markup Language: PGML*, submitted by Adobe Systems Incorporated, International Business Machines Corporation, Netscape Communications Corporation and Sun Microsystems Inc., 03 April 1998.
- *Vector Markup Language: VML*, submitted by Autodesk Inc., Hewlett--Packard Company, Macromedia Inc., Microsoft Corporation, Visio Corporation, 13 May 1998.
- *WebCGM*: submitted by the Boeing Company, Council for the Central Laboratory of the Research Councils (CCLRC), Inso Corporation, Joint Information Systems Committee (JISC), Xerox Corporation, 19 August 1998.

The standards community frequently tasks the OGC with reviewing proposed standards with regard to their applicability within the geospatial domain. Nevertheless a suitable transfer mechanism for communicating vector-based images of geospatial data stubbornly refuses to emerge.

3.3 Interfaces and Messages

If one wishes to reach a large audience using the Internet, it is necessary to work with a wide range of client configurations. It is not practical to make too many assumptions about the browser configuration of a casual user that makes an inquiry, the result of which includes a geospatial element. When the requirement is for a very thin client that makes the minimum of demands on resources on the client browser, a raster image remains the best way of communicating a map over the Internet.

However there are a number of techniques for making life more pleasant for the more frequent user of geospatial data over the Internet. Downloadable controls and applets can provide improved ergonomics and client side interaction. A vector based description of geospatial data rendered at the server, allows a client to fuse input from multiple servers, provide instant feedback from embedded tags and incrementally change the display during update. Rendered data need not represent the full information content of a feature, for example a rendered line may have been filtered based on the display scale to remove unnecessary vertices.

The experiences of many GIS vendors would suggest that the ability to communicate a structured result representing a map is crucial. This practical use of structured results would appear to represent the most likely outcome of the 'interface versus messages' debate.

Interfaces provide a high-level and familiar model to program against. However methods defined in those interfaces may return sizeable, low-level data structures (messages) as a result.

It might be noted that the unwritten assumption of the OpenGIS Simple Features Specification for CORBA was that CORBA objects are remote. This was in contrast with OLE/DB where typically the COM objects are local, although they may be representing data from a remote database. Consequently the CORBA specification includes elements which are designed to reduce the number of communications between client and server. The primary way of doing this is to take advantage of CORBA's ability to return structures and sequences as method results. These are returned 'by-value', in contrast to the more usual approach of returning a reference to another remote object.

Specifications are required in two areas; the interfaces themselves and the structures that are returned. Of course Internet based technologies for transmitting machine readable, structured data have already emerged, for example XML. One way to close the gap between specifications written for different DCPs is to require them to pass structured data using platform neutral standards. In such cases the requirement on a body like the OGC is to establish how they are used to encode standard geospatial concepts.

4 Interoperability

4.1 Interoperability at the Client

Most attempts to define 'interoperability' require much more space than is available here. However it is worth noting that one can identify a number of different ways in which interoperability might be realized. Whereas we may aspire to one approach, the initial implementations of OGC specifications may only operate at another

If there is a standard interface through which one client can interact with a number of different servers, then one has achieved a useful level of interoperability. For example one might have written a generic display tool that retrieves geospatial data from a number of servers and displays the result as a continuous, multi-scale map. One might be able to select a feature that originated on one server and use its location as input to a spatial query on data from another server.

All this can be achieved with an approach that requires the client to explicitly manipulate multiple server connections. The individual servers are not aware of one another. Thus some tasks (like a spatial join between data from two different servers) needs to be either handled, or at least coordinated, by the client. For example one might wish to find all towns that contain a railway station where data for towns and railway stations exist on different servers.

Most commercial solutions today implement 'interoperability at the client'. Although there are a number of products that automate the coordinate multiple servers, most of these derive from the requirement to manage transactions across distributed servers.

4.2 Interoperability between Servers

The example query described above required the coordination of multiple servers to obtain a result. One might imagine a 'super server' designed specifically to coordinate requests amongst multiple servers, without any need to store data of its own. A client of such a server would be relieved of many of its tasks. However truly distributed servers need to be 'aware' of one another in the absence of a coordinating client. A good example is the handling of feature relationships.

For the purposes of this discussion, a feature relationship is a pairing of pointers that allow two features to be 'related'. While the features involved live on the same server this can be achieved using techniques specific to the implementation on the server. However, if there is a need to establish a relationship between features on different servers, it is necessary for both servers to share some mechanism for feature identification[14]. Frequently the solution involves an addressing model in which a feature identifier is expanded into a sequence of 'scope identifiers' and identifiers within those scopes. Systems employing such an approach must also be able to deal with the distinction between a feature not existing and the server on which it lives being unavailable.

4.3 Interoperability between DCPs

The desire to a see a generic solution to the problem of providing interoperability between different DCPs, most notably COM and CORBA, is one that faces any vendor that cannot afford to rely on a single platform. As clear water opens up between the evolving Java platform and the place for Java in the Windows environment, it is difficult to imagine that this is not a problem that will afflict increasingly many people. Because the OGC is platform neutral, it is a problem that they face also. Clearly it is beyond the scope of the OGC to define a solution, but they monitor the situation and inform their members.

5 Conclusions

The OGC has set itself the task of developing specifications that it can bring to the market place with the promise that the GIS vendor community can deliver against those specifications. By removing the risks of vendor specific solutions and elevating the level of interaction between software components, it hopes to make the task of exploiting the geospatial data (that many organizations have access to) both easier and more common. In setting itself the dual challenges of providing technically sound specifications in a commercially realistic manner, it lays itself open to tensions readily identified with by GIS vendors. Consequently the OpenGIS Consortium provides a rare forum for identifying, discussing and, where possible, resolving these issues. This paper has sought to give an indication of the range and importance of these issues.

References

1. The Open GIS Consortium: http://www.opengis.org/.
2. Object Management Group: http://www.omg.org/.
3. Sun Microsystems: The Source for Java Technology, http://www.javasoft.com/.
4. Cetus Links: Object-Oriented Language: Java / General, http://www.objenv.com/cetus/oo_java.html.
5. Object Management Group: The Common Object Request Broker Architecture (CORBA), http://www.omg.org/corba/beginners.html.
6. Microsoft: Component Object Model (COM), http://www.microsoft.com/com/default.asp.
7. World Wide Web Consortium: http://www.w3.org/XML/.
8. The Open GIS Consortium: The OpenGIS® Implementation Specification, OpenGIS® Simple Features Specifications for OLE/COM, CORBA and SQL, http://www.opengis.org/techno/specs.htm.
9. National Institute of Standards and Technology (NIST): Federal Information Processing Standards Publication 193, 1995 February 3. http://www.nist.gov/itl/div897/pubs/fip193.htm.
10. Manola, F.: Towards a Web Object Model, http://www.objs.com/OSA/wom.htm.
11. Internet Engineering Task Force (IETF): http://www.ietf.org/
12. World Wide Web Consortium: http://www.w3.org/.
13. World Wide Web Consortium: A new interest in vector graphics for the Web, http://www.w3.org/Graphics/Activity.
14. Sargent, P.M.: Feature Identities, Descriptors and Handles, This volume.

Evolution of EO/GIS Interoperability towards an Integrated Application Infrastructure

Günther Landgraf

European Space Agency, Directorate of Application Programmes
Remote Sensing Exploitation Department, ESRIN
Via Galileo Galilei, I-00044 Frascati (Roma), Italy
guenther.landgraf@esrin.esa.it

Abstract. Combination of data and services (i.e. 'Interoperation') is one of the key concerns to develop a geographical information business network, in particular for near real-time information derived from satellite-based earth observation. In this logic the current situation of the satellite ground segments is not satisfactory and will require a conceptual improvement. Various international standardization and harmonization activities, like CCSDS OAIS, CEOS CIP, FGDC GEO, OGC OpenGIS, ISO TC/211 are identified as the basis for interoperability, but will need to evolve to an 'integrated application infrastructure' that organizes service interoperation, thereby building production chains for specialized high-value application products. Within the area of application programmes the European Space Agency has a vital interest to support this convergence and is investing in research projects to support European industry to participate in this process.

1 Introduction

The world of geospatial information is currently undergoing a rapid and profound evolution. There is the perception that time is mature for the development of a new industrial branch, with the opportunity to further develop the public sector and in parallel create a profitable private sector allowing beneficial cost sharing for both.

For three out of the five main technologies involved – geographical information systems, remote sensing, geopositioning, telecommunications and distributed computing – the space sector can contribute to effective solutions. Therefore the European Space Agency has a natural interest to explore the technological needs for the deployment of this sector.

This is in line with a general re-orientation of the Agency's work: while technology research has to remain the basis, there is the wish to enhance the Agency's efforts towards a more application and industry-oriented approach, better exploring and preparing the potential commercial exploitation of technology investments. This is particularly true for the sector of remote sensing, where an increase of attention towards the actual exploitation of data and their use for applications – as opposed to a genuine technological focus on the space segment – will be required.

Application-oriented data exploitation is characterized by the need to combine data from different sources, possibly involving services of many different providers. By definition this requires 'interoperability' in its largest sense, i.e. the technical possibility to combine data and services from different sources into 'something new' (which can be a new product or a new service). In the context of the present paper service interoperability is extended to any function potentially required for data exploitation. This starts from resource identification or 'advertisement' (directory), includes services that are a traditional focus for interoperability activities like the catalogue (including metadata search and ordering), but extending to new functions to be standardized in the area of archive access, formatting and processing services.

Traditionally interoperability is achieved by 'translators', implemented (and maintained) with more or less effort, achieving mapping between different models with more or less loss of information. This is a reasonable approach if the number of data and services providers is small. If this number is too high, the number of necessary translators grows exponentially with a consequent collapse of this approach.

The alternative solution is interoperability by 'standardization'. All players commit to offer their data and service according to a standard, either by implementing it directly or by translating to it. A federation of such providers being able to exchange data and services according to a well-defined standard is forming a de facto 'integrated interoperable infrastructure', which can exchange data and services in a performant and cost-efficient way. This is the basis for the development of an 'Integrated Application Infrastructure', which organizes the 'interoperation' of this federation by implementing the application-specific workflows on the top.

2 The Current Situation and its Limitations

In the past, civil Earth Observation satellites were strongly technology driven programmes, without pretending that the high investment for the development, construction and launch as well as the cost of operation of the satellite could ever be directly recovered. For what regards the data exploitation, ESA budgets for production activities for the supported satellites (ERS, Landsat, JERS, NOAA, SeaStar, IRS, Nimbus) were limited to correction and geocoding of the acquired images, without provision for efficient coordination of further application-oriented exploitation.

Furthermore development budgets were allocated with a focus on the single satellite mission, closing each single satellite programme is a world of its own, with its proprietary terminology, data representation and services. This was acceptable for the scientific exploitation of the carried instruments, but prevented the propagation of these powerful information sources to really operational use in applications. Scientists are used to adapt to specific terminology, formats and access mechanisms and develop temporary workarounds for their specific experiments. Typically an experiment or research activity is carried out once or for a limited time period, so there is no permanent load of additional cost from difficult access to data or the high effort for combining them.

This is different for the applications sector. The additional time required to access and merge data with different access methods and heterogeneous formats can considerably reduce the value of a product for applications, where real-time or near real-time information is needed. Furthermore any "overhead" cost for building, maintaining and operating translators may be a decisive element if a "product" can be offered at a price which is acceptable for the market or not.

It is therefore evident that an innovative approach is needed to perform the step from the current situation of 'satellite-focused' programmes towards an fertile environment for the growth of new applications and business sectors.

3 The Need for an Interoperable Production Infrastructure for Spatial Applications

A long production chain needs to be organized to arrive from the raw material ("single source products" like satellite observations as provided by the satellite operators) to the application service provider, who can finally offer the extracted application-specific "product" – rich in domain-specific information but small in size – in a suitable way for end users. It will not be sufficient to extract this information from a single-source product, but it has to be combined with other geospatial information, e.g. other satellite, airborne or ground-based observations, maps and GIS data.

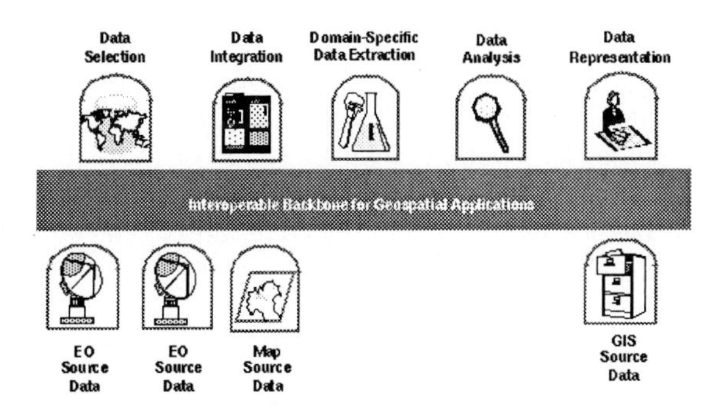

Fig. 1. A 'typical' application production chain to deliver Earth Observation information to an end user, integrating it with map and GIS data. The required data and services are attached to the interoperable backbone, allowing quick and standardized access. The governing 'active object' which organizes the interoperation workflow is not shown.

There are many 'production steps' involved to achieve a final application product, usually falling in one of the following categories:

Data Selection: Given the enormous amount of data acquired, rapid analysis if a new acquisition contains information useful for a specific domain (e.g. an oil slick, an iceberg, a fire, etc.) is an essential activity to prioritize the processing of commercially valuable information. New efficient and very specialized algorithms will be required to perform this step.

Data Integration: Merging of different data – usually from different sources – is an important step towards a "semi-finished" product which then requires only a more simple and rapid step of extraction of all relevant information. This activity requires particular skill and familiarity with a multiplicity of data characteristics. It will be impossible to deploy such a service in a cost-efficient way, if lower-level providers furnish their data according to 'local' standards, implying the costly need to adapt the application to every single country, country-internal area and even township. The lack of interoperability – and consequently of interoperation – is probably one of the most prominent reasons why a 'data integration industry' barely exists today. Local markets are too small to guarantee satisfactory return on investment, and for the global market the cost and time required for all necessary adaptions is prohibitive.

Domain-specific Data Processing and Information Extraction: The step of reducing the data to the aspects relevant for an application domain is relatively simple, but requires a combination of domain-specific knowledge and expertise in data contents and format.

Data Analysis: In many cases the end user will not be interested anymore in an image representation, but only in an extracted summary information (e.g.: crop forecast). This kind of analysis usually requires large domain- and remote-sensing specific expertise and highly specialized algorithms.

Data Representation: The final step to be executed by the specialized application service provider is to assemble all relevant information and convert it into a format suitable for the specific application user. Depending on the domain, this can be an electronic map, an email message, a fax, a database update, etc. Again, this potential business sector faces problems similar to the data integrators. High-level application products serve very specific needs and therefore have a limited end user community. For this reason the initial investment to develop the service needs to be limited by a standardized interoperable infrastructure, with this latter one becoming a mandatory prerequisite for profitable deployment of this business sector.

Depending on the application domain all these steps will be executed by different constellations of service and value-added providers, with different workflows putting the various steps above in sequence. Typically many of these steps will be repeated to achieve the final product after passing various levels of intermediate products.

We must not have too high expectations on the price that a broad consumer market is willing to offer for a 'piece' of geospatial information. Therefore it is important that the cost of initial investment and reproduction for the value-added service industry is as limited as possible, still allowing an attractive profit margin. Where the geospatial information has a high cost of initial investment (e.g. satellite images) the price can only be kept down if this investment can be distributed between a high quantity of sold products – a goal which is only achievable if it is used by a vast number of end users in multiple different applications. The 'high price' policy applied in the past has by far not succeeded to cover the initial expenditure and today we can already observe a decrease in the price at which remote sensing products are commercially offered.

The big advantage of the geospatial industry is that all 'raw materials' can be made available in digital electronic format, so we don't need highways and trucks to ship them from one participant to the other, but 'only' a network with sufficient bandwidth. Consequently the time to 'ship' the 'assembly components' between the 'factories' can be reduced to a neglectable factor. Making full use of state-of-the-art technologies in distributed computing and telecommunication a geospatial production chain can be built in accordance with the 'virtual enterprise' paradigm, cutting production costs by an order of magnitude with respect to traditional approaches.

What is really essential is to have an interoperable network that allows access to data and services for their manipulation at sustainable cost. This basic infrastructure has to consist of enhanced and cheap network capabilities, together with an EO/GIS application standard protocol. With the availability of such an interoperable backbone, each 'provider' in the geospatial sector can 'attach' his data and/or services at marginal costs, consequently obtaining the relevant profit derived from their usage. Only the availability and easy accessibility of products of a certain type and level will enable "business" for value-added providers of even more specialized higher-level products, targeted to specific application use.

For what regards the network part there are other sectors who will drive the development. The multimedia market including digital TV also requires this kind of infrastructure and the market forces behind this domain are an order of magnitude bigger than ours. It can also be expected that some public infrastructure investment will occur to accelerate the new markets. The EO/GIS application standard protocol on the other hand is proprietary to our domain and we have to invest into it by ourselves.

As the final vision of a 'growing' infrastructure, for each single application domain it can be expected that the availability and easy accessibility of specific subsets of data and services will be the 'critical mass' for the commercial deployment of this application – which itself is not only a consumer of other geospatial services and data, but is a new service itself, that can be attached to the network (potentially being part of the 'critical mass' for even new, today not imaginable services). Applying this recursively, a production infrastructure for spatial applications can dynamically build up an 'information value tower', providing more and more suitable products for an ever-growing community of end-users.

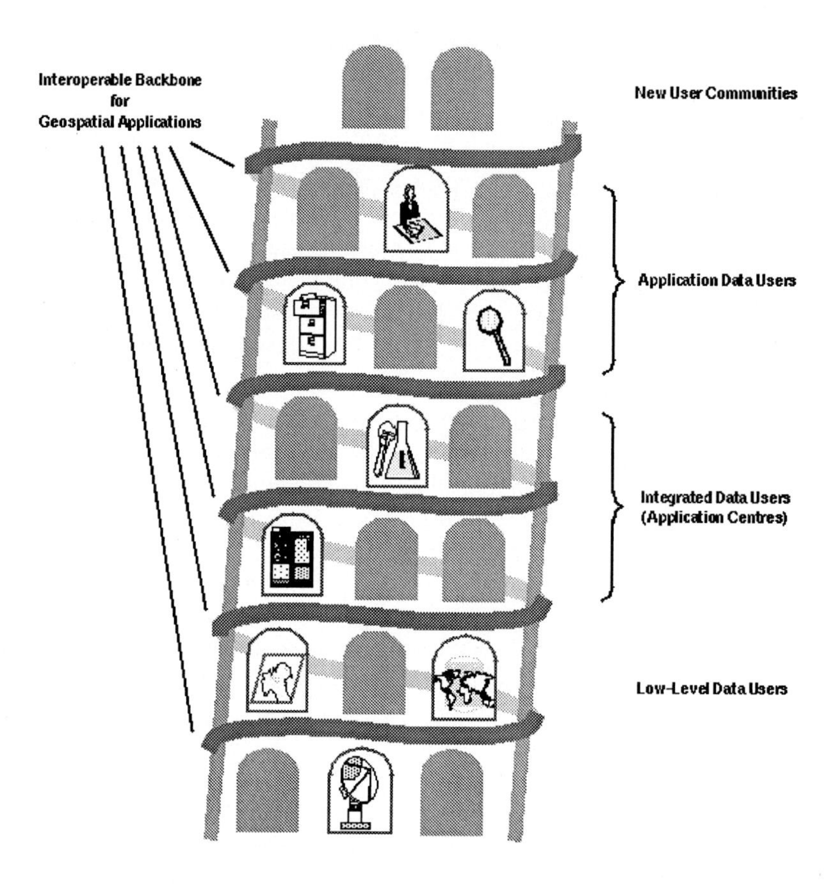

Fig. 2. The availability and easy accessability of a certain level of products and services is the "critical mass" for the commercial deployment of a new layer of products, satisfying new user communities. The total of user communities for higher level products can be expected orders of a magnitude larger than for lower-level products.

4 Ongoing Interoperability Activities and Trends

Many different bodies – with different focus and level of abstraction – are currently working in parallel on the subject of interoperability. The ones considered hereafter are only the subset, which is currently at least monitored in some way be the European Space Agency. The present contribution does therefore not claim to be complete.

4.1 CCSDS

The Consultative Committee for Space Data Systems (CCSDS) is an organization officially established by the member space Agencies. It meets periodically to address data system problems common to all participants and formulate sound technical solutions.

Of particular interest is the reference model for an Open Archival Information System (OAIS) [9], developed in response to ISO TC20/SC 13.

The OAIS model contains an in-depth analysis of data, metadata and services, including archival, metadata query, ordering and retrieval.

4.2 CEOS Activities

CEOS – the 'Committee for Earth Observation Satellites' [8] – is an international platform to discuss issues related to satellite-based remote sensing. Its "Working Group Information Systems and Services" (WGISS), is structured into various subgroups, which achieve their technical work through tasks teams. Of particular interest are the activities of the 'Access' subgroup.

The CINTEX Task Team and IMS. The 'Catalogue INTeroperability Experiment' (CINTEX) is a historical milestone for gaining experience on the issue of interoperable catalogue access. This activity was mainly sponsored by NASA, who contributed the IMS client and the IMS gateway, which has been customized by various satellite data providers. The IMS network implements the second level of interoperable federation identified in the CCSDS OAIS model, i.e. interoperable query and metadata retrieval via a global node which can distribute a query to multiple local archives. The IMS client basically allows formulation of such queries and visualization of metadata information by a single client.

While the level of syntactic interoperability is satisfactory and IMS is continuing to be enhanced, CEOS decided, that the level of standardization should be improved by a more formally specified protocol to better exploit information derived from different sources. This would implement the Interoperable Catalogue System (ICS) as third and fully functional level of an interoperable OAIS federation, i.e. including standard ordering and dissemination mechanisms.

The Protocol Task Team and CIP. The 'Catalogue Interoperable Protocol' [1] further standardizes metadata and services related to interoperable access to satellite-based remote sensing data. It's initial version (CIP-A) was mainly developed by the European Space Agency, whereas the European Union and NASA are the prime technical contributors to the successive CIP-B version. As main innovation CIP introduced the concept of structuring data into hierarchical 'collections' with a two-level search approach - first for the collections of potential interest and successively inside these collections. Furthermore interoperable ordering with the related problem of user management and authentication was addressed and an abstract model for the specification of order options agreed.

For what concerns the technical foundation, CIP was decided to be implemented as application-specific profile of Z39.50. This had the big benefit of being able to inherit existing standards and software for search and retrieval, and allowed harmonization with the FGDC standard for Digital Geospatial Metadata [3], which had been defined for the geographic community.

The drawback is the difficulty to extend a protocol like Z39.50 – focused on search and retrieval – to other services in a natural way. These inherent limitations lead ESA to the conviction that the next generation of a CEOS interoperable protocol should be based on CORBA, which meanwhile has become an acceptably mature standard.

The protocol task team is closely observing the progress of ISO TC/211 [5], with the target to even better unify the CIP and Geospatial profiles on this basis. On the other hand there is considerable interest in the OpenGIS activities, which has lead to a proposal to the OpenGIS Consortium (OGC) for the "Catalogue Services Request for Proposal". The building of this proposal which maps the CIP services to XML was a mainly NASA-driven initiative and aims to put the rich experience of the CEOS Protocol Task Team at the disposition of the OGC.

Apart from the various specifications the following CIP software is currently becoming available:

- the INFEO middleware produced by the European Union within the CEO programme;
- the CIP-ODBC gateway produced by the European Union, allowing to connect local databases to the CIP federation network;
- the CIP client produced by the European Space Agency [6].

4.3 The FGDC GEO profile

The U.S. Federal Geographic Data Committee has developed a Z39.50-based interoperable protocol standard for Digital Geospatial Metadata ("GEO") [3] in parallel to the remote sensing community. A registry of GEO databases is kept at the FGDC and is known as the Clearing House. This standard is also successfully used in the 'CEOnet' (Canadian Earth Observation Network) where it is implemented by the ISITE software, a public domain Z39.50 software package supporting the FGDC metadata standard.

However, seemingly also FGDC had felt that more extensive services required a different base protocol and have prominently sponsored OpenGIS demonstration activities.

4.4 The OpenGIS Consortium

While all other initiatives are government-driven, the OpenGIS Consortium (OGC) [4] is an industry-driven trade association. From the achievements in the last years it can be concluded, that the level of effort put into this standardization activity is an order of magnitude greater than in the others. The correctness of the initial focus on a CORBA-based infrastructure has been confirmed by the growing maturity and acceptance of this distributed computing solution, even if OpenGIS has elevated its abstraction level to be compatible also with other distributed computing platforms as DCE and DCOM and languages like SQL and JAVA. However, excessive opening could turn out expensive and dangerous for the standardization effort and will require a strong control of the OGC Technical Committee to avoid counterproductive 'pushing' of members for their own base technologies.

Extrapolating the current trend, OpenGIS has to be judged as a very promising platform to specify the baseline of an interoperable EO/GIS infrastructure.

4.5 ISO/TC 211

TC211 [5] is the Technical Committee tasked by ISO to prepare a set of standards related to geographic information. There is a common line in all other standardization efforts to modify any current standard along the lines given by ISO. However, the ISO focus is rather on data, while the other standards cover also practical implementation aspects.

5 Activities of the European Space Agency

As indicated earlier, the European Space Agency is placing more emphasis on application-targeted activities, as opposed to the past mostly science-related approach. For the development and deployment of applications two main ground segment activity lines need to be envisaged:

- The investment in research activities in the application domains themselves, i.e. special processing, algorithms, etc.
- The creation of a basic infrastructure that can be reused by multiple applications.

This latter infrastructure will enormously profit from successful international standarization efforts and it can be anticipated that it would not even be possible without them. Data and service providers are globally distributed, and many of them under national competence. In particular in the European scenario standardization is a must for the backbone infrastructure required for the deployment of extensive application-oriented data exploitation. Apart from the participation in international activities, many pilot projects have been initiated to unify the services provided for the single satellites into an interoperable framework:

- The CIP client [6] serves as demonstration and test tool for the CIP infrastructure on one hand and for the potential of an interoperable custom client capable of locally manipulating the retrieved data on the other hand. The CIP client is completed since December '98 and available via the Web.
- The Multi-Mission User Services (MUIS) project is developing a multi-mission distributed infrastructure, providing access for all ESA supported Earth Observation missions with a unified service concept including product and sensor guide, service directory, inventory, browse, on-line ordering, archive access (including post-processing). The initial release MUIS-A provides ESA's "earthnet online" service [7] which can be accesses at "earthnet.esrin.esa.it". Initially it was planned to use the CIP protocol as internal standard, but the speed of the international standardization activity could not meet the requirements of the project. Thus, a MUIS abstract model and interoperable protocol (GIP) was developed, and the results were partly fed into the CIP activity.
- CICCIA (Catalogue Interoperability through CORBA compliant Infrastructures and Architectures) will analyze current specifications in the area of interoperable catalogues and define an implementable CORBA-based specification. This activity has to be seen in support to the MAAT activity, but will also be the "working horse" for Agency participation in CORBA-infrastructure oriented standardization activities.

While the above projects are targeted towards the provision of interoperable services, other pilot projects have been undertaken to implement the "interoperation" of these services with an application-oriented view:

- ISIS (Interactive Satellite Image Server) has implemented a first prototype of a possible application infrastructure, focussing on oil-spill detection.
- The PATHEMA project implements a CORBA-based demonstrator system for generating multilayer thematic products and will generate detailed requirements on functions for data archiving, access and manipulation.
- RAMSES (Regional earth observation Applications for Mediterranean Sea Emergency Surveillance) is a major experimental project co-financed by the European Commission and industry, involving multiple countries in the Mediterranean bassin. It will provide a CORBA-based demonstration infrastructure for an interoperable production chain, resulting in a major acquisition of practical experience with the detailed problems in practically setting up an integrated application infrastructure.
- The MAAT (Middleware Architecture for distributed earth observation Application Systems and Tools) project will analyze the end-to-end architecture of a CORBA- based application infrastructure and will specify the required objects and their interfaces. This study will essentially try to capture all experiences gained from earlier prototypes, the MUIS GIP protocol, and the

CIP activity, aligning them with the OpenGIS concepts. As a result a detailed abstract specification for an interoperable production infrastructure for spatial applications is expected.

6 Conclusion

There is good reason for optimism that the 'critical mass' for the deployment of a new large business area is achieved. However, we have to clearly understand that we are still in a pioneering phase. We are making big steps forward, but on the more detailed scale there are big problems waiting to be resolved and sometimes we will even need to take a step back. There are areas where much work has been done – e.g. for all services related to catalogues – while other essential ones are only at the beginning of being resolved. These include e.g.:

- standardization at data level, e.g. cross-calibration just to mention one very basic problem related to remote sensing;
- data access methods including reformatting;
- data processing functions.

For a long time we will still have to live with 'islands of interoperability', small areas where the vision already works, while other areas will still have to undergo a conceptual re-thinking due to new facts and problems popping up. It will be important to maintain the current strong and positive collaboration of all involved bodies to abbreviate the construction period by pulling all together in the same direction.

References

1. Best, C., et al.: Catalogue Interoperability Protocol (CIP) Specification – Release B: CEOS/WGISS/PTT/CIP-B,Issue 2.4. CEOS WGISS – Protocol Task Team (1998)
2. Best, C., Hayes, L., Nebert, D., Reich, L., Percival, G., Hilton, J., Smith, S.: Alignment of CIP and GEO Profiles. Published in: Strobl, J., Best, C. (Eds.), 1998: Proceedings of the Earth Observation & Geo-Spatial Web and Internet Workshop. ISBN: 3-85283-014-1
3. Z39.50 Application Profile for the Content Specification for Digital Geospatial Metadata or "GEO", 22 April 1997, Federal Geographic Data Committee
4. Buehler, K., McKee, L.: The OpenGIS Guide. From `http://www.opengis.org` (1998)
5. ISO TC211 Geographic Information/Geomatics `http://www.statkart.no/isotc211/`
6. Gorman, M., Ceciarelli, A., Brownlee, Simon: The CIP Client. From "The ESA CEOS home page" (1998) `http://ceos.esrin.esa.it`
7. Landgraf, G., Fusco, L.: Earthnet online: The ESA Earth Observation Multi-Mission User Information Services start the Operational Phase. In: Gudmandsen, P. (ed.): Future Trends in Remote Sensing - Proceedings of the 17th EARSel Symposium. A.A.Balkema, Rotterdam Brookfield (1997) 5156

8. Committee for Earth Observation Satellites - http://ceos.ccrs.nrcan.gc.ca
9. Reich, L., Sawyer, D, et al.: Reference Model for an Open Archival Information System (OAIS), CCSDS 650.0-W-4.0, White Book, CCSDS (1998)

Feature Identities, Descriptors and Handles

Philip Sargent

Laser-Scan Ltd.
Cambridge CB4 0FY, UK
Philip.Sargent@computer.org
http://www.laser-scan.com
http://www.sargents.demon.co.uk/Philip/

Abstract. Finding the "right" geographic feature is a common source of interoperability difficulties. This paper reviews the issues and discusses how *persistent feature identifiers* can be used to support relationships and incremental updating in dispersed inter-operating information systems. Using such identifiers requires common definitions for concepts such as "scope" of datasets and identifier namespaces. This work extends current understanding in the Features Special Interest Group of the Open GIS Consortium (OGC).

1 Introduction

Compared with the requirements of an individual analyst, operational use of geographic information in a multi-user, multi-organisation application, adds significant new requirements in data maintenance, data transformation, lineage tracking, schema maintenance and metadata update [1, 2].

These additional functions tend to involve several datasets with specific relationships, e.g. one dataset may be a prior version of another. We then require some way of tracking individual features across those datasets. The standard example is where some attributes of a feature are under the update authority of a different organisation from other attributes; but somehow all parties must agree that they are referring to the correct feature even if many (or all) the attribute values change.

A study of feature identity presupposes that we will be dealing with moderately persistent real world objects which are observable as distinct entitites (at least for a while): entities that exist long enough to be worth naming and talking about. Thus this paper is firmly placed in the "object" rather than the "field" tradition of GIS, with the proviso that some of these objects may have indistinct boundaries [3] and may be temporary, e.g. sandbanks, storms and forest fires.

This paper attempts to review conceptual structures which may underpin future interoperability standards. File data formats have a relatively short useful life compared to the life of the data they transport and "standard" function interfaces have even shorter lives, but the data model has a much longer life: almost as long as that of the data itself.

2 Background

2.1 Unfulfilled Needs

Current treatments of feature identity do not adequately support common user needs of incremental publishing (serial updates) and value-adding; where a user of a spatial dataset wishes to add more information to a published dataset, and yet retain synchronisation when that data is updated [1, 2].

2.2 Conceptual Data Model

It is necessary to outline a conceptual data model in which to frame the spaces and domains under discussion. The following is a simplified version of the Open GIS Consortium (OGC) conceptual data model [4]:

Real World: the entire world in objective reality
Conceptual World: the observed subset of the real world
Geospatial World: a categorisation and classification of that subset
Dimensional World: the classified entities with metric representations and spatial reference systems, but not yet represented in any software system
Project World: the entities in a logical schema defined by a particular information community
Software Worlds: a set of representations of the entities in an overlapping set of increasingly capable software systems with defined schemas

In this paper a "feature" will be taken to be a *software representation* of a real-world object [4], e.g. a lake, road or city, which can have associated with it a number of attributes, some of which are *geometric representations* ("geometries"), i.e. shapes with locations. (Note that this definition differs from a commonly understood meaning where a feature *is* the geometric representation in a spatial reference system.) Thus a school is represented as a feature with one associated complex geometry which is the set of polygons representing the floor plans of the buildings and another that is the boundary of the site.

If we examine the conceptual data model sketched out above, we will see that unique labelling can only be done for discrete objects which are already a categorisation of a subset of the real world.

2.3 Labelling the Real World

The first hurdle to overcome is the disbelief of those who think that suggesting the use of feature identifiers means that we must label and index everything in the real world. That is clearly infeasible and for almost all purposes we cannot assume that such an index exists.

However there are organisations which do maintain unique identifiers for a great many types of real world objects, e.g. road bridges are numbered within the jurisdiction of a local government's civil works department, telegraph poles are labelled by the local telecommunications organisation, and every computer Ethernet card has a unique number burned in to it during manufacture (we know they are unique even if we don't know where they are). These labelled real world objects are almost invariably man-made or even entirely man-imagined, e.g. land parcel identifications, for the simple reason that natural objects usually permit less precise delineation, e.g. is an estuary part of the coast or the riverbank? This indeterminacy is of several distinct types and has been discussed in detail in a recent conference [3].

The types of feature identity that will be introduced later in this paper must be able to interwork with these pre-existing labelling schemes that are maintained by a variety of organisations with very different construction grammars and quality control standards. A key point is that the labels have to be maintained: mistakes happen and must be corrected, incorrect numbers are applied to real objects and real labels are misrecorded in software systems. Thus the real world label must be related to, but distinct from, any geographic information system's *feature identity*. For these reasons it is clear that real world labels can never substitute completely for some purposes, even though such labels probably provide more added value than other types of feature identity.

2.4 Practicalities

Practical interoperability has to take account of pre-existing data which may have been constructed using entirely different semantic principles. In the case of feature identity, the task in hand is not so much to provide support for those existing dataset collections which provide sophisticated persistent identifiers as to provide mechanisms whereby datasets without persistent identifiers can nevertheless offer useful services as if they did. For example, a good standard should not preclude the possibility that identity is constructed using a key based on one or more attribute values (as used in many RDBMSs), but neither should it mandate such an approach since many other GISs have a concept of identity which cannot be adequately represented by that approach.

The process of standardisation introduces its own oddities and restrictions which are not present in either commercial or academic research software. In addition, the goals themselves are different from most academic work in GIS. University research tends to look for proof that an innovative technique is feasible, elegant and efficient, irrespective of its compatibility with existing systems, whereas standardization research has to bring the bulk of the existing commercial implementations with it. Thus introducing new concepts has to be done with great care:

- find the *minimum* number of new concepts required
- consider whether these concepts need to be reified as software objects;
- if so reified, whether these objects should be named, and
- if so named, over what namespace the names should be unique.

A "namespace" is a concept which has been found increasingly useful in the design of computer languages. The C++ language recently elevated it to be an integral part of the language, and the Python language is fundamentally designed around it [5].

Thus while a concept may be universally agreed as being a useful aid in structuring a problem area, it can be nevertheless be productive avoid naming or reifying it because that can then avoid an entire harmonisation argument.

For those objects (concepts) we are going to name individually we then consider how we might "get one" and what we could do with it once we've got it, i.e. what other object might be a factory for it, and what operations we might want to perform on it or with it.

Example The technique of reducing conflict by avoiding the reification of a concept is demonstrated if the OGC concept of 'Feature Type".

The OGC abstract specification defines the concept of *feature type* which specifies the attributes (properties) that a feature of that type can have, but the OGC Simple Features for SQL specification avoids making that concept into a manipulatable object and instead each *individual feature* can be queried as to what attributes it supports. This avoids introducing a new type into the function interface and avoids introducing a new namespace, the list of names of feature types, with its own uniqueness constraints. The savings in elapsed time to produce the standard and to test proposed implementations for conformance more than makes up for the slight inconvenience of not having a standard for *feature type* itself (which could be introduced into later standards if absolutely required).

2.5 Dataset

A "dataset" has an obvious meaning when a simple GIS organises its persistent storage as files. However, large and complex GIS applications involve databases and possibly a large number of files for import, distributed update, etc. Thus we need a tighter definition or a different concept entirely.

The OGC uses the term "Feature Collection" to mean an object which represents a collection of features but which also may support its own attributes. The fundamental operation on a feature collection is to make available (in some way) the features in the collection. Feature collections many be permanent, e.g. a dataset, or transient, e.g. the result of a query. The metadata of a dataset thus become attributes of a feature collection. Compatibility with other standards for metadata *content* can then cause problems because some which come from bibliographic communities allow "repeated fields" with different data, whereas most GIS data models for features (and an OGC feature collection is a variety of feature) allows only "name = value" semantics.

The existing OGC "Simple Features" specifications do not need to go into details such as defining the *feature type* of a *feature collection* [6] but the OGC Abstract Specification suggests that any feature could belong to several *feature collections* at once.

2.6 Schema

Many of the queries that one wishes to perform on a dataset logically should be queries addressed to the dataset's *schema*, e.g. questions as to what *feature types* the dataset holds, what attribute names and types are defined for each *feature type*, etc. If we consider a set of related datasets, e.g. a set of versions, then we can see that the schema itself has a broader existence than each individual dataset and might be better identified with a "scope" (Sect. 3.2).

Clearly if versions of the "same feature" may be found in several different datasets, it should have some of the schema in common between them (but datasets may not necessarily the same internal structure: versioning should be able to cope with restructuring, e.g. of a directory tree [7]).

Consider for a moment the case where we have a real world *feature description*, in this case the different feature representations of the real entity may have *nothing* in common: for example, the London suburb "Richmond" appears as a feature in both the London Tube map and as part of the UK postal code coverage, but these have nothing in common and it is sensible to consider them as two different features (software representations).

Initially it would be simpler to just assert that the schemas must be identical in all datasets in the same scope where a feature identifier may be used. However we must bear in mind that serious geographic information applications are approaching 24 hour – 365 day operation, so some allowance for dynamic schema evolution will certainly be needed in the near future.

2.7 Lineage and Metadata

An important relationship between datasets which are intended to share feature identity scopes is "lineage": the history of data. Lineage should be described in the metadata of a dataset: "the currency, accuracy, data content and attributes, sources, prices, coverage, and suitability for a particular use" [8].

For feature identity management we need something more precise than the rather loose semantics and grammar, defined only in natural language descriptions, which are commonly used in metadata descriptions. If a dataset is composed of several persistent *feature collections* then each could contain its own metadata, and in the limit of granularity, every individual feature could contain metadata on how it was constructed and under what conditions its attribute values were originally measured.

There are geographic-specific metadata protocols and systems [9] as well as names and types [8], but the recent enormous growth in non-geographic metadata protocols such as RDF [10] implies that the GIS-specific protocols will have a short life before they are merged into the mainstream.

3 How Many Varieties of Identity?

3.1 Descriptors and Handles

We have the concepts *feature* and *dataset* (*feature collection*). We think we need the concept of *feature identifier*, but some thought will show that we need two such concepts: a feature *descriptor* and a feature *handle.* A *descriptor* is some way of specifying a feature from "outside the system", by listing some sufficiently unique combination of attribute values, where the list of attributes will be highly application dependent. If an external agency maintains real world labels, then a single attribute value may be sufficient: but from the software designer's point of view the uniqueness of such labels cannot be entirely relied upon. A feature *handle*, however, is a concept that is "inside" the software system and which is required to have quite tightly defined uniqueness properties which are enforced by the software itself.

Feature handles should generally be considered "opaque" and it might not even be sensible to think of them as "values" at all. Some handles might be string of text which is a query (in any language) which is sufficient to retrive the feature itself.

3.2 Scopes

Scope is a dataset management and metadata issue. A scope is a unit within which updating management can occur and probably the level at which schemas are defined. A scope is a collection of software and data in which a *feature handle* has meaning. If we are going to insist on uniqueness, we must define scope.

There are two interpretations for scope, the first broader than the second:

Meaning Within a scope, a *feature handle* has meaning.

Reachability A scope is a "domain of reachability" in that a reference (a *feature handle*) from a feature to another feature can "reach" another, e.g. to imply a relationship.

A scope to be used for update could be larger than a domain of reachability, i.e. a system may allow references *from* a feature to another only within a part of the scope in which that reference (feature handle) has meaning, but might allow corrections to be made to anything it knows about. (Existing examples are systems which allow topological relationships only between features in the same thematic layer.)

Even at our limited current state of knowledge, we can probably suggest that our lives will be simpler if we propose that a feature handle has meaning only within *one* scope, and that the handle itself contains the means to uniquely identify that scope.

How we identify scopes and how we define the function which evaluates a *feature handle* to produce (on demand) that feature or how we evaluate the handle to produce some kind of *scope handle*, is another matter.

3.3 Universal Identifiers: URIs, Monikers, GUIDs etc.

The problem of unique object identifers has been tackled by several other software industries before geographical information users became interested or aware of the issue.

Existing distributed computing platforms all offer their own solutions to the problem, but the objects in these cases are responsive "live" software processes, not "dead" geographic features hidden inside proprietary software systems and not accessible to dynamic enquiries. These "live" objects include Microsoft COM objects, CORBA nameservices, Inter-Language Unification (ILU) "String Binding Handles" (SBHs) and on-going work to develop Internet Service Location protocols [11, 12]. Some of these object identifiers include type fingerprints and version information as well as providing uniqueness and persistence.

The object identifiers from the bibliographic and World Wide Web communities [10, 13–15] are more what we require for geographic features though these too are evolving towards a "live" web-object way of operating.

3.4 Names or Addresses?

It is important to understand that names are not the same as addresses. They are conceptually distinct and some schemes implement them distinctly.

- An identifier is a name with particular persistence and uniqueness properties.
- A *naming scheme* is a system for creating these names.
- A name is *resolved* to an address by a *resolution service.*
- An *object server* uses the address to retrieve the named object.
- If you have a name, you need a *registry service* to tell you what resolution services are appropriate for it.

A practical advantage of the separation of names and addresses is that an address can be ephemeral even if the name is permanent. By separating resolution as a separate service, a name can outlast its originating organisation [16].

3.5 Mechanisms

When discussing any kind of object identifier problem, many people want to get straight in to discussing whether their favourite implementation mechanism will do what is required. There are basically two mechanisms for creating unique identifiers and the second comes in two main types:

1. Pseudo random generation
2. Federated hierarchical organisations
   (a) where the same grammar is used by the subsidiary authorities
   (b) where each subsidiary authority defines its own subsidiary adressing scheme

Pseudo-random generation is the surprisingly effective technique of generation a large random number from some local source of entropy, e.g. timing of keystrokes on a keyboard or a short analysis of network traffic coupled with an absolute clock value. The probability of two separately generated identifiers being the same can be reduced to arbitrarily low levels by ensuring that the number is large enough and the entropy unbiased enough [17]. Microsoft's COM GUIDs use this method.

Federated naming schemes use a unique central authority which administers some top-level prefixes which it distributes to a number of other authorities, each of which adds something sequentially to the prefix and distributes further.

Conceptually, the identifer in a federated system gets longer and longer as the depth of the subsidiary tree gets deeper, but in practice it is quite possible to work within a maximum defined length so long as this is increased universally every few decades. The Internet Protocol (IP), Domain Name Service (DNS) and Ethernet card numbering systems work like this. Some naming protocols put an explicit depth on the tree.

The subsidiary naming systems may be subject to separate international standards, e.g. the *service type names* of the Service Location Prototocl (SLP) are registered with the Internet Assigned Numbers Authgority (IANA) [12].

The World-Wide Web architecture assumes that resource identifiers (URIs) are identifiable by their *scheme name* which determines the subsidiary naming scheme. This applies to all URNs and URIs (including URLs) [15]. A URN in absolute form consists of:

`<scheme> : <scheme-specific-encoding>`

where the *scheme name* usually contains usually only lowercase letters and digits. The scheme name identifies the naming service and, implicitly, the resolution service, which would be used to resolve the identifier [14, 15]. A subset of schemes use a common generic syntax:

`<scheme> ://<authority><path>?<query>`

The familiar "`http:`" URL uses a DNS *machine-name* (and optional port number) as the authority. Note that the *query* option means even the `http:` scheme can be extended to arbitrary encodings using cgi scripts and "?" parameter separators. The *Handle System* architecture can be defined with a scheme name "`hdl:`". Thus a particular document has the persistent name `hdl:cnri.dlib/february96-urn_implementors` which (currently) resolves to the address: `http://www.dlib.org/dlib/february96/02arms.html` [16].

Given the existence of one universal naming system for organisations (DNS), there is no great reason to invent and maintain another. In the past, individual industries have had to organise their own hierarchical organisation naming systems, e.g. the International Article Numbering Association (EAN) which assigns manufacturer identification numbers for barcodes for retail goods and much else besides (`http://www.ean.be`).

Today the naming systems for extensions to multimedia email formats, Java packages, URIs and no doubt much else are "piggy-backed" onto DNS. Thus if an individual has a personal website `http://www.sargents.demon.co.uk` he can be sure that he can create a unique name for his collection of Java utilities by calling the package `uk.co.demon.sargents.utils`.

It has been suggested that feature identifiers should use their geographic location and feature type as part of their unique identity. Unfortunately this runs into so many problems with different resolutions, projections, datums, accuracies and Feature Class Codes that this approach is not now being seriously followed.

4 What do we need Identifiers For?

Having reviewed a bewildering array of identification mechanisms, nearly all under active development, we can see that we really do need some clearer idea of what we want identifiers for and how we want to use them:

1. When doing updates, we need to be able to determine if the supplied identifier in the update refers to a feature in the original dataset to decide whether to update it or to create a new one.
2. When comparing versions, we want to be able to find the previous version of a feature from the current version and vice versa.
3. When doing some work outside a GIS, we want to be able to make a reference to specific features in a published dataset which persists for some indefinite time into the future.
4. Within some defined scope (usually the "same" dataset), we want to assert that a relationship of a certain type exists between two features and we want this relationship to persist when any feature collection containing the related features is copied to another dataset.
5. When we copy a feature collection in its entirity, it would be useful if there were some simple relationship between the feature handles in the two copies, e.g. differing only in some prefix, so that access in both directions were quick and easy.

We introduced feature descriptors and feature handles earlier; but how can we use them ?

- A feature *descriptor* is useful only for finding and then acquiring a specific feature. It has no other purpose and since feature descriptors come from "outside" the system expressed in a variety of types of language there can be no guarantee that two different descriptors will not produce (resolve to, evaluate to, return) the same feature. In which case the two (different) descriptors have "equality by reference".
- A feature *handle* will also produce a feature, but some *scopes* will *also* define that only identical feature handles can return the same feature. This means that "equality by value" implies "equality by reference".
- There is also the intermediate case where a feature descriptor may be defined by a global scope which provides a "re-phrasing" service such that two different descriptors of specified types could be cast to a canonical form in which "equality by value" does then imply "equality by reference".

What do we mean when we say a feature *descriptor* "returns a feature"? There are two possibilities:

1. we get a lump of binary data encoded in a "well-known format" which contains all the feature's attribute data and sufficient references to a schema to be able to decode the names and types of the attributes,
2. we get a feature *handle* which incorporates the identification of the scope in which is is valid.

Whereas when we ask what happens when a feature *handle* "returns a feature", we must mean that we get the binary lump.

Assuming that we used the descriptor to make a query on some third-party indexing service, we get a handle but we must then find the actual dataset repository. We require some naming scheme so that we can use the handle to then obtain information about the scope object itself (the dataset).

Each scope may have its own coding function which it uses to evaluate the handle and to return the feature itself (in a well-known binary format).

Some scopes may publish their coding functions and allow independent access into the dataset, e.g. a directory tree of files in well-known formats, others may maintain their own integrity and require access using opaque handles using a private coding function.

4.1 Mechanisms to Declare Scopes

The URI and RDF mechanisms described earlier (Sect. 3.5) are specifically designed to be applied to old software systems inherited from a previous age ("legacy" systems, though strictly speaking they should be called "heritage" systems). Thus so long as an existing geographic data repository is not still being updated, it can be *annotated* by setting up a URI which contains a subsidiary naming scheme and RDF descriptions of the metadata and schemas, and a subsidiary URI offering a unique identifier system for the individual features.

5 Putting it all Together

We have clearly seen that we need some type of "repository" which manages multiple feature collections (datasets), which maintains scopes, which can respond to queries about schemas, which can evaluate feature handles to return features and which can be identified and located from part of a feature handle's observable value. We need such a repository for each "project" on which several people are working; incorporating several "lines of effort" at the same time. We have also seen that we need something (else?) which can evaluate feature descriptors and, if valid, return a feature handle. This latter service could make use of RDF and other metadata services and protocols.

It is suggested that the repository manager be identified with a URN and that feature handles also be legal URNs but where the initial part of the scheme-specific encoding is the URN of its manager.

It is not necessary that the repository manager actually be "on line" at any time: the important characteristics are soley the *persistence* and *uniqueness* of the identifiers. If desired, these could be maintained by an entirely manual process consisting of paper forms and authorised signatures as currently used in many file-and-directory-based GIS archives. (The probability that an organisation will want to participate and yet refuses to register with DNS is assumed to be neglible.)

There are places for randomly generated identifiers, e.g. when generating new feature handles in disconnected remote sites, or generating short locally unique strings for storage as feature attributes [17]. However, since we need some kind of federated system for relating repositories which are going to exchange data anyway, it makes sense to use that for the primary architecture. We should also remember that not every feature necessarily needs to be issued an identifier, especially in inherited systems, and that identifiers do not necessarily need to be held "in" the dataset as attributes on the features.

The types of relationships, e.g. version relationships, between the different feature collections making up a collectively-managed "scope" could usefully be partially standardised [7] to encourage a software component market. The types of relationships which already exist in ad hoc, manually managed systems are varieties of "source" data related to "working" datasets and eventually "published" data. Those GISs which provide version services have better defined semantics for the narrower domain of version control.

6 Conclusions

Doing a decent job of a "simple" matter of proposing standard ways of constructing feature identifiers thus turns out to involve interelated aspects of dynamic schema discovery, metadata granuality, formal version control semantics, distributed/replicated unique naming systems and a lot more besides. Despite the temptation to thow up our hands in horror, there does seem hope that very tightly-defined and narrowly focused *feature handles* may yet provide some usable functionality which is worth the effort of implementation.

Acknowledgements

The paper is a personal viewpoint written while the author was a Visiting Scientist at the European Commission Joint Research Centre at Ispra, Italy. Notable contributions were made by Adam Gawne-Cain (cadcorp), Grant Ruwoldt (Bentley Systems, Inc.), Yaser Bishr [University of Münster], Adrian Cuthbert (Laser-Scan Ltd.) and David Arctur (Laser-Scan, Inc.) in the Features Special Interest Group of the Open GIS Consortium Technical Committee's Core Task Force. Thanks are also due to the OGC staff who facilitated our work and showed that they understood what we were doing and why.

No part of this paper should be construed as any intention by OGC to produce proposals for standards in this area. However, this is an area in which ideas are developing rapidly [18] and the reader is recommended to investigate the current position by looking at the OGC website at http://www.opengis.org.

References

1. Hair, D., Timson, G., Martin ,E.P.: Feature Maintenance Concepts, Requirements, and Strategies, Version 3.0 May 28, 1997, Published by U.S. Geological Survey/National Mapping Division.
2. Arctur, D., Hair, D., Timson, G., Martin ,E.P.: Issues and prospects for the next generation of the spatial data transfer standard (SDTS). Int. J. GIS **12** (1998) (4) 403–425
3. Burrough, P.A., Frank, A.U., (eds.): Geographic Objects with Indeterminate Boundaries. GISDATA2. Taylor and Francis, London (1996) ISBN 0-7484-0386-8 (series editors Ian Masser and François Salgé)
4. Open GIS Consortium: The OpenGIS® Abstract Specification Topic 5, The OpenGIS® Feature, http://www.opengis.org/public/abstract.html.
5. Lutz, M.: Programming Python. O'Reilly & Associates, Inc., Bonn Cambridge Paris Sebastopol Tokyo (1996). ISBN 1-56592-197-6. http://www.python.org.
6. The Open GIS Consortium: The OpenGIS® Implementation Specification, OpenGIS® Simple Features Specifications for OLE/COM, CORBA and SQL, http://www.opengis.org/techno/specs.htm.
7. Kaler, C.: Versioning Extensions to WebDAV (Web Distributed Authoring and Versioning), Internet Engineering Task Force (IETF), Work in progress — Internet Draft August 6, 1998, Document draft-kaler-webdav-versioning-00.txt in http://www.ietf.org/internet-drafts/, see also http://www.ics.uci.edu/~ejw/authoring/
8. ISOTC 211WG 3, Geospatial data administration, Part 15: Metadata, http://www.statkart.no/isotc211/wg3/wg3welc.htm.
9. Committee on Earth Observation Satellites: CIP - Catalogue Interoperability Protocol, http://lcweb.loc.gov/z3950/agency/profiles/cip.html.
10. Manola, F.: Towards a Web Object Model, http://www.objs.com/OSA/wom.htm.
11. Inter-Language Unification (ILU) Concepts, ftp://ftp.parc.xerox.com/pub/ilu/2.0a12/manual-html/manual_1.html.
12. Guttman, E., Perkins, C., Veizades, J., Day, M.: Service Location Protocol, Internet Engineering Task Force (IETF), Work in progress — Internet Draft, July 30, 1998 Document draft-ietf-svrloc-protocol-v2-08.txt in http://www.ietf.org/internet-drafts/, see also http://www.ietf.org/ids.by.wg/svrloc.html.
13. Bray, T., Hollander, D., Layman, A.: Namespaces in XML, World Wide Web Consortium (W3C) Working Draft 18-May-1998 http://www.w3.org/TR/WD-xml-names.
14. Berners-Lee, T, Fielding, R., Irvine, U.C., Masinter, L.: Uniform Resource Identifiers (URI): Generic Syntax, Internet Engineering Task Force (IETF) Request For Comment (RFC) 2396, August 1998, http://info.internet.isi.edu:80/in-notes/rfc/files/rfc2396.txt.
15. World Wide Web Consortium: Naming and Addressing: URIs, http://www.w3.org/Addressing/Addressing.html.
16. Sun, S.X.: Handle System: A Persistent Global Name Service — Overview and Syntax, Internet Engineering Task Force (IETF), Work in progress — Internet Draft July 16, 1998, Document draft-sun-handle-system-01.txt in http://www.ietf.org/internet-drafts/, see also http://www.handle.net.

17. Leach, P.J., Salz, R.: UUIDs and GUIDs, Internet Engineering Task Force (IETF), Work in progress — Internet Draft February 4, 1998, Document `draft-leach-uuids-guids-01.txt` in `http://www.ietf.org/internet-drafts/`, see also `http://www.ics.uci.edu/~ejw/authoring/`
18. Bishr, Y.A.: A Globally Unique Persistent Object ID for Geospatial information Sharing, This conference: Interop'99, The 2nd International Conference on Interoperating Geographic Information Systems, Zurich, March 10-12, 1999.

A Global Unique Persistent Object ID for Geospatial Information Sharing

Yaser A. Bishr

University of Münster
Institute for Geoinformatics
Robert Koch Str. 26-28
D-48149 Münster, Germany
bishr@ifgi.uni-muenster.de

Abstract. Achieving interoperability between geospatial applications is a challenging research issue that attracts the attention of a growing number of scientists. By interoperability we mean that users of geospatial information systems can share their information in a distributed and heterogeneous environment. In such a distributed environment establishing relationships between geospatial objects require a mechanism to provide a persistent and globally unique object identifier, GUOID. In this paper we argue that given the reasons for the need of GUOIDs, an object is not required to be stored with a GUOID in local databases. Instead, a mechanism to provide a GUOID is only required when objects are outside the address space of the local database.

1 Problem definition

Object ID is an essential component in object technology for distributed processing. The main thrust of this paper is to attempt to resolve the issue of persistent object identifiers. This issue was realized during the recent OpenGIS(tm) meetings and is currently being investigated. We use the term *Object* to refer to a computational instance of an abstract data type that has an identifier and data, and provide services accessed through interfaces. The object identifier, ID, can either be locally or globally unique, temporary or persistent. Global uniqueness, GUOID, is used here in the unrestricted sense, that is an ID is unique everywhere, all the time.

An object ID as referred to in this paper does not refer to *key* or *foreign key* attributes in the relational database sense. Furthermore, If two objects have the same GUOID it does not necessarily imply that they refer to the same real world object. It implies that the two computational instances are the same. An Object ID is an opaque identifier which is understood and resolved by the underlying system. It is more like a handle which is a variable in which the system can store meta information about an application and some object used by the application.

OID can be temporary or persistent, locally or globally unique (more about persistent vs temporary OID is in section 2). The system presented here is called the globally unique object identifier system, GUOID. A persistent GUOID has several functions in a distributed heterogeneous environment. In general we need persistent object ID for:

1. Version management and comparison when an object is retrieved by client(s) and later updated at the source's database. A mechanism is needed to allow the client(s) to be notified and have reference to the new version.
2. Back tracking updates: sometimes, e.g., for copyright, a client would like to know when and who has originated a retrieved or updated an object. For example an object X is sent to a client who updates or modifies it and stores it as X1. Later, X1 is sent to another client who also modifies it and stores it as X2. It is needed to have reference from X2 to X at any time. Even the more challenging situation is that X2 can acquire reference to X without any prior knowledge of X and without persistently storing any reference to it.
3. Maintain a correct reference to objects even if location is changed. In many cases information sources might change the address of their physical data storage, or even the whole organization might move to another city. In this situation a GUIOD schema should be independent of physical addresses.
4. maintenance of complex objects created from primitive distributed ones. This is perhaps one of the ultimate goals from interoperability and distributed geoprocessing. In an interoperable environment we usually don't need to keep a copy of the primitive building blocks of the complex objects that we have defined in our application. We only need to keep reference to them. There are several reasons to that. For instance, The application only requires the complex object and there is no practical use from its primitives. Another example is when the complex object is required for mission-critical operations and the latest update of its primitives is always needed.

In this paper we present a mechanism to generate global unique object identifiers, GUOID. The mechanism maintains the autonomy of the database management systems and allows them to generate their local unique object identifiers when the object is local, and converts it into a global persistent identifier when the object is posted to outside world..

Section two reviews the different activities and research efforts. In this section efforts are classified according to the type of the OID service they provide, i.e., temporary, persistent, and/or unique. The GUOID system is introduced in section 3. Section 4 introduces a non exhaustive list of cases of interactions between information sources and clients. The cases help to demonstrate the strength of the GUOID. The paper is then concluded in section 5.

2 Temporary vs Persistent vs Globally Unique OID

Temporary OIDs are assigned to objects which are created by a server object on behalf of a client object and are destroyed at the end of the session, or at the end of the originator's life. On the other hand, OIDs which persist after the end of a session between a client and a server are called *Persistent OID*, e.g., serialized objects in Java. Current paradigms of digital objects mostly support short life cycle that does not outlast the session time, as will be shown in section 2.1.

Not all persistent or temporary OID are necessarily globally unique. Current DBMS provide OIDs services that are unique within the DBMS address space. The address space can either span one machine if the DBMS support single user environment, or span several intranet or internet machines if the DBMS support multi-user environment. However, these solutions are not designed to solve the problem of OID when applications need to reference or retrieve objects that are distributed in more than one DBMS. In this case, a globally unique object ID, GUOID, is essential, which is the focus of this paper.

2.1 Temporary OID

ODBC is a good example of temporary OID implementation. The Open Database Connectivity ODBC implements OIDs as handles. ODBC uses three types of handles: *environment handles, connection handles, and statement handles* [5]. The connect and the statement handles are managed under the environment handle. The relation between the environment handle and the connecting handle is one to many. Similarly the relation between the connection handle and the statement handle is one to many. The environment handle is the global context handle in ODBC. It redirects the scope of the ODBC engine to the underlying connection and statement handles. Every application that uses ODBC starts off with the function that allocates the environment handle and finishes with the function that frees the environment handles.

The connection handle manages all information about the connection. A connection handle is used to keep track of a network connection to a server, i.e., a session, and for all routing of function calls across the network. The statement handle is used for all processing of SQL statements. Each statement handle is associated with only one connection handle. When ODBC receives a function call from an application and the call contains a statement handle, it uses the connection handle stored within the statement handle to route the function call the correct DBMS.

2.2 Persistent OID

The common object request broker architecture, CORBA, provides an interesting approach to persistent OID. The model is called persistent object service, POS cite6. POS provides a comprehensive interface architecture to handle the relationship between its different components. As will be shown later, our GUOID model has the POA in its heart. Therefore, we will summarize POA model in the sequel.

In POA, a persistent identifier, PID, is intended to allow a CORBA object to have a persistent reference to its data, e.g., attribute values, in a database. As shown in Fig. 1, POA has four components: persistent identifier, persistent object, , persistent object manager, and persistent data service.

The PID identifies one or more locations within a database that house the persistent data for an object. The ID is generated in a string representation. A client can create and initialize a PID, and associate it with object to be used in a persistence model.

The Persistent objects, PO, are objects that can store and retrieve their own data. In other words, a CORBA object must have a PID to be able to store its data. A PO can connect and disconnect from a database, and store, restore, and delete its data from a database. The persistent object manager, POM, acts as

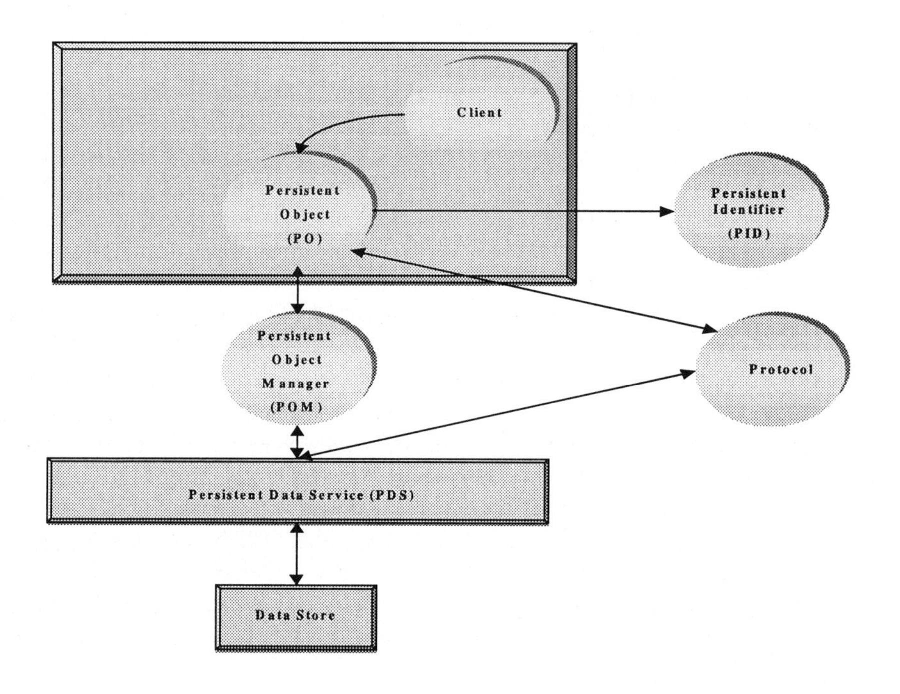

Fig. 1. The components of the POS, after Hoque 1998

a switchboard, routing requests from PO to the correct data store. Because the POM acts as a router for a client, the client is shielded from the specific data storage and retrieval mechanisms used by a given database.

The persistent data service, PDS, acts as the mediator between a persistent object and its associated database, once the database is located by the POM. Because databases vary in their functionality, the approaches used to accomplish PDS to database interaction are usually different.

2.3 Globally unique OID

The Internet community has adopted the term Uniform Resource Name, URN, for a name that identifies a resource or unit of information independent of its location. URNs are globally unique, persistent, and accessible over the network and identify resources of information. For example *uni-muenster.de* is a valid URN. A uniform resources locator, URL, identifies the location of an instance of the resource identified by the URN. For example, uni-muenster.de/index.html is a valid URL syntax [10].

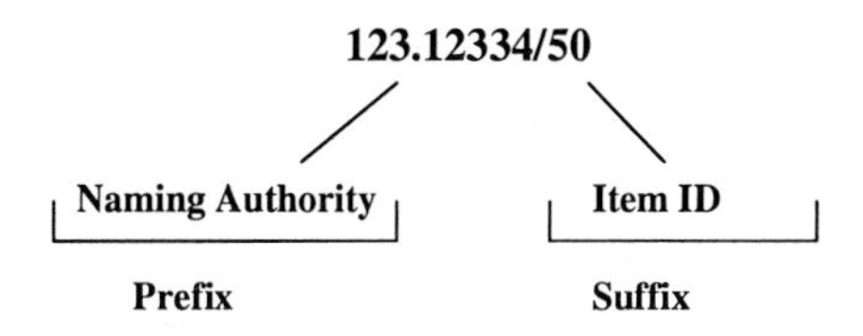

Fig. 2. Syntax of the handle system

Several proposals have sprung out of the URN initiative. The digital object identifier, DOI, initiative launched in October 1997 by the international Digital Object Identifier Foundations aims at providing a framework to manage intellectual content, e.g. literature, and the rights which accompany that content, such as access rights and copyright [1, 2]. Khan, et al. 1995 have introduced a framework for distributed digital object services. The framework provides a schema for naming, identifying, and invoking digital objects [8].

The handle system is a system for assigning, managing, and resolving persistent identifiers, known as "handles," for digital objects and other resources on the Internet. Handles can be used URNs CNR, 1998, Sun, X., 1998). The handle system is currently presented as an Internet-Draft to the Internet Engineering Task Force (IETF) [7].

The Handle System includes an open set of protocols, a namespace, and an implementation of the protocols. The protocols enable a distributed computer system to store handles of digital resources and resolves those handles into the information necessary to locate and access the resources. Perhaps one of the main advantages of the handle system is that the associated information can be changed as needed to reflect the current state of the identified resource without changing the handle, thus allowing the name of the item to persist over changes of location and other state information.

Every handle in the handle system is defined in two parts: its naming authority, otherwise know as its prefix, and a unique local name under that naming authority, otherwise known as its suffix as shown in Fig. 2. The Handle System protocol mandates UTF-8 [11] as the only encoding for any handles specified in the protocol packet.

3 Discussion

We recognize that the temporary object ID approach is not a disadvantage. The two techniques, persistent and temporary IDs, serve two different purposes. Temporary OIDs in general are more practical and efficient when reference to objects is only needed during a client-server communication session, and no further reference is required when the session is terminated.

The CORBA POS model provides a comprehensive protocol to locate and retrieve objects that have a persistent ID. However, it does not guarantee that an ID is globally unique in the unrestricted sense. The POS can guaranty that the PID is unique within the space of a client-server session. On the other hand, the handle system provides a practical protocol and naming scheme that ensure a globally unique and persistent object ID. The main disadvantage of this approach is that it violates the autonomy requirements in distributed processing and enforce its schema on the underlying information systems. It also suggest that each object is stored with its unique ID, which adds overhead on the local database that have to guarantee each time an object is created that its ID globally unique. This in fact does not allow the application to implement its own independent unique ID schema. In some case applications need to have an interpretable ID for other purpose.

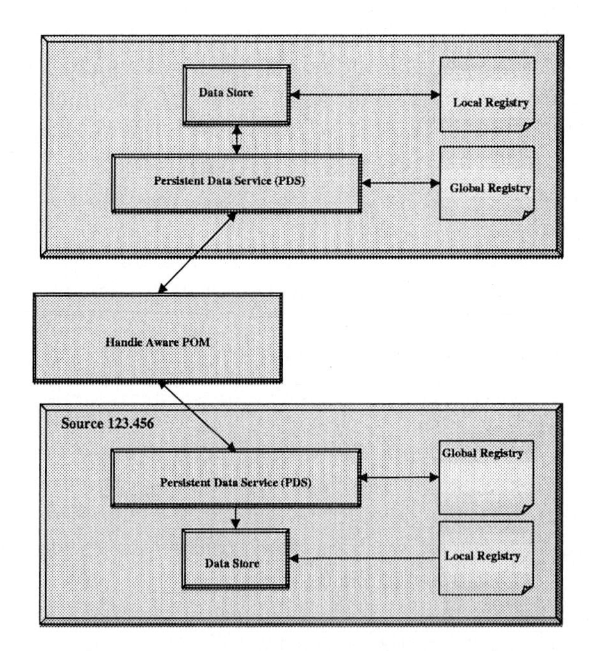

Fig. 3. Components of the GUOID system

In the next section we will present the GUOID system and attempt to tackle the issues we have raised here. We first introduce an overview of the system. We will then examine several use cases to demonstrate how the system would behave.

4 An Overview of the GUIOD system

As shown in Fig. 3, the core of the GUOID system is based on the POS and the handle system. In the GUOID system and similar to the Handle system a source address has two parts a prefix and a suffix. Each source has one prefix and can have several suffixes depending on the objects to be communicated with the outside world. The main characteristics of the GUOID is that the system does not enforce an OID schema on the local data store. Instead, the GUOID replaces the local OID and assign it to merely any outgoing object.

Communicating parties should adopt and be able to resolve the handle system mentioned in section 2.3. This is done cooperatively between the PDS and the local data store. As shown in Fig. 3, the POM is responsible for routing the objects to the correct client and source. The PDS resolves the GUOID and maintains the relationship between the prefix and the GUOID in a global registry file. The data store resolves the suffix part of the GUOID, and maintains the relationship between the local ID and the suffix in the local registry file. Each source or client maintains its own global and local registry files. The registry files maintains only a reference to the incoming objects not the outgoing ones. The information about the relationship between the local object ID and the GUOID is stored in two registry files as shown in Fig. 4. This mechanism of registering the relationship between GUOID and local OID allow the source to know who received its objects and allow the client to know who sent the object at any given time. This is different from the Handle system which enforces the underlying local database to save the objects with their globally unique handles.

Information sources who want to provide the GUOID service should maintain a local unique object ID in their local database. The local ID should not be changed so often and should be persistent as long as possible. In case that the source supports versioning, a registry of the relationship between the ID of the original version and that of the subsequent ones should be maintained.

In this reminder of this paper we introduce some use cases to help us understand the system behavior. The following cases assume that there is a client and an information source, or simply a client and a source.

Case 1: Searching for an object for the first time, e.g. data mining The GUOID does not support this case. This is in fact intuitive and should not be supported since no prior knowledge of the object is known to the client. GUOID is un-interpreted octet string, compatible with the handle system, and therefore is not involved in the search process. The GUOID is not intended to replace logical identifier of objects, e.g., keys, foreign keys and other attributes meaningful attributes in the database sense.

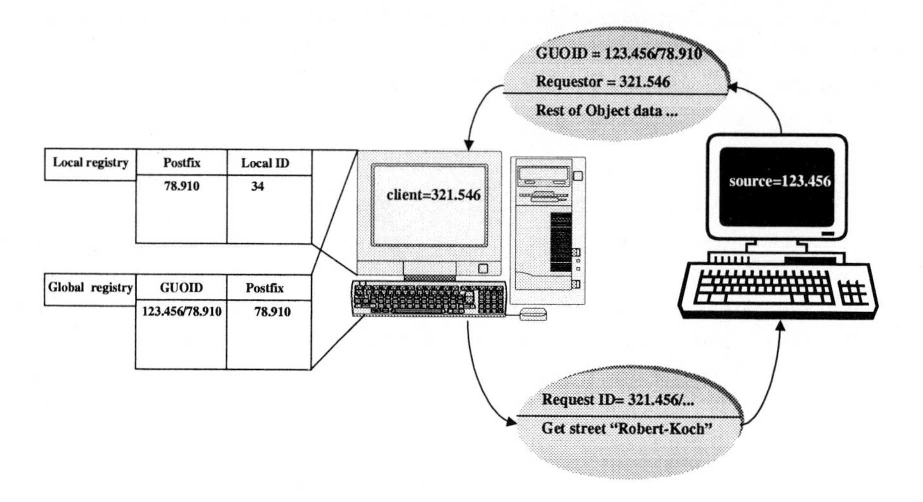

Fig. 4. Interaction between client and source

Case 2: Object is retrieved for the first time As shown in Fig. 4, a request, or a query, is sent by the client to the source. The request has a reference to the client's address. If the requested object is found, it is then sent to the client, based on the address that was associated with the original request. In the GUOID system, each outgoing object has two identifier attributes. The first attribute has the full GUOID address of the source. The second attribute has the prefix address of the client that "tells" the object where to go. The structure and the interaction of the global and the local registry files are also shown in the figure.

Case 3: Object is retrieved and later referenced for update or retrieve further information
The client has an old version, or require extra information of the object. Fig. 5 shows a flow chart of the operations involved. The local registry file of the client has a reference to the object's suffix. First the local registry file is searched for the suffix. When found it replaces the object's local ID and then sent to the PDS. The PDS in turn searches the global registry file for the prefix that corresponds to the suffix and then sends the object to the corresponding source via the POM. At the source, the PDS truncates the prefix and sends the object to the local data store. The local data store updates the object. The reverse of the process is then repeated and finally the object is stored at the client.

Case 4: Object is retrieved from a database, then original object is updated at the source In this case the client requires to be notified of any updates on the retrieved objects. A broadcast mechanism can be designed such that the source sends an update alert on the network. The alert includes a list of GUOIDs of the updated objects. The PDSs on the network may receive

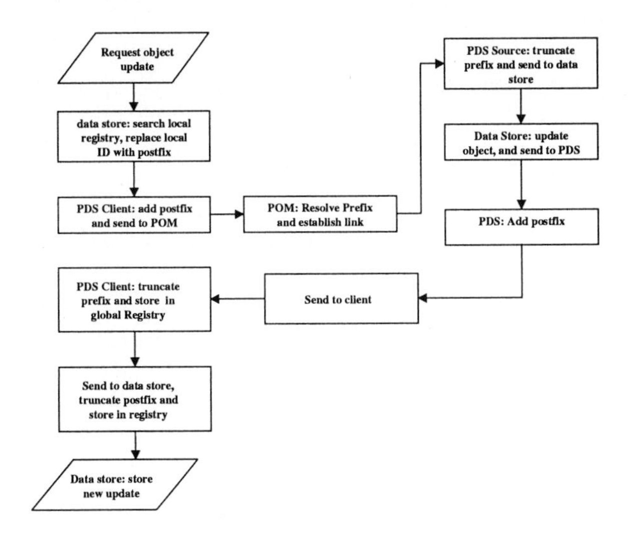

Fig. 5. Operations involved in object update in GUIOD system

the alert and search their global registry files against the GUOID list. If a match is found case 3 is repeated. Otherwise, the PDS can ignore the alert.

Case 5: Object is retrieved and then sent to another client This is similar to the situation mentioned in the list point number 2, section 1. In this case an object is retrieved not from the original source and the client need to reference the original source. Since any outgoing object will have its original GUOID assigned to it, the receiving client will always have a reference to the original source. Case 6: Local ID of the original object is changed Although not recommended, same broadcast mechanism mentioned in case 4 can be applied. The client then receives the broadcast alert and updates the global and the local registry files.

Case 7: Complex object is created from primitive ones that are stored remotely

The GUOID system does not require the data store to explicitly store the object. The client may choose to store only reference to the primitive objects and supply to the application remote operations to access the remote primitives.

5 Conclusions

The crux of the OID problem is to ask the right questions. Instead of only asking the classic question

"How to assign a globally unique OID". Three questions should be asked instead: "Why do we need a GUOID?". we have tried to answer this question in section 1. The second questions is *"Do we really need to have our local database store its objects with a globally unique object ID?"*. Sections 2 and 3, showed that we only need globally unique object IDs when communicating an object with the

outside world. Based on the answer of the first two questions the third question was " how to assign ID to object such that they appear to be globally unique to the outside world, and at the same time maintain the application autonomy and appear unique to local applications.

We have introduced the GUOID system which is a hybrid between the CORBA model and the handle system. The GUOID system goes one step further and provides a mechanism answer the third question using registry files.

Only the most basic elements of the GUOID infrastructure are described herein. These elements constitute a minimal set of requirements and services that must be in place. One restrictive point of the GUOID system is it assumes an ideal world of no system crashes, no error exceptions, etc. Further investigation for fault tolerance and error recovery is required. Finally, the use cases are not meant to be exhaustive, further investigation is required to cover larger spectrum of cases and implementation tests.

Acknowledgements

The author would like to thank the OpenGIS consortium for their partial support of this work.

References

1. Arms, W. "Key concepts in the architecture of the digital library ". D-Lib Magazine July 1995.
2. Arms, W., Blanchi, C., Overly, E., "An Architecture for Information in Digital Libraries". D-Lib Magazine, February 1997.
3. Berners-Lee, T., Masinter, L., McCahill, M., et al., "Uniform Resource Locators (URL)", RFC1738, December 1994. http://ds.internic.net/rfc/rfc1738.txt
4. CNR, 1998. The Handle System. http://www.handle.net
5. Geiger K. (1995). Inside ODBC. Microsoft Press. ISBN 1 55615 815 7.
6. Hoque R. (1998). CORBA 3. IDG books, Forster City, CA 94404.
7. Internet Draft – Handle System: A Persistent Global Naming Service - Overview and Syntax (submitted to the IETF 14 Nov 97; Updated 16 Jul 98). http://www.handle.net/draft-sun-handle-system-01.html
8. Kahn, R. and Wilensky, R. A framework for distributed digital object services. D-Lib Magazine, May, 1995.
9. Sun, X. 1998 "Internationalization of the Handle System - A Persistent Global Name Service"; a paper presented at the 12th International Unicode Conference in Tokyo, April 1998.
10. The URN Implementors. "Uniform Resource Names, a Progress Report". D-Lib Magazine, February 1996.
11. Yergeau, F., "UTF-8, A Transform Format for Unicode and ISO10646", RFC2044, October 1996. http://ds.internic.net/rfc/rfc2044.txt

Implementation Architecture for a National Data Center

Steve Ramroop and Richard Pascoe

Department of Information Science, University of Otago, Dunedin, New Zealand.
sramroop,rpascoe@infoscience.otago.ac.nz,
http://www.otago.ac.nz/

Abstract. In this paper, the authors' main focus is presenting the implementation architecture of Ramroop and Pascoe's [9] conceptual model for data integration in a Geographic Information System (GIS) environment. The model is briefly discussed and extensions to the model are made. These extensions are also applied to the notation Ramroop and Pascoe [9] used to denote the entire data integration process. The model consists of a *Data Center* and *Data Agencies* at a national level. Details of the actual processing steps followed within the Data Center is discussed and the general architecture design is described. The implementation of the concept is modular starting with the first service called the Selector Broker. Multiple tools are being considered for further investigation.

1 Introduction

Over the past decade the use of Geographic Information Systems (GIS) has increased by a vast majority of professionals and non-professionals. One of the major problems that the development of GIS applications is facing today is data acquisition, Devogele et al. [3].

Large quantities of data sets are available as hardcopy and softcopy. At the National level, such data sets are viewed as independent repositories which are maintained by the various organisations that collect the data sets.

When integrating data, specific difficulties are encountered. Abel et al. [1], stated that the difficulties encountered by an application developer are the presence of differing data models (such as the network model, the relational model, and so on); different Application Program Interfaces (API); and semantic clashes. These difficulties all vary from one data set to another which hinder the immediate realization of a solution to the integration problem. On-going research in this area is described by: Ramroop and Pascoe [9], Abel et al. [1], Devogele et al. [3], Pascoe [7], and so on.

Ramroop and Pascoe [9], presented a conceptual model for National integration of geographic data. In their model, the entire process of integrating data sets is modelled starting with the user formulating a query, to the final destination data which is transformed into an acceptable format. Their model reflects the typical National scenario where organizations are autonomous because of their varying mandates. The need for organizations to co-operate and inter-operate together motivates the sharing of data and ultimately the development of federated databases.

In this paper, Ramroop and Pascoe's [9] model and notation for representing the overall process are discussed and extended in Sect. 3 and Sect. 4 respectively. The main focus in this paper is to address the architecture of such a system which is discussed in Sect. 5.

2 The Virtual Data Center

The model addresses the problem of data integration by making use of a *Virtual Data Center*. Ramroop and Pascoe [9], defined a Virtual Data Center as:

> *The virtual processing center through which data sets from Data Agencies are located, transformed, and delivered to other Data Agencies who have requested data sets to satisfy their needs.*

Data Agencies will have one or more users and will provide the software necessary for participating as a member of the Virtual Data Center. The Center is made up of a number of services which are executed in a specific sequence starting with the user query. These services are specific with specific operations. The general details of these services and their operators are discussed in Sect. 3.

The Virtual Data Center can be accessed by all users whose interest lies in integrating data. Similarly, Abel et al. [1] presented the design of their view of a Virtual GIS for distributed spatial data processing in heterogenous environments. Their system, however, is aimed at extensibility and scalability through the distribution of the processing load. Their system's architecture comprises global frontends and several backend component systems through which the processing load is distributed.

In this paper, the architecture of the Virtual Data Center is presented by explaining the processes involved when integrating data sets at the National level. Ramroop and Pascoe [9], indicated that, the process of integrating data sets, involves, in no particular order:

- selecting data sets appropriate to the GIS application;
- transferring data sets into a common data format (without unnecessarily loosing any information); and
- merging data sets initially collected at various level of details (such as graphic databases with varying map scales).

These processes are representative of the many task needed to be performed and considered when integrating data sets. In Ramroop and Pascoe's [9] conceptual model the Center's processing features are used as a basis for further research.

3 National Conceptual Model

The Conceptual Model representing a National data integration strategy is shown in Fig. 1. The model comprise of two components: the *Virtual Data Center* and the *Data Agencies*. Each Data Agency continues to operate separately as they did in the past by creating, storing, and using geospatial data sets. However, when the need for other data sets arise, then the services of the Virtual Data Center is accessed.

The model indicates the processing of geographic data sets which are stored at the Data Agencies using various data schemas defined by the multiple GISs available. The storage of such data sets are usually a collection of related files that stores the geographic locations and attributes of each feature. Therefore, these files are usually large and their contents are variable when compared to other data types (for example customer data from banks) which are readily manipulated and restructured to be used in other applications.

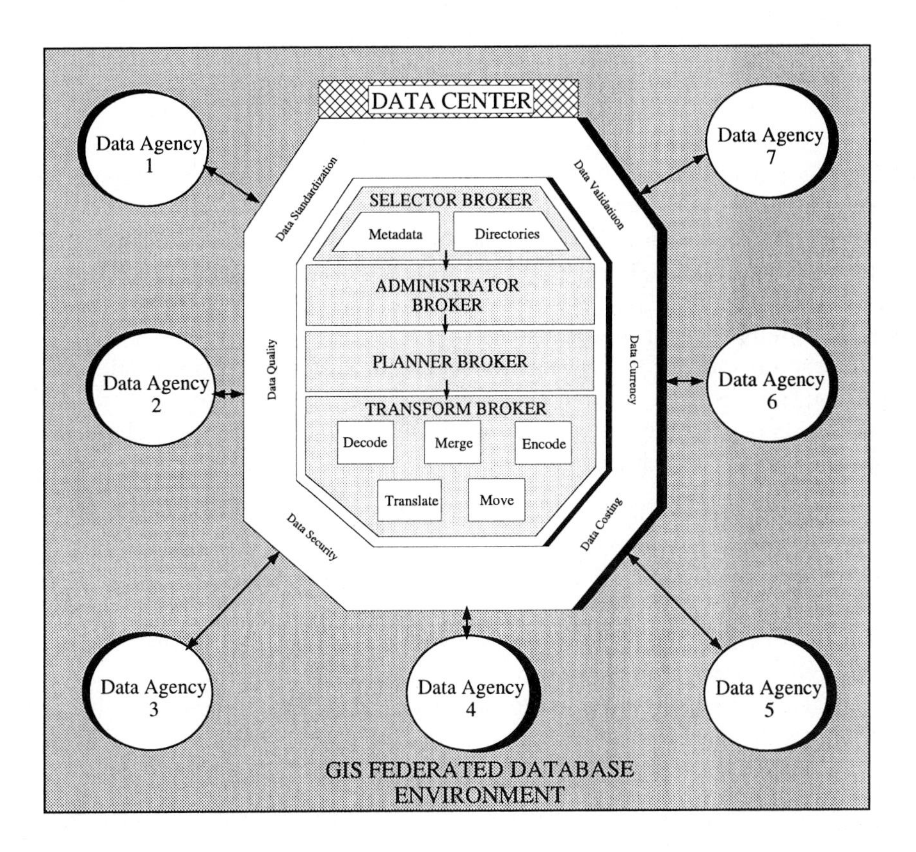

Fig. 1. National Conceptual Model

3.1 Brokers

The services of the Virtual Data Center is facilitated by the use of *Brokers.* Ramroop and Pascoe [9], defined a Broker as:

> *The agent responsible for executing specific processes associated with data integration within the Data Center.*

The Brokers divide the generic process of geographical data transfers into smaller, specialised tasks which are performed by different modules. There are four Brokers, three of which were defined by Ramroop and Pascoe [9]:

- a Selector Broker for finding, selecting, and prioritizing data sets appropriate to user requirements;
- a Planner Broker for optimising the data transformation process overall; and
- a Transform Broker to perform the functions of the encode, decode, move (communicator), translate, and merge modules for combining data sets.

A fourth, an Administrator Broker for managing legal implications is defined below. Combining the use of these Brokers enables Data Agencies to select, access, and merge multiple data sets from other Data Agencies satisfying their needs.

Administrator Broker. At the National level, Data Agencies have spent large sums of money to capture data (Peel [8], Johnson et al. [5], Dueker and Vrana [4], Abel et al. [1], and so on). Although money is spent throughout the development of any GIS implementation, 80% of the total cost is during the initial data capture stage, Aronoff [2]. Therefore, using data available from other Data Agencies is a potentially cheaper alternative.

Laurini [6], commented on the aspects of administration of geographic multi-database systems. He indicated that consideration should be given to encourage the creation of an inter-organisation protocol. This includes the mechanisms to deal with problems such as:

- *copyright*: Who is the owner of the data?
- *access rights*: not all end-users are granted permission to retrieve or to use any kind of data everywhere, therefore some limiting access rights must be defined;
- *difficulties during prototype implementation*: indeed during the integration procedure some sites can crash then, who is responsible?
- *results property*: by mixing information issued from different databases or sources then, who is the owner?
- *accounting problems*: what is the cost of the data and who is paid?

Such a protocol would be the corner stone upon which the Administrator Broker will be developed to address.

Issues of cost for data and other legal administrative concerns. The *Administrator Broker* is defined as:

The agent responsible for managing all legal transactions associated with the sharing of data sets.

The Administrator Broker is responsible for:

- requesting legal procedures from Data Agencies corresponding to the selected metadata;
- informing the user of the legal procedures to be followed; and
- ensuring that all legalities are done before the metadata is transferred to the Planner Broker.

Typically, the Administrator Broker is responsible for ensuring that all legal procedures (such as payments, copywrite, intellectual property rights, royalties, security, and so on) are considered before processing is allowed to continue. Therefore, the Administrator Broker ensures that the relevant Data Agency authorities are informed of the pending transaction with the user.

Having included the notion of the Administrator Broker, the notation presented by Ramroop and Pascoe [9], will now be extended to include the Administrator Broker.

4 The Notation

Ramroop and Pascoe [9], defined a notation (Table 1), for describing the transfer of data from one GIS to another as a sequence of transformations. The notation defines a number of operators. Each either contributes to the transformation of data values from a representation required by one GIS to that required by another GIS, or changes the location where data is stored.

A generic transfer is denoted by:

$$\mathcal{Q} \overset{select}{\rightsquigarrow} \{\mu\} \overset{admin}{\longrightarrow} \{\mu_a\} \overset{plan}{\longrightarrow} \tau(\mathcal{S}_*) \overset{*}{\rightarrow} \mathcal{D}$$

where

$\mathcal{Q} \overset{select}{\rightsquigarrow} \{\mu\}$ denotes the set of metadata $\{\mu\}$ reported back to the user by the Selector Broker in response to the user query $\mathcal{Q}$

$\{\mu\} \overset{admin}{\longrightarrow} \{\mu_a\}$ denotes the selected set of metadata processed by the Administrator Broker to ensure that all legalities are done, to produce an approved metadata set $\{\mu_a\}$

$\{\mu_a\} \overset{plan}{\longrightarrow} \tau(\mathcal{S}_*)$ denotes a function τ representing the transformation processes to be performed on each data set $\{\mu_a\}$ by the Transform Broker

$\tau(\mathcal{S}_*) \overset{*}{\rightarrow} \mathcal{D}$ denotes data sets $\mathcal{S}_*$ are transformed using either one or more of the decode, translate, encode, merge, and move operations into the final destination data set(s) $\mathcal{D}$

Notation	Associated term	Definition
$\mapsto$	A data *transfer*	The transfer of a data set from one representation to some other representation.
$\xrightarrow{translate}$	A data *translation*	The transformation of a data set to conform to a different conceptual, implementation, or physical schema.
$\overset{move}{\rightsquigarrow}$	A data *movement*	The physical movement of a set of values from computer system x to computer system y.
$\xrightarrow{*}$	Many data transformations	One or more data translations or data movements.
$\mathcal{Q}$	User query	User request specifying multiple criteria $(\alpha, \beta, \gamma \ldots)$
μ	Metadata	Information associated with each data set.
$\overset{select}{\rightsquigarrow}$	Select operator	A process which selects and prioritized metadata.
$\xrightarrow{admin}$	Administrator operator	A process which addresses the legalities associated with the sharing of data.
$\tau()$	Planner Function	A function representing the optimal sequence of data transformations.
$\xrightarrow{plan}$	Plan operator	An operator which selects the order and sequence of data transformations.
$\xrightarrow{decode}$	Decode operator	An operator which transforms data values in a file format into equivalent values in memory.
$\xrightarrow{encode}$	Encode operator	An operator which transforms memory data values into equivalent data values in a file format.
$\parallel \xrightarrow{merge}$	Merge operator	An operator which combines two or more data sets into one data set.

Table 1. Notation used for describing data transfer at a National level

5 System Architecture

To best understand the overall architecture of the entire system, the processing steps are identified in Fig. 2. Generally, the data integration process is initiated by the user while the actual integration of data is done by the Virtual Data Center. Once a query is submitted, the services of Brokers are all dependent upon the other starting with the Selector Broker.

Apart from the final output of the destination data, there is another exit from the Virtual Data Center. Such a loop-hole occurs at the Administrator Broker. This occurs if the Administrator Broker reports back to the user with legalities indicating that the selected data is not accessible, (for whatever reason), then the user will be given the opportunity to make another selection or quit. In cases where the Selector Broker reports back to the user with metadata indicating that some data sets are in analogue form, then further processing within the Center will cease for such data set(s).

The system architecture of the entire Data Integration System at a National level is divided into two infrastructures. One is associated with the storage of metadata at the Data Agencies, and the other is associated with the Virtual Data Center.

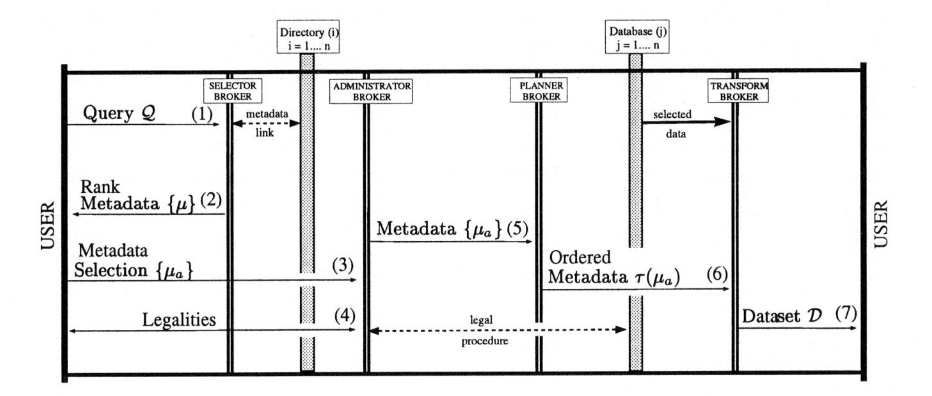

At the National level there are 'i' metadata directories and 'j' databases. When integrating data there are roughly seven steps. They are as follows:

1. The user sends a query either with single or multiple criteria to the Selector Broker.
2. The Selector Broker reports back to the user with the metadata that matched the query, however, prior to reporting back to the user the following sub-processes are executed:
   - heterogenous directories are searched for matching metadata; and
   - the resulting metadata is ranked as a percentage of the criteria.
3. The user selects the metadata corresponding to the desired data set(s) which is sent to the Administrator Broker.
4. The Administrator Broker flags the respective Data Agencies containing the desired data set(s), requesting the legal procedure to follow with respect to acquiring copies of the data. This information is sent to the user. The user ensures that all legalities are done and then submits an acceptance code to the Administrator Broker.
5. Once the code is accepted by the Administrator Broker, the metadata is sent to the Planner Broker to order the transformation processes needed to be performed on each data set.
6. The Planner Broker orders and sends the transformation recommendations to the Transform Broker.
7. The Transform Broker copies the selected data set(s) from the federated databases and applies the respective transformation on each data set according to the recommendations made by the Planner Broker. The final destination data set is then sent to the user.

Fig. 2. Processing steps used for National Data Integration

5.1 Infrastructure of the Data Agency

Each Data Agency is responsible for storage and maintenance of their own metadata sets. For the purposes of this research heterogenous databases and directory systems are used to store metadata. A variety of standards for metadata is being defined. Examples are: the Australia New Zealand Land Information Council (ANZLIC) Metadata XML/SGML standard [1], providing guidelines for the content of geographical metadata; Content Standard for Digital Geospatial Metadata (CSDGM) [2], developed by The Federal Geographic Data Committee [3]. To implement the Selector Broker, the ANZLIC standard is used as the basis for defining and storing metadata.

---

[1] `http://www.environment.gov.au/database/metadata/anzmeta/`

[2] `http://www.fgdc.gov/Metadata/ContStan.html`

[3] `http://www.fgdc.gov/index.html`

The ANZLIC metadata standard is utilized in Australia and New Zealand. This standard is a working standard and for the purposes of avoiding the issues related to other standards, the ANZLIC standard is assumed to be more than sufficient for this research and fulfill the requirements needed to perform the services of all four brokers. Other additional assumptions for each Data Agency is that they would all follow the ANZLIC standard to store metadata; they would have a fast and reliable Internet access; they would allow the execution of permission to run the infrastructual software; and they would have a commitment to the overall Data Integration concept embodied by the National Data Center.

5.2 Infrastructure of the Data Center

The Virtual Data Center is owned by all Data Agencies which are inter-connected via the Internet. The Center consists of the software that implements all of the infrastructural software at each Data Agency and the software of protocols needed for interconnecting these agencies. The Center's software is stored and managed by designated Data Agencies. The Center is platform independent which is made accessible through an applet embedded within an HTML (Hypertext Markup Language) document. When a user accesses the Center, a copy of the applet is copied into the memory of the user's computer where the processing is done.

6 Test-bed Implementation

The implementation of the Center is being researched in a phased basis, starting with the implementation of the Selector Broker since this is the input to the Center. A test-bed is developed in order to test various strategies associated with each Broker. A number of strategies will be defined and each will be evaluated in terms its effectiveness and compared with the others.

The tools being used to implement the test-bed are those which are commonly used by the GIS community for example object oriented programming, relational databases, object databases, and object-relational databases.

6.1 Selector Broker

The Selector Broker is built using Java. The Java Database Connectivity (JDBC) provides the ODBC connection interface to SQL (Structured Query Language) databases. Therefore, Java programs can access data in almost every SQL database including Oracle, Sybase, DB2, SQL Server, Access, FoxBase, Paradox, Postgres, SQLBase, and XDB. However, tests are being done using metadata stored using Access, Dbase, ObjectStore, and Postgres database management systems.

The Selector Broker uses metadata stored at each Data Agency. The services of the Selector Broker includes:

- searching for metadata based upon the criteria stated in the user query;

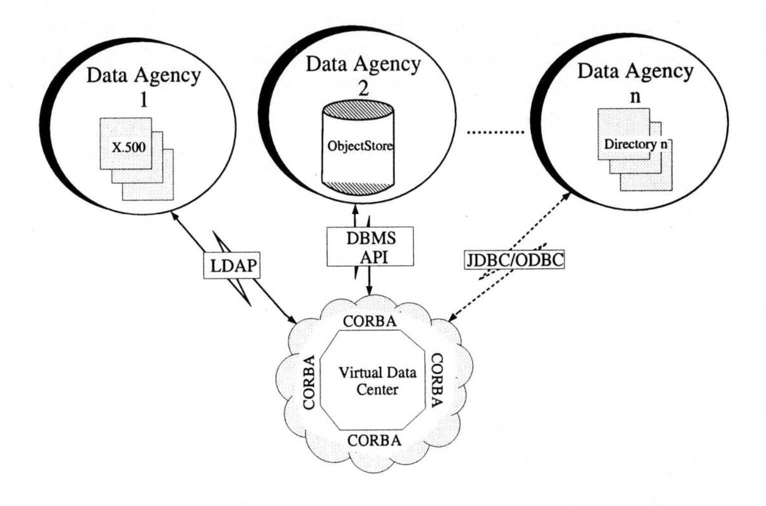

Fig. 3. Tools used for selecting metadata

- prioritizing the metadata results as a percentage of the criteria; and
- reporting back to the user with the results.

The approach for developing the test-bed for the Selector Broker is one in which the user has access to the Center through an applet which accesses the metadata stored in relational, object, and/or an object-relational databases. The tools facilitating the functions of the Selector Broker is shown in Fig. 3.

The query input to the Selector Broker is built using the metadata standards as defined by ANZLIC. All possible criteria for selection is listed on a Graphical User Interface (GUI). This is necessary because users would be informed upfront of the type of metadata stored and the criteria available to build the query.

In the process of writing the infrastructual software associated with the Selector Broker, access to ObjectStore meta database is made possible using CORBA ORB (Object Request Broker). To do this, CORBA is used to describe all available services, components, and data. Using CORBA's IDL, the metadata repository stores the interface specifications of each object an ORB recognizes.

7 Summary

In this paper the concept of data integration at the National level was addressed by extending the research of Ramroop and Pascoe [9]. Their model and notation for a Virtual Data Center was briefly presented and extended by introducing the Administrator Broker and its operator into the notation. The data integration processing steps was identified which assisted in the design of the overall Center architecture.

The infrastructures for the Data Agencies and the Virtual Data Center were then presented concluding with the tools used to define a test-bed for the Selector Broker. Further research is being done with regard to the tools to be used to execute the roles of the other Brokers.

References

1. Abel, D. J., Ooi, B.C., Tan, K. and Tan, S.H.: Towards Integrated Geographical Information Processing. International Journal of Geographic Information Systems, (1998), vol: 12, no: 4, pp: 353-371.
2. Aronoff, S.: Geographic Information Systems : A Management Perspective, (1993). WDL Publications, Ottawa, Canada.
3. Devogele, T., Parent, C. and Spaccapietra, S.: On Spatial Database Integration. International Journal of Geographic Information Systems, (1998) , vol: 12, no: 4, pp: 335–352.
4. Dueker, K. J. and Vrana, R.: Systems Integration: A Reason and a Means for Data Sharing. Sharing Geographic Information, (1995), pp: 149–155. Editors: Onsrud, H. J. and Rushton, G. Center for Urban Policy Research, New Brunswick, New Jersey.
5. Johnson, B. D., Shelly, E.P., Taylor, M. M. and Callahan, S.: The FINDAR Directory System: A Meta-Model for Metadata, (1991), pp: 123–137. Metadata in the Geosciences. Editors: Medyckyj-Scott, D., Newman, I., Ruggles, C. and Walker, D. Group D Publications Ltd. United Kingdom.
6. Laurini, R.: Spatial Multi-Database Topological Continuity and Indexing: A Step Towards Seamless GIS Data Interoperability. International Journal for Geographic Information Systems, (1998), Vol: 12, No: 4, pp:373–402.
7. Pascoe, R.T.: Data Sharing Using X.500 Directory. First International Conference on Geoprocessing, (1996), pp: 689–698. Spatial Information Research Center (SIRC), University of Otago, New Zealand.
8. Peel, R.: Unifying Spatial Information. Mapping Awareness, July, (1997), vol: 10, no: 8, pp: 28–30.
9. Ramroop, S. and Pascoe, R.T.: Notation for National Integration of Geographic Data. 10th Annual Colloquium of the Spatial Information Research Centre (1998), pp:279–288. Editor: Peter Firns. Spatial Information Research Center (SIRC), University of Otago, New Zealand.

Architecture Considerations for Advanced Earth Observation Application Systems

Hermann Ludwig Möller[1], Marcello Mariucci[2], and Bernhard Mitschang[3]

[1] European Space Agency, Directorate of Application Programmes,
hmoeller@estec.esa.nl
Formerly ESA-ESRIN, Italy, since January 1999 affiliated to ESA-ESTEC, The Netherlands.
[2] Universität Stuttgart, Institut für Parallele und Verteilte Höchstleistungsrechner (IPVR),
Marcello.Mariucci@informatik.uni-stuttgart.de
[3] Technische Universität München, Department of Computer Science,
Bernhard.Mitschang@informatik.tu-muenchen.de

Abstract. Application systems in the earth observation area can be characterised as distributed, platform-inhomogeneous, complex, and cost intensive information systems. In order to manage the complexity and performance requirements set by these application scenarios a number of architectural considerations have to be applied. Among others the most important ones are modularization towards a component architecture and interoperation within this component model. As will be described in this paper, both are mandatory to achieving a high degree of reusability and extensibility at the component level as well as to support the necessary scalability properties. In our paper we refer to the state of the art in earth observation application systems as well as to a prototype system that reflects to a high degree the above mentioned system characteristics.

Key Words: Distributed Information Systems, Earth Observation Systems, Applications, Interoperability, Middleware, CORBA

1 Introduction

Since the early '70s an increasing number of satellites orbit our planet and make observation data related to sea, land, and atmosphere available globally. The data is used in support to a number of applications; the best known might be the daily weather forecast satellite maps and animations shown on the TV news programmes. Fleets of new satellites will produce about 1TByte of new data every day and soon the amount of data collected within a single year will equal the size of all acquired data of the last 25 years. With the increase in observation platforms also the number of applications is increasing. For example, since the early '90s, Europe has been exploiting its European Remote Sensing Satellite

[1] The contribution of the author is related in particular to his work during a one year research fellowship assignment at NASA/GSFC

ERS-1 and -2, e. g. for sea ice monitoring, oil-pollution monitoring or in support to disaster management. Earth observation (EO) satellites represent an investment of several hundred million ECU per space craft. In order to justify such investment, which still is largely public funded, new emphasis has been given in recent years to the development of ground segments with focus on exploiting the data streams received from the satellites for specified applications also beyond scientific use. The present paper focuses on architectural considerations of application-specific information systems for data exploitation.

2 Earth Observation Application Data and Information Management

EO systems distinguish between a space segment, comprising the space craft and associated command and control systems for flight operations, and a ground segment, comprising facilities for data acquisition, processing, archiving and distribution. In the following focus is given on the ground segment and in particular ground system elements that may help to relay observation information to its specific use in end-user applications. EO Ground Systems are defined by means of three levels:

- 'Data Level' (DL): large scale infrastructures primarily operated by space agencies and satellite operators as data providers (DP), handling the data acquisition from the EO satellite and the data handling for standard data processing and archiving;
- 'Information Level' (IL): infrastructures primarily operated by Value Adding Companies (VAC) and scientific institutions (SI) for creating higher level application specific information, e. g., through thematic processing, and used by Value Added Resellers (VAR) for distributiong such information;
- 'End-User Level' (UL): user access infrastructure, interface and local infrastructure serving scientists, governmental and commercial users, and the educational sector.

Traditionally, DL infrastructures interface directly with the user segment on UL, serving a multitude of user requests through a single system architecture. Search and order of standard data products are the main functions externally accessible in such systems. Figure 1 illustrates now the additional level, at present in prototyping stage, the IL. The IL constitutes an additional layer, interfacing upstream with the DL for the provision of standard EO products, and downstream interfacing with the UL for application-specific user access and distribution. The IL is not one single system but comprises a multitude of smaller systems each serving a different user domain and each interfacing the DL level separately. Some IL functions, in particular data ingest may be provided by space agencies and data providers in general. Other IL functions, e. g. thematic, application-specific processing, may migrate towards VAC or SI. Selected IL functions may be shared within a cluster of individual IL systems. IL functions such as archiving of thematic products may be covered by DP, thus reaching from DL into IL functions. End-users (EU), depending on the service support required may be associated either with an SI, a VAS or a VAR; even in some cases EU may be registered with more than one information service supplier.

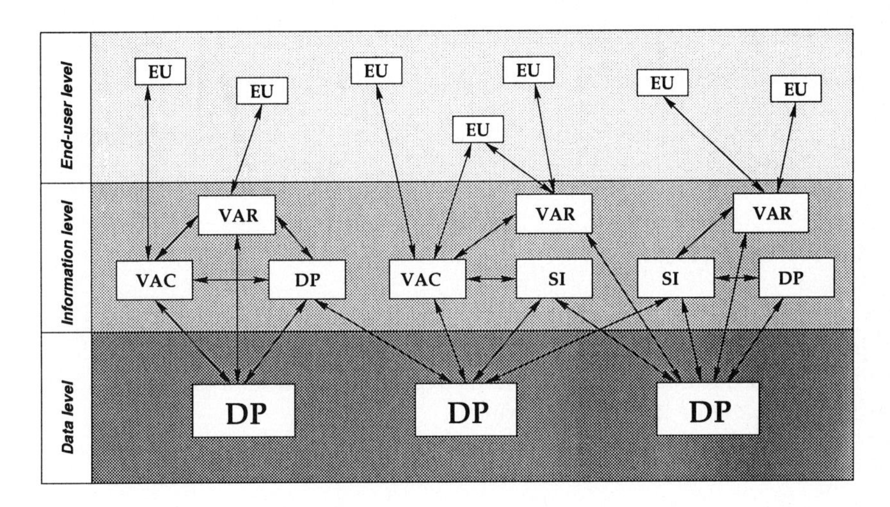

Fig. 1. Information Level

Prior to focusing on the IL, its functions, associated interoperation, technology issues, and federation concepts, an estimation of EO data volumes and access statistics shall help to understand the magnitude of the problem present day DL are facing. Estimates in the domain depend significantly on a number of assumptions, e. g., user behaviour and market evolution, and on the range of satellites considered, e. g., geostationary and low-orbiting satellites, meteorological and environmental satellites. However, the numbers provided will help in a first approximation to determine a number of requirements for the design of the IL infrastructure and IL system concepts.

2.1 Data Volumes

EO Ground Segments handle high volumes of data, both in terms of archived historical data and in terms of newly ingested data. Estimates of the total volumes considering all major LEO observation constellations world-wide point at more that 300 TByte of archived data in over 20 million inventory records, varying between a few kBytes and more than 2 GByte in size. Ingest of new EO data is estimated in the range of 1 TByte per day, with a corresponding user pull above 2 TByte in the next future. The actual user pull will depend on how successful the data exploitation will be performed. One important enabling element for such exploitation and global data usage are interoperable IL systems.

Existing Archives (1970's-present)			Number of Users	**Daily** Ingest from Satellites as from 1998/99	**Daily** User Pull as from 1998/99 (media&on-line)
Total Data Volume	Inventory Entries	Single Data Item Size			
> 300 TByte	> 20 Million	kByte – > 2 GByte	> 20000	500 GByte – > 1 TByte [a]	> 2 TByte

[a] Numbers derived from NASA, ESA and EUMETSAT estimates

Table 1. EO Information Systems - Volume Estimates - World-wide

2.2 Access Statistics

As an example, based on earlier NASA estimates for US based systems, Figure 2 provides estimates of category and number of users, and nature of electronic access to EO data and information, as expected for state-of-the-art DL systems becoming available in the near future.

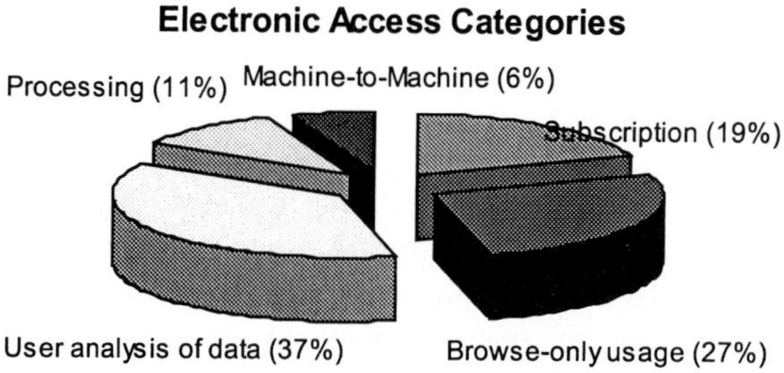

Electronic Access Categories

Processing (11%) Machine-to-Machine (6%) Subscription (19%) User analysis of data (37%) Browse-only usage (27%)

Fig. 2. EO Information Systems – Electronic Access Categories

According to the example, the scientific community with an estimated >12000 users globally represents the largest part of the user population. It is also within this community that 'user analysis of data' at UL is most prominent with 37%, as the scientific usage of data often requires in depth analysis of particular phenomena related to a data set. User requests leading to 'processing' represent another significant usage type and may include application specific processing of data, e. g., extraction of thermal fronts from sea surface temperature data along coast lines for detection of fishing grounds. Such thematic processing represents an estimated 11% of the accesses. It is also within this type of usage that user access may result in subsequent machine-to-machine interoperation hidden to the user, e. g., transfer of a particular data set from a remote archive to a thematic processing system.

A different utilisation is related to the users in the educational community, estimated at 70000 – 200000 users. Usage is focused on browse-only (27%), e. g., investigation of thumb-nail images without further data processing or analysis, and subscription services (19%), where users register for data provision related to specified locations, times, or events, e. g., the weekly image of their hometown.

3 Information Level Federation Concepts

3.1 Information Level

At present, DL systems need to handle very diverse usage profiles in a single system. However, government use and in particular the percentage of commercial usage are expected to grow significantly in the coming years as distinguishable application domains. VAC and SI emerge for serving these new user market segments. They will require automation. E. g., a high percentage of 'user analysis of data', today performed by experts at UL after consultation of DL, may need automation to serve these new user communities. To do this effectively, systems will need to be capable of handling application specific usage in a way today only available for expert users, for example a scientist with the knowledge where to find the data and which processing to apply in order to obtain desired information. Prototyping of such systems, constituting the forerunner of IL systems, is currently in progress in a number of selected application domains.

First results have shown that individual IL functions may be applicable to more than one application domain and system. E. g., advertisement of available services, which is today rather limited because services are already known to the scientists user group, becomes a challenge across IL systems in order to attract new governmental and commercial users to the information services. Interoperation of participating components for advertisement, storage, processing, workflow, and access is becoming essential. Proper partitioning of applications into a federation of individual IL systems is important in order to achieve the scalability and performance required by individual user groups. I.e., distribution and interoperation of data and functions and machine-to-machine interoperation is becoming an issue. Table 2 provides a comparison of complexity between individual data level systems vs. information level systems.

Individual IL systems distinguish themselves from large-scale DL systems in that they typically respond to the needs specific to a scientific discipline or particular application, e. g., flood monitoring or urban planning. Unlike satellite operator's large-scale data systems managing multiple TBytes, these systems are mostly concerned with a more limited amount of data, e. g., multiple GBytes, specific to a defined usage. The number of individual systems is expected to be one order of magnitude higher for IL than for DL. DL systems interoperate primarily internally, whereas IL systems may develop different degrees of interoperation in smaller clusters. User interfaces, rather uniform for DL may become application specific on IL, reflecting well-identified requirements of much smaller and defined user groups. This will lead to complementary interfaces to

	DATA LEVEL	INFORMATION LEVEL
Nature of System & Service	Universal ('query from hell')	Specific to application
Typical Data Volume per system/service	TBytes	GBytes
Number of individual systems/services globally	< 10	> 100 (estimate for 2005)
Level of interoperation	Primarily internal to individual system, Inventory external	Interoperation in a number of clusters (estimate > 20 globally) to varying degree depending on type of federation
User Type	Mostly scientific	Science, commercial, education/wider public
Number of users per system/service	> 10.000	±10 government and commercial ±50000 education/public
Processing requirements per system/service	Complex, across many scientific disciplines	Known with application, e. g., one algorithm per system
Access/Dissemination requirements	Universal	Tailored to user type

Table 2. Complexity: Data vs. Information Level

the general purpose data search and discovery interfaces dominant in today's Earth Observation User Information Systems on DL [9].

The emergence of open, network-centric middleware, in particular the Object Management Group's Common Object Request Brokers Architecture (CORBA) [11, 16, 10], and the wide availability of advanced communication networks, in particular the Internet, provide the essential elements for the required underlying infrastructure for distributed IL services, operating in a federation of individual systems.

3.2 Various Federation Scenarios

DL system developments inherently provide a high level of interoperation in a distributed environment as they are typically implemented as single large-scale projects, e. g., ESA's ENVISAT Payload Data Segment (PDS) [13], or NASA's Earth Observing System Data and Information System (EOSDIS) [3] with its Data Model and Distributed Information Management (DIM). However, this interoperation is merely internal, i.e., interoperation options for external data providers and different DL systems either assume the adoption of the internal standards of a given DL system or external interoperation is limited or non-existent. The Committee on Earth Observation Satellites (CEOS) has made an attempt to achieve interoperation of DL systems for a limited functional scope, i.e., for queries of large scale inventories, and has specified the Catalogue Interoperability Protocol (CIP) [1].

In contrast to the above, IL systems are expected to be developed in a multitude of smaller development projects. As individual projects may not have the resources to develop and serve all required service functions internally, a market opportunity for externally available, interface-compatible system functions and services beyond catalogue services exists. IL systems and services provided primarily by VAC and SI would orchestra in a federation characterised by

- The *level of distribution* of individual information service functions, like thematic image processing, and its allocation under the responsibility of different players in the service market, e. g., data providers, VAC, SI.
- The *level of interoperation and reuse* between different systems and distributed information service functions, like common advertisement services or sharing of a common data archive between IL systems under the responsibility of VAC or SI.

Four different degrees of federation are proposed in the following, ranging from a 'Non' Federation, with distribution and interoperation only within the DL, to a 'Full' Federation configuration where most functions are migrated to IL and distributed with a high degree of function interoperation.

'Non'-Federation The 'Non' federation (see Figure 3) represents the *state-of-the-art* in EO information systems and in principle reflects today's DL-only configurations, i.e. without IL. All functions typically are provided through a single distributed DL system (per world-region) appearing to the user as a single system. DL sub-systems, e. g., for processing and archiving, interoperate according to a single DL-internal schema, called the common DL schema.

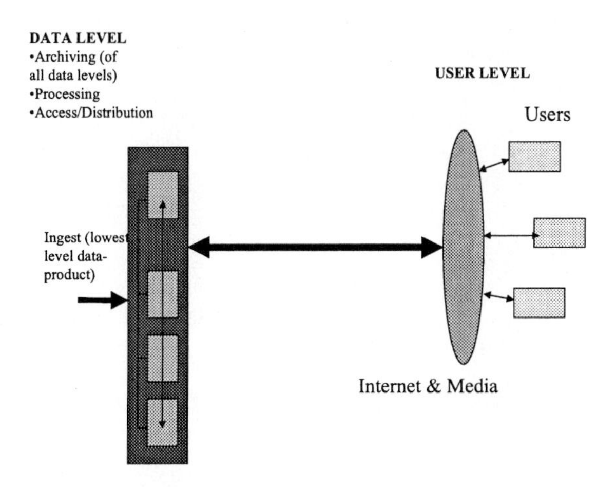

Fig. 3. 'Non'-Federation

This schema distinguishes a *user view* from a *conceptual data description* and a *physical data description.* The user view is a logical description that represents the end-user perception of the data. The conceptual data level describes the data assets, illustrating relationships among classes, e. g., defining the data input/output relations per sub-system, and specifies the attributes of the data. The physical data level refers to the platform dependent representation of the data as implemented using commercial database management systems. The common DL schema is essential for DL-internal interoperation. Different DL may offer catalogue interoperability services based on an agreed schema subset or a derived standard, limited to their inventory subsystems. Reuse of components within the DL is maximised. Partitioning of the DL into application-specific information system components is however strongly limited by the fact that individual components are typically designed to handle a broad range of different information. Therefore, DL systems as such are not easily adaptable or sizable to application-specific systems as defined in the IL.

'Processing' Federation The 'Processing' federation depicted in Figure 4 shows thematic processing as first functions migrated into a thereby created IL. Application specific information is typically generated under VAC or SI responsibility. Some VAC or SI may decide to share processing functions, i.e., an algorithm may be executable within a cluster of VAC IL systems. User access and distribution has not changed and is still achieved through the DL system. The IL typically builds on the DL schema. Although the 'Processing' federation improves the adaptability of the system to application-specific requirements and allows a first separation of service functions, its scalability and its reuse potential are still limited by the underlying DL.

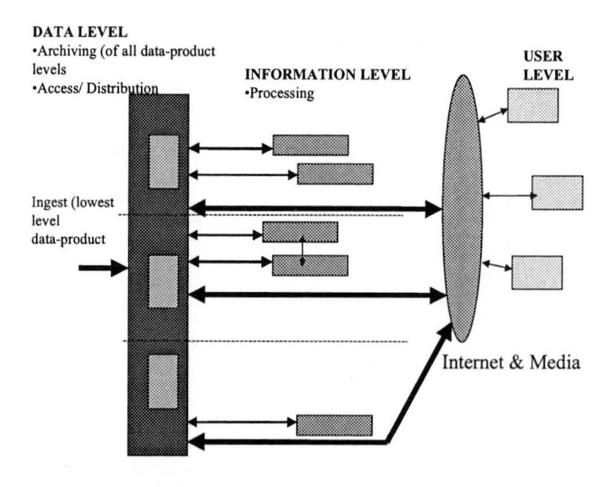

Fig. 4. 'Processing' Federation

'Processing and Access/Distribution' Federation The 'Processing and Access/Distribution' federation (in Figure 5) is an essential step towards a federation allowing the VAC or SI to identify and distinguish itself visible to the user. The IL may be defined by different overlapping IL schemata, which for archival functions may still be those of the associated DL. A cluster of a few IL systems may decide to operate through a single user interface and provide a common user request management, e. g., forwarding of user requests and context information between their services. However, access functions are expected to serve as distinguishing feature for individual IL systems and will only show a limited level of interoperation. This interoperation may focus on directory and trading services for the advertisement of available IL services across a cluster. Interfaces to GIS for the provision of non-EO information as complementary data may be part of a VAC or SI offering. Standard data product archival is still performed at DL, management, storage and archival of thematically processed data may however be allocated to the IL. IL sub-system components, e. g., ingest module, may be reused from one IL system to another. With most functions migrated to the IL, the level of modularization of service functions, the system scalability, its reuse potential and adaptability are high. *Prototype systems* reflecting such a federation concept are currently being demonstrated (see chapter 4).

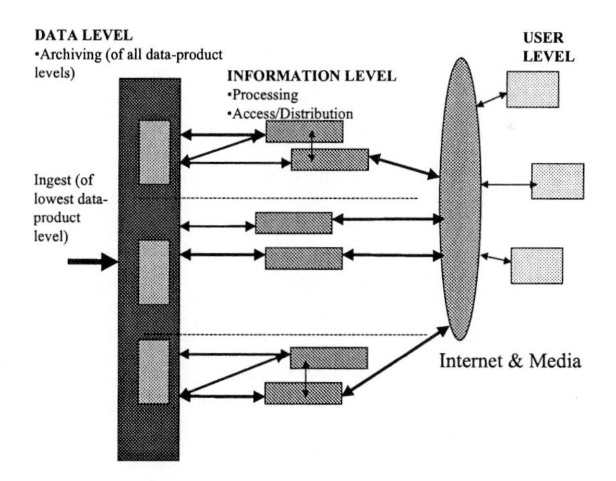

Fig. 5. 'Processing and Access/Distribution' Federation

'Full' Federation The 'Full' federation depicted in Figure 6 reduces the DL to the long-term archive for lowest level data products. Other archival functions have been moved into the IL segmented according to geographical regions and applications. Together with interoperable interfaces for archive services within a cluster, the number of interoperable, related processing services is expected to increase. User access/dissemination is expected to remain with a lower level of interoperation. Interoperation with traditional DL systems is no longer an issue for most VAC or IS as this is dealt with within the IL. The lack of any common IL schema maintained outside the IL provides a challenge to the interoperation within the IL. Leading federation members, e. g., those comprising the archival function, may provide a reference data model and serve as architectural model. Level of reuse, scalability, modularization, and adaptability is comparable to the 'Processing and Access/Distribution' federation.

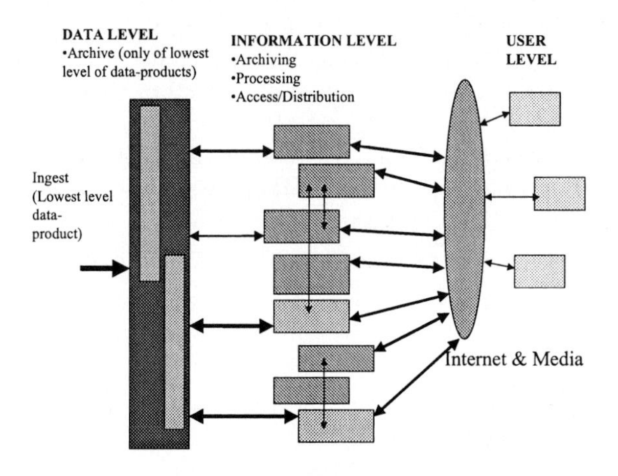

Fig. 6. 'Full' Federation

Outlook on Federations Depending on the success of DL systems in the different world-regions and the pressure from application markets, the degree of federation in the longer run may be different between regional markets. It may be established between the 'Processing and Access/Distribution' federation and the 'Full' federation, which most likely remains merely a long term perspective. Much will depend on the way in which new enabling technologies will be inserted into the development process. Leading IL service and system developments may set the pace for the federation, or a co-ordinating working group may provide recommendations for the IL.

3.3 Technology Considerations

During the last years system engineering and software design for federated information systems have undergone tremendous technological changes. Modern architectures are based upon components with well-defined interfaces and behaviour. This step enables reuse, extensibility, and scalability of systems or system components. Adoption of new technology or application requirements can be done by exchanging single components only. Furthermore, each module can be developed by separate companies with different expertise. In order to enable the usage of third-party software published without source code, several component models have been defined, e. g., Microsoft's DCOM or ActiveX [15], IBM's DSOM [7], OMG's CORBA [10], etc.

Considering the great Internet wave starting in the early 90's, component-based design became even more important. Companies want to establish so-called virtual enterprises, accessing resources of their partners and offering e.g. online order facilities for their customers. Thus there was an increasing need for standardized components which can interoperate with other ones via Intra/Internet, no matter which programming language or operating system is used. The success story of Java and its well defined components started. Though Java offers several ways for component interoperabilty, it is still a particular programming language. In order to be extensible w.r.t. any kind of system aspects it is not appropriate to define interfaces of components in a specific and single programming language. For example, in independence w.r.t. interface definition can be easily achieved by means of CORBA's Interface Definition Language (IDL). It is independent of programming languages, but mappings exist or can be developed for any language as needed. In addition, arbitrary components can interact through ORBs and interfaces, and the behaviour of basic components are already standardised by the OMG.

Though we have just detected a suitable component model, federated systems generally raise another requirement: Data models of each participating partner have to be compatible. CORBA does not offer mechanisms to resolve this issue, but federated database technology (e. g., as provided by so-called database middleware like IBM's Data Joiner [5]) together with standardization endeavours (e. g., OGIS) may help. The different federation scenarios presented before differ in their degree of distribution of functions and degree of standardisation. From the database point of view, these differences translate to system transparencies for data storage and data access. 'One-stop-shopping' is the idea that a user can issue an information service request to a (logically) single system and is freed from possible query decomposition into multiple sub-queries to distributed data sources, issuing those queries, and perhaps finally integrating the results to be presented to the user. For higher level information it may be required to know how this information was derived. The ability of the system to support such questions is the issue of pedigree or data provenance. Clearly, less control over data and metadata model, and an increased degree of distribution make the necessary transparencies more difficult to be achieved.

4 A 'Processing And Access/Distribution' Federation Prototype

In order to prepare for the challenge described above, a number of prototype developments have been initiated, e. g., the Earth Science Information Prototypes (ESIPs) [4] in the USA, or European projects funded by ESA and the European Commission. A European prototype, the Interactive Satellite Image Server (ISIS) project [2, 6], and its successor project RAMSES, aim at defining and validating IL interoperation with emphasis on the distinct definition, distribution, and interoperation of functions such as data ingestion, processing, cataloguing, and user access as well as their implementation and validation on a common system backbone. More information on this prototype can be found in [14].

4.1 Distributed Functions

Figure 7 depicts the various IL functions identified and illustrates their interconnection through a common bus based on CORBA technology. The User Client interfaces refer to the UL, whereas the Data Ingest interfaces refer to the DL. This architecture has been applied by three IL system developments, each for a selected application: detection of fishing grounds, urban planning, and oil pollution monitoring. All components are defined through a well defined CORBA IDL interface. A high level of reuse of components for standard data ingest, image processing, and catalogue has been achieved between the systems. Although client and workflow functions are application specific, different modules on sub-component level are reusable also, e. g., image display and animation.

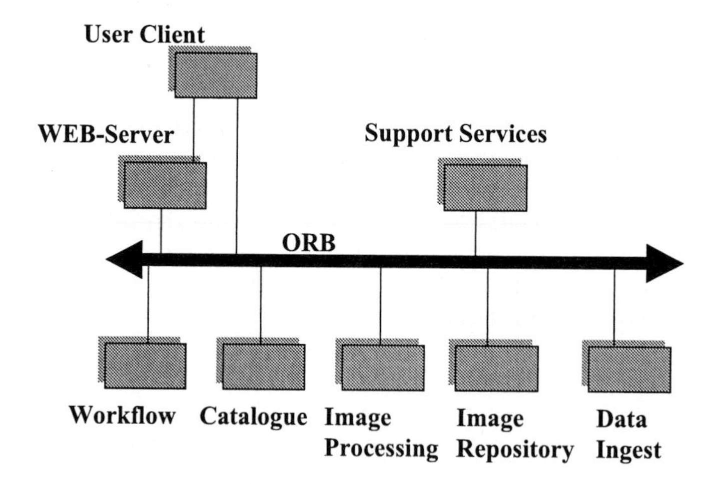

Fig. 7. Information Level System Components

Figure 8 presents the interactions between the different functions for a given application, here the monitoring of oil pollution through the analysis of radar imagery. The UL client's interface to IL functions is through a workflow module which has a priori knowledge of the user, its application, and the associated relevant catalogue entries and processing algorithms.

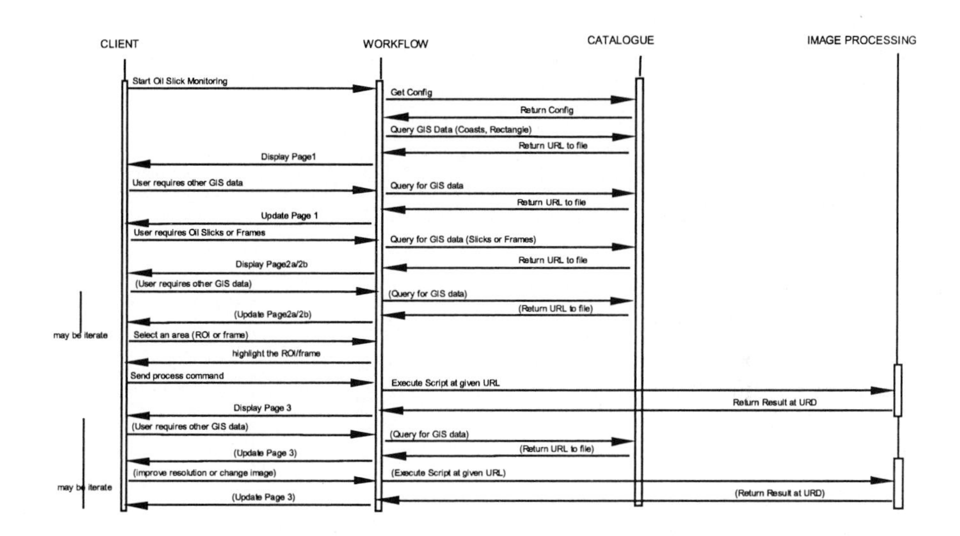

Fig. 8. Example: IL internal Interaction between IL components

The workflow manages the client requests and the catalogue queries, e. g., it automatically retrieves the user relevant catalogue information at the start of the user session. In the example it triggers the processing function once the user has confirmed a pre-selection of a suitable data set. Such data set is identified by the catalogue based on a number of parameters provided by the user. It is displayed as a vector or sub-sampled image on the user screen for selection.

The prototype makes use of the Internet InterORB protocol (IIOP) for client access and has been implemented based on Orbix products. The OpenGIS simple feature specifications are under consideration for interfacing external GIS for the provision of complementary, non-EO data. At present, this data is stored as vector data inside the catalogue component.

4.2 Application of CORBA Services

A number of commercially available CORBA service are suitable to be directly applied in support to earth observation data and information services. An example is the Trader service [16] which can be used as a directory function where different catalogues register with a description of the nature of available data sets, e. g., European radar imagery, ERS-2 available at the ESA-ESRIN catalogue. A catalogue query may thus identify and be routed to the adequate catalogue site without running a full query on all sites (see Figure 9).

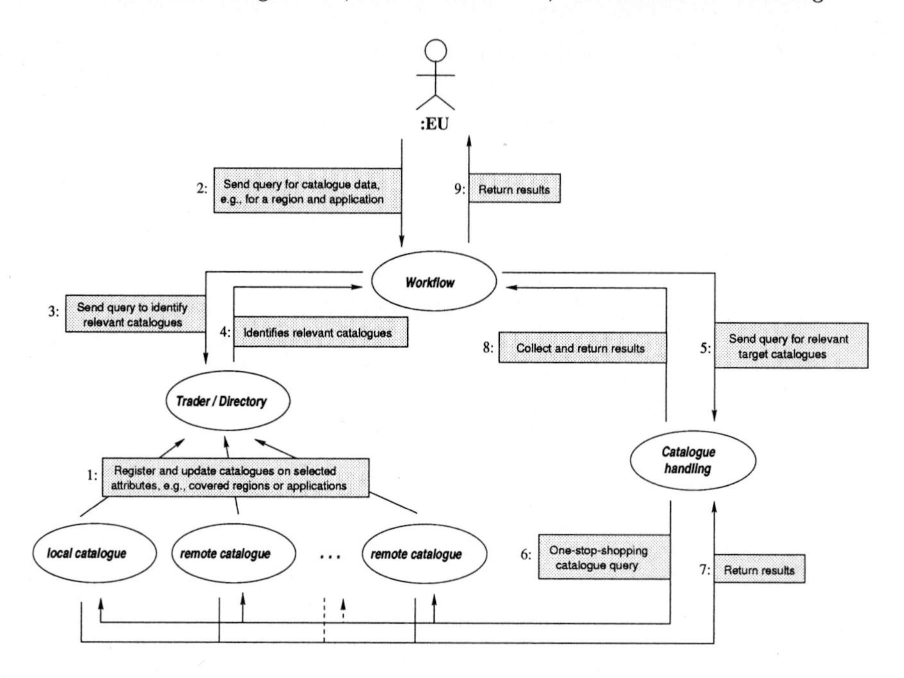

Fig. 9. Directory Service and CORBA Trader Service

Another example is the Event service [16] as illustrated in Figure 10. Well specified events are triggered by the ingestion of new data sets into the system archive. Events specification may include the source of the data, information on its applicability for selected applications, or a first indication on geographical coverage. End users (EU) may register at the subscription service for a sub-set of such events, and in conjunction with a message box will be notified by the system in case such event occurs. The same events may be used for logging and accouting, or may act on workflow functions to automatically trigger processing functions on the newly ingested data.

5 Conclusion And Future Issues

The 'One-size-fits-all' approach of today's state-of-the-art EO information systems, i.e. large-scale DL systems, may be adequate for the handling of standard EO products. But it leads to overly complex solutions in view of the multitude of emerging, very different user communities and application domains. It also risks not to meet the adaptability requirements resulting from the future role of VAC, SI and VARs, which demands a higher degree of independence to distinguish their service offering. An additional system layer, the IL, may provide the adaptability needed, balancing the complexity of individual, smaller systems against an adequate level of interoperation among such systems in a federation. Prototype demonstrations indicate that the evolution of the Internet,

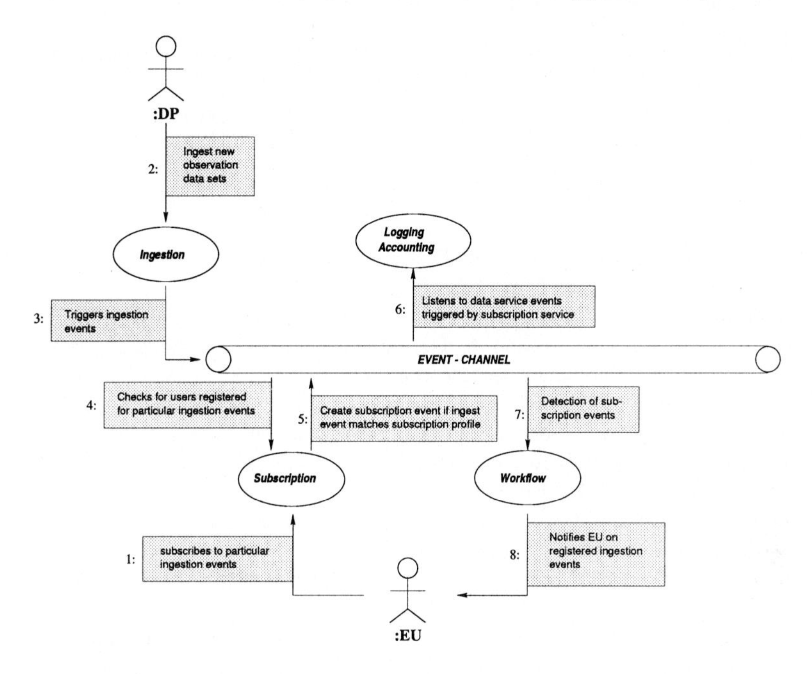

Fig. 10. Data Subscription Service and CORBA Event Service

together with the already cheaply available processing and storage power, and the emergence of open middleware standards, provide the technological basis for federations of IL systems yet to develop.

A number of activities are underway worldwide with space agencies and related organisations to advance along this line. In particular, the authors are involved in the development of a 'Processing and Access/Distribution' prototype [14] and the ESA author has prepared a modelling study for better mapping CORBA and EO IL systems [MAAT] which initiated recently. This shall help to perform system verification in a real case of an application and user scenario (oil pollution monitoring) and in optimising the EO service component model to make best use of CORBA.

References

1. Triebnig, G., et al.: Catalogue Interoperability Protocol Specification (CIP) - Release B, CEOS Working Group on Information Systems and Services, 1997.
2. Moeller, H.L.: European Earth Observation Application System Prototype, CODATA - Conference on Scientific and Technical Data Exchange and Integration, Bethesda, MD, USA, December 1997.
3. NASA Earth Observing Data and Information System EOSDIS, `http://spsosun.gsfc.nasa.gov/New_EOSDIS.html`.
4. Creative and Innovative Working Prototype Earth Science Information Partnerships in Support of Earth System Science, NASA CAN-97-MTPE-01, May 1997.

5. IBM Co.: Data Joiner: Administrator Guide and Application Programming, IBM Co., San Jose, 1997.
6. Moeller, H.L.: The Interactive Satellite Image Server Project, International Society for Photogrammetry and Remote Sensing - ISPRS Workshop 'From Producer to User", Boulder, Colorado, USA, October 1997.
7. Lau, Chr.: Object-Oriented Programming Using SOM and DSOM, Wiley & Sons, 1995.
8. Object Oriented Environment for Distributed Facilities, Earth Observation Applications Architecture, ESA Statement of Work, June '98.
9. Landgraf, G.: The ESA Multi-Mission User Information System (MUIS) and their evolution, Earth Observation & Geo-Spatial Web and Internet Workshop '98, Salzburg, `http://earthnet.esrin.esa.it`.
10. Orfali, R., Harkey, D., Edwards, J.: Instant CORBA, Wiley Computer Publ., 1997.
11. Object Management Group: The Common Object Request Broker Architecture: Architecture and Specification, Revision 2.2, `http://www.omg.org/corba/corbiiop.htm`, Upd. February 1998.
12. Fischer, P., Gorman M.: Generation of Mutilayer Thematic Products for End Users - A WWW Interfacing System for Integration earthborne GIS and spaceborne EO Information, Earth Observation & Geo-Spatial Web and Internet Workshop '98, Salzburg, `http://earthnet.esrin.esa.it`.
13. ESA: Envisat-1, Mission & System Summary, ESA Publication, March 98.
14. Regional earth observation Application for Mediterranean Sea Emergency Surveillance, Project Description, July 1998, `http://tempest.esrin.esa.it/ramses`.
15. Sessions, R.: COM and DCOM, Wiley Computer Publishing, 1998.
16. Siegel, J.: CORBA: Fundamentals and Programming, Jon Wiley & Sons, 1996.
17. Sellentin, J., Mitschang, B.: Data-Intensive Intra- & Internet Applications - Experiences Using Java and CORBA in the World Wide Web, in: Proceedings of the 14th IEEE International Conference on Data Engineering (ICDE), Orlando, Florida, February 1998.

A Spatial Data Infrastructure for the Hindu Kush – Himalayan Region: Some Institutional and Political Issues of Interoperation

Peter Bitter

ICIMOD
GPO Box 3226
Kathmandu
Nepal
peter@icimod.org.np

Abstract. This paper describes the institutional and political context of a spatial data infrastructure for the Hindu Kush – Himalayan region. It then outlines the role and present activities of the International Centre for Integrated Mountain Development (ICIMOD) in general and its Mountain Environment Information Service (MENRIS) Division in particular. Some pragmatic steps to build a regional spatial data infrastructure that are envisaged by MENRIS are discussed. Emphasis is being put on metadata, standards definition and generation of regional key data sets at 1:250'000 scale.

Key references. Spatial data infrastructure, Metadata, Hindu Kush – Himalayan Region

1 Background

The Hindu Kush – Himalayan Region, which is the primary area of activity of the International Centre for Integrated Mountain Development (ICIMOD), covers an area of more than 4 Mio sq. km and habits a population of approximately 150 Mio people. The region is made up from 8 different countries or parts thereof (Afghanistan, Bangladesh, Bhutan, China, India, Myanmar, Nepal, Pakistan). Most of the area is sparsely populated and, due to the very limited agricultural and industrial potentials, plagued by rampant poverty. Due to the topographic difficulties and the remoteness from bigger population centres, the infrastructure is weak as well.

There is an increasing number of *trans-boundary concerns* in the region: for instance the issues of environmental degradation, such as forest depletion and soil erosion, but also poverty and migration, are increasingly being recognised to be of a regional rather than purely national domain. Moreover, there are some proven and more suspected highland - lowland interactions (such as the causes of flooding in Bangladesh, [1]) which often lead to emotional debates and political tensions in the absence of good, publicly available data and established facts. Finally, there are the questions regarding the utilisation of trans-boundary resources such as rivers, which gather increasing political sensitivity as the resources become scarcer.

Unfortunately, the relations among some of the neighbouring countries in the region are not as good as they could be. Consequently, there is no effective institution to tackle the issues of trans-boundary resources at the political level, like it had been established in other parts of the world (for instance the Mekong Committee, the Rhine Forum, or the Alpine convention, to name but a few).

Considering the political situation, it is not surprising that data on territory and natural resources, like maps or hydrological records, have traditionally been highly sensitive material most of the regional countries. In some of them, large and medium scale maps are still completely off limits despite the fact that we have been down the age of reconnaissance satellites for quite some time. The few bits and pieces that are available are often not comparable from one country to another, or they are extracts from global datasets like the Digital Chart of the World or IGBP's Global Land Cover [2] data. These datasets are typically of a 1:1mio scale and have limited suitability for mountain areas with their high variability of conditions.

Finally, the mountain areas are marginal border zones for the bigger countries like China, India, Pakistan or Bangladesh, and consequently enjoy a relatively low priority on the national political agenda.

To sum it up: A factual demand for an infrastructure of spatial data at medium to small scales (1:100'000 - 1:250'000) on mountain areas in general and the Hindu Kush- Himalayan region in particular can be taken for granted [3]. They would be helpful to investigate issues of trans-boundary resource use and environmental change in a region that, inter alia, is origin of much of the water that sustains about 600 million people.

In the absence of a regional political body, the main users of such an infrastructure will be:

- Scientific institutions like ICIMOD itself, institutions concerned with global climate change, or Universities in the region
- International Organisations like UNEP or FAO
- Donor and development organisations (to target development interventions)
- National Government Institutions (as far as they lack more accurate data)

However, due to the mentioned security and political concerns, the conditions to establish this infrastructure are not favourable.

2 The Role of ICIMOD

Out of widespread concerns about environmental degradation and poverty in the region, the International Centre for Integrated Mountain Development has been established in 1983 as a forum for research, scientific exchange and documentation in and on the region as well as advisory services. This has notably been the first and - to the author's knowledge - so far only institution with an explicit focus on this region. The founding fathers were the development co-operation agencies of Switzerland and Germany, while UNESCO and His Majesty's Government of Nepal stood patron. Over the years, the support has gradually expanded to include about a dozen different donors now. The eight countries of the region are Members of ICIMOD; they delegate representatives into the Board of Governors. Through the Board, but perhaps even more through manifold contacts with scientific institutions and individuals, the centre is firmly anchored in the region. However, it has to be noted that ICIMOD's role is a strictly scientific one, and it has to avoid contentious issues. For instance, ICIMOD should not publish any maps that depict national boundaries to avoid being dragged into one of the many unsettled boundary conflicts between the neighbouring countries.

Internally, ICIMOD is structured into three thematic divisions (Mountain Farming Systems, Mountain Natural Resources, and Mountain Enterprise and Infrastructures). The thematic divisions are supported by a Documentation, Information and Training Service, the Mountain Environment Natural Resources Information Service (MENRIS), and the Administrative, Financial and Logistical Service.

ICIMOD's conceptual approach has gradually evolved from a project- to a programme approach, aiming at stronger thematic and regional integration. This is being reflected in the 4-year Regional Collaborative Programmes that have been taken up in 1995 and 1999.

3 MENRIS - Past and Present

The Mountain Natural Resource Information Service (MENRIS) of ICIMOD has been established in 1991 with initial support from the Asian Development Bank and UNEP- GRID. The objectives have been and continue to be to [4]:

- establish a network of nodal agencies in the regional member countries and serve as a resource centre to them
- develop a database on geomorphology, soils, land use, vegetation and related factors through remote sensing techniques
- develop mountain-specific applications of GIS
- facilitate the application of GIS and RS and the use of the MENRIS database by the nodal agencies for environmental and natural resources planning, management and monitoring
- improve the co-ordination of regional and project-related mapping and the monitoring of projects introduced in the regional member countries by various international organisations

In the first years, the main focus has been on installing the necessary hard- and software and on getting acquainted to the technology in ICIMOD itself. This has been followed by a phase of capacity building in the region: A substantial training programme for professionals has been established gradually and complemented by short seminars for managers and policy-makers, and hard- and software has been supplied concurrently to partner institutions in the region. Further, a series of case studies has served as demonstration examples and for training purposes. More recently, these case studies have started to evolve into support of 'real-world- applications' of GIS in partner institutions.

However, comparatively little has been achieved in terms of a comprehensive geographic database of the region. The case studies referred to above remained essentially patchwork confined to particular project areas, and to date there is little that MENRIS can offer in terms of homogenous regional spatial data which can not be offered by other institutions as well. But there are some rays of hope: Since Nepal has adopted an open data policy earlier, MENRIS has been able to build an extensive digital database of Nepal at 1:250'000 scale. Bangladesh and Bhutan have joined this open data policy recently and it seems that MENRIS will be able to populate the regional spatial data base with data from those countries as well. Perhaps most significantly for the region as a whole are the very encouraging news that have come from India recently: The Government has set up an national task force on information technology which recommended, inter alia, that the Survey of India makes the existing digital topographic data at 1:50'000 and 1:250'000 scales available to the public at no cost and without copyright restrictions. However, the Defence Ministry still has to overcome its reservations [5].

Considering the more mature status of MENRIS itself, the improved institutional capacities in the region, and the gradual easing of data restrictions, it is felt that it is a good time now to earnestly pursue the creation of a regional spatial data infrastructure. Some of the steps that have been envisaged in MENRIS are discussed below, together with some issues of interoperability that will arise in the process. The primary objectives of these activities are:

- to increase the availability and accessibility of relevant geographic data on the region
- to enhance the exchange of geographic information within the region

The activities broadly fall under one of these categories:

- capacity building
- facilitation of data exchange
- generation of regional key datasets

Interoperability of GIS in the Hindu Kush-Himalayan region will by and large mean the facilitation of data exchange, which is no small achievement, given the political and institutional context. Therefore the next section will focus mainly on this activity.

4 Planned MENRIS Activities 1999 - 2002

4.1 Capacity Building

The substantial capacity-building programs that have already been started in the previous programme will continue under the Regional Collaborative Programme

1999-2002 (RCP-2). However, it is hoped that the already existing curricula and training materials and the increasing availability of qualified staff in the partner institutions will gradually ease the ICIMOD's burden in this regard. The increasing prevalence of standard computers in government and academic offices will also gradually reduce the demands to supply such equipment. However, demands for software and special equipment (digitizers, plotters) will remain high.

4.2 Facilitation of Data Exchange

Metadata Server. A substantial amount of geographic information on the Himalayan region has been compiled by many institutions, development co-operation projects, and individual researchers. To date, most of it exists in analogue form, but there is also a growing number of institutions and projects using GIS facilities to compile their own databases. The problem is that this valuable information is hardly accessible, especially after the end of the respective projects. Moreover, it can be extremely cumbersome to retrieve ancillary information; even such basic things as the projection system of a map are often unknown.

To improve the access to existing and new geographic data, MENRIS tries to take a lead to provide metadata services to the user community in- and outside the region. This has also been one of the recommendations of the Space Informatics Seminar 1996 [6] which was held in Kathmandu.

In a first step, it is planned to document all the MENRIS data holdings. In a second phase, other existing data on the region shall be documented as well. This would not mean that ICIMOD actually holds that data, it just provides a pointer to the holding agency. It goes without saying that we are again dependent on the co-operation of our regional partner institutes and the many researchers outside the region. Finally, it is also envisaged to make this catalogue accessible through Internet.

The Metadata server shall document the following types of geographic data:

- interpreted GIS datasets on topography, geology and soils, land cover, hydrography, transportation, administrative units, settlements, socio-economic statistics
- raw and geo-referenced satellite images which have been acquired by MENRIS or one of the partner institutions
- air photographs
- possibly also paper maps on themes as above (not decided yet)

In addition to that, the metadata server shall contain some general reference information like national mapping systems (geodetic datum, projection, sheet indices), satellite frame references, locations of GPS base stations, satellite receiving stations, addresses of institutions, etc.)

However, in order to be included in the metadata server, a dataset should fulfil certain minimal conditions:

- Some degree of comprehensiveness in terms of area coverage and completeness: While it will be difficult to define an exact limit as to what shall be included, it clearly makes no sense to spend time on documenting trials and extremely local datasets.
- The dataset must be accessible, at least under certain conditions that must be spelled out clearly. There is no point in documenting data that will not be released by the holding agency under any circumstances.

The metadata server shall include a menu-driven graphical user interface which allows simple geographical queries like: "what data on landuse exists for my region of interest?" or: "is my area of interest completely covered by a particular satellite scene?" This will be done by providing reference data (administrative boundaries, hydrography, topography) from existing global datasets, and the footprints of the datasets which are documented. The user will be enabled to select the reference data that he wants to display, and this will be filtered according to the currant map scale. Then he can select an area of interest and query the metadata base according to data type (as above), keyword (e.g. geology / landcover / topography etc.), scale, date, etc. and display the metadata of the query results.

Development and Promotion of Standardised Applications. It is also planned to develop a number of standardised applications that can be adopted by institutions in the region. This would improve the comparability of geographic information in the region. The focus will be on applications that produce information relating to phenomena that are a primary concern in large parts of the region:

Land cover mapping and monitoring from satellite images A system to incorporate DTMs, agro-ecological zonations and previous land cover maps into a digital classification shall be developed. A primary requirement to ensure compatibility with other datasets is a precise rectification, including terrain distortions. In addition to the land cover map according to the standards as described in 3.1, the approach should also yield some measure of reliability as output.

Inventorying and monitoring of glacier lakes Glacier lakes, or actually the risk of their sudden outburst (GLOFs), pose a serious threat to many settlements and infrastructures like hydroelectricity plants in the region. The use of satellite images for monitoring them has been demonstrated successfully [7] it is now a matter of standardising the methodology and promoting it.

Biodiversity mapping and monitoring The preservation of the region's richness in biodiversity is a growing concern [8]. Yet there is not much known as to where exactly the 'hot spots' are that deserve particular attention. Some methods to use satellite images to assist in the mapping of biodiversity have been developed elsewhere [9], but need to be adapted to the extreme variability of the Himalayas.

4.3 Development of a Regional Geographic Database

Adaptation of Standards. With regard to the envisaged regional geographic database, standards are required to ensure the compatibility of the data. To save time and effort and to achieve interoperability with the 'outside world', existing standards shall be adopted as far as possible and modified only where absolutely necessary.

Standards are required primarily in three fields:

– Topographic base data

  In order to produce a homogenous topographic database of the region, the individual elements should be clearly defined (e.g. what is considered a highway, what is a secondary road). Such standards already exist and could be adopted without much change. However, one has to bear in mind that the topographic database will be compiled from very different primary sources with their own inherent standards which will not always easily translate, and the various national mapping agencies will show little inclination to change their own standards in the near future. Thus the pragmatic short-term solution will be to find out what is the smallest common denominator. However, a more proactive role can be taken with regard to the purely technical aspects, like data format, digitising accuracy, etc.
  Thus the issues (or inhibitors) of interoperability are largely semantic ones, as far as topographic data standards are concerned. This is not surprising, given the fact the region is composed of a multitude of very different cultures, which attach very different meanings and conceptualisations to seemingly simple features like 'forest' or 'river'.

– Metadata

  The situation looks more promising with regard to metadata standards. Since this is a relatively new topic, there is not much of a legacy to be carried with, and a more forward-looking approach can be taken. In view of existing tools to create metadata entries, a preliminary decision has been made to adopt the FGDC metadata standard [10]. This would also allow interoperation with metadata parsers that are based on the much more confined DIF standard [11]. However, further consultations with the partner institutions will be required, and it has to be accepted that many fields of records on existing geographic data will remain blank, because the relevant information can not be retrieved any more.

– Land cover classification

  The idea here is to develop a land cover classification scheme that can be implemented with reasonable accuracy by digital classification of satellite imagery. Since the envisaged scale is 1:250'000, imagery of medium to low resolution, such as IRS-WIFS or NOAA AVHRR shall be used. The emphasis will be on relatively frequent monitoring of large areas rather than detailed land use mapping for the purpose of planning; hence the precision of the classes will be limited. What is important is 'upward compatibility' to global land cover classifications, like those of IGBP.

Generation of Homogenous Key Datasets. ICIMOD is also trying to establish a homogenous regional spatial database at a scale of 1:250'000. This scale is larger than existing global datasets which are mostly at 1:1mio scale or smaller, and maps of this scale are gradually being released by some of the regional countries. The database shall contain data which play a key role in ICIMOD's and other institution's research on mountain development and mountain environment. Some elements are:

- administrative units (mainly as a base to visualise statistical data)
- transportation network
- hydrography: rivers and hydrological records (as far as available)
- a Digital Elevation Model (DEM) - an essential item in almost all mountain-related GIS and RS applications: Modelling of soil erosion, slope instability, acceptable land use intensity, hydrological flows, but also precise processing of satellite images all require a DEM of suitable resolution.
  The GTOPO30 model of the USGS is currently the only model that covers the whole region. However, NASA/JPL are planning to acquire new InSAR data through a shuttle mission in 1999 and produce a global DEM of 1" resolution [12]. It is understood that a reduced-resolution version (3") will be made available at nominal cost, which will be an enormous benefit to ICIMOD and the region. The resolution of 3" would also allow the computation of derivatives (slope, aspect) at sufficient accuracy for the envisaged scale of 1:250'000.
- Land cover

  The Indian IRS WiFS data of 188 m resolution seems to offer a good compromise between resolution and manageability of the amount of data, and it would fit neatly into the regional database of 250 000 scale. However, the only two spectral bands pose a limit with regard to interpretability.
  First trial classification have indicated that water bodies, snow, barren land, sparse vegetation, forests, rainfed and irrigated agriculture can be differentiated. It is hoped that the use of auxiliary data such as agro-ecological zonations and DEMs can improve the classification. On the other hand, the combination with satellite land cover data will also help to evaluate the agro-ecological zonation.
  Major problems are the availability of satellite images (both technically and logistically), and the sheer impossibility to do any ground truth studies at this scale.

- Inventory of biodiversity and protected areas

- Inventory of glacier lakes

Since the staffing and funding situation of MENRIS is extremely constrained, a collaborative approach has to be taken. This means that most of the actual work to build the databases has to be done by the partner institutions in the region. The standards and standardised methodologies as listed above are expected to help in the process, but the role of ICIMOD will be limited to one of a catalyst. Moreover, it has to be kept in mind that ICIMOD does not have a mandate to enforce anything on the partner institutions - it has to rely on the goodwill of the partners and, to some limited extent, on the 'co-ordinating power of money' it can disburse under various programmes.

Possible applications of these data sets range from priority area selection for social development through policy information on land use, infrastructure, energy and natural resources to climate modelling. Potential users are the thematic divisions within ICIMOD itself, other national and international research institutions, and international organisations like FAO, UNDP, UNEP, with their regional committees, but also NGOs and Donor organisations. Last, but not least, national Governments and sub-national authorities can make use of the data. In order to enable them to do so, it is essential that the data be available at nominal cost.

5 Conclusions and Perspectives

The demand for good, reliable and homogenous geographic data on mountain areas has been established clearly in scientific circles and international organisations. However, in the case of the Hindu Kush - Himalayan region there is at present hardly any corresponding drive from the governing political institutions. The prevailing security concerns have traditionally been inhibiting the creation of and dissemination of geographic data.

As those restrictions are gradually easing, more qualified manpower becomes available, and an increasing number of local and national geographic databases are being built in the region, it is felt that the time is right to start building a regional spatial data infrastructure for the benefit of scientific and international organisations. In the absence of large budgets and grand designs, a pragmatic approach shall be taken by MENRIS/ICIMOD to integrate existing pieces and improve their accessibility. It is envisaged that MENRIS will assume a clearinghouse function for the region mainly by providing metadata services and continued networking of professionals in the region.

A limited number of generic datasets shall be created from remote sensing images.

References

1. HOFER, T. (1998): Floods in Bangladesh: A highland-lowland Interaction?, Geographica Bernensia G48, University of Berne
2. USGS EROS Data Center (1992): Global Land Cover Characterization. http://edcwww.cr.usgs.gov/landdaac/glcc/glcc.html
3. UNCED (1992): Agenda 21, Chapter 13. United Nations Conference on Environment and Development, Rio de Janeiro 1992. http://www.fao.org/waicent/faoinfo/forestry/Mountain/ch13txt.htm
4. PRADHAN, P. & BITTER, P. (1998): Geographic Information Systems and Remote Sensing in the HKH Region. Proceedings "Mountains 2000 and beyond"; International Conference on Sustainable Development of the Hindu Kush-Himalayan Region
5. GIS@DEVELOPMENT (1998): IT Task Force Recommendations, that can make the GIS really happen in the country. GIS@Development, July-August 1998
6. UNCRD (1997): Space Informatics for Mountain Resources Management. Proceedings of the second Space Informatics Seminar for Sustainable Development; United Nations Centre for Regional Development
7. MOOL, P.K. (1995): Glacier lake outburst floods in Nepal. Journal of the Nepal Geological Society; Vol. 11 (Special Issue)
8. ICIMOD (1998): Biodiversity management in the Hindu Kush-Himalayas. Newsletter No. 31
9. MYINT, M. (1996): The use of remote sensing data for inventory on biodiversity of national parks: A case study of Glaungdaw Kathapa National Park in Myanmar. PhD thesis; Asian Institute of Technology
10. FGDC (1998): Content Standard for Digital Geospatial Metadata (version 2.0). FGDC-STD-001-1998, http://fgdc.er.usgs.gov/metadata/contstan.html
11. NASA (1998): Directory Interchange Format (DIF) Formal Syntax Specification v6.0. http://gcmd.nasa.gov/software_docs/dif_syntax_spec.html
12. JPL (1998): Shuttle Radar Topography Mission (SRTM). Jet Propulsion Lab, http://www-radar.jpl.nasa.gov/srtm/

Interoperability for GIScience Education

Karen K. Kemp[1], Derek E. Reeve[2], and D. Ian Heywood[3]

[1] National Center for Geographic Information and Analysis
University of California, Santa Barbara, CA USA
kemp@ncgia.ucsb.edu
[2] Department of Geographical and Environmental Sciences
The University of Huddersfield, UK.
[3] Robert Gordon University
Aberdeen, Scotland

Abstract. The advent of interoperating GISs has many implications for education. While there are certainly important issues to discuss with regards to additions to the curriculum which address the technological and institutional impacts of interoperating GISs, this paper focuses on the theme of interoperability *for* education. Interoperability provides a context for the development of shareable education materials which in turn allow for collaborative education in a field in which rapid technological developments are making it difficult for individual instructors to stay up-to-date with both the science and the related technologies. Such collaborative education initiatives raise many issues, both technical and institutional, but a number of existing projects provide some basis for rapid developments. An international effort to create an infrastructure for the development and distribution of interoperable, shareable GIS education materials is described.

1 Introduction

In December 1997, the National Center for Geographic Information and Analysis (NCGIA) and the Open GIS Consortium (OGC) convened an international conference and workshop on Interoperating Geographic Information Systems (Interop'97). Topics addressed at Interop'97 included the current state of research in related disciplines concerning the technical, semantic, and organizational issues of GIS interoperation; case studies of GIS interoperation; theoretical frameworks for interoperation; and evaluations of alternative approaches [4]. Arising from these discussions about GISystems interoperation was an awareness that interoperation might have important implications for GIS education.

Many of the measures of the success of interoperation identified at Interop '97 are specified as measurable changes in the content of GIS courses. This suggests that GIS education may become an unwitting accomplice in the move to interoperation. However, an alternate view may be that GIS education will become a fortunate beneficiary. The vision of interoperating GISs foresees ubiquitous GIS and the corresponding necessary pervasive spatial thinking and awareness. The same vision also acknowledges that success in interoperability means that there are many things which will no longer need to be learned. How must GIS education change with interoperability?

There are two perspectives to consider in this context: 1) Interoperability and GIS education, and 2) Interoperability for GIS education. While the first of these perspectives is an important growing theme for GIS educators [6], this paper focuses on the second perspective – interoperability *for* GIS education. The motivation for this interest comes from a recognition that GIS educators in the private and public sectors are faced with both an opportunity and a dilemma. As the GIS vendors move to open systems which can be integrated with many traditional operations, the use of spatial data and analysis will become widespread throughout business, government and education. Hence the need for GIScience education is expanding rapidly. However, at the same time, rapid changes are occurring in both GIS technology and the structure of higher education. These shifting foundations make it impossible for individual GIS educators to stay on the leading technological edge where their students need them to be. Collaboration in education is now essential.

Given the urgency of these issues and the need to begin considering the education community's response, an international workshop on Interoperability for GIScience Education (IGE'98) was organized soon after these issues were first discussed at Interop'97. This workshop was held in Soesterberg, The Netherlands on May 18-20, 1998 [8]. This paper examines the issues raised at this meeting and outlines various existing and new activities in the context of these discussions.

2 The Opportunity

GI and its associated technologies are migrating outward from the specialist niche markets in which they have been embedded over the last 20 years [1]. This means that a greater number of individuals are going to need to work with the technology in their everyday lives. Eventually this interfacing will be seamless, as users are able to perform high tech spatial tasks via intuitive interfaces. However, that points lies some time in the future. In the meantime the education community will need to provide a broad based education strategy to deal with this growth in demand.

Many educators in both the public and private sector are already responding to this challenge in their own individual ways by providing:

- Web resources such as the Virtual Geography Department and the NCGIA Core Curricula.
- Flexible education programs such as those provided by distance learning (e.g. UNIGIS),
- Virtual learning centers such as the Western Governors' University and ESRI's Virtual Campus.

3 The Dilemma

However, all of this is being done against a background where:

- A significant percentage of GI knowledge, particularly as it relates to the technology, becomes outdated within less than 6 months.
- New GI products, services and ideas are appearing at a rate beyond any one individual's ability to keep track.
- It is impossible for an individual educator to stay at the technological leading edge in their field and to keep their learning materials up-to-date.
- The model of higher learning is changing from a traditional, one-time-through university education experience to a flexible lifelong learning environment.
- Mature, busy students are demanding effective and efficient learning opportunities.
- Many students are no longer satisfied with the talk and chalk approach to university education.
- Professionally designed education products now compete against traditional one-off materials.
- Central support for traditional education institutions is shrinking while for-profit education institutions are beginning to compete for the growing number of mature students.
- An increase in demand for just-in-time education is apparent in both academic and industry settings.
- Concern is increasing over how the quality of educational GI programs can be maintained in light of decreasing budgets and rising student demand.

The aim of the Soesterberg meeting was to explore how the GI community can work together to develop an Interoperable or Open environment in which educators can exchange resources and add value to these resources for use in their own unique educational settings while at the same time retaining intellectual (and commercial) copyright. Can such an enterprise provide a framework for collaborative education which allows GIS educators to stay on the leading edge of both the technology and the changes happening in higher education? Both technical issues, such as metadata, data formats and technology, and educational/institutional issues related to collaborative education and sharing of resources need to be considered.

4 What does Interoperable Education mean?

While the use of the term interoperability within the context of education may be misleading, we have found that it is a useful shorthand for describing the need for creating materials which are shareable and can have multiple uses in various contexts from traditional instructor-led teaching to independent self-directed learning. Certainly, in most cases, materials will not need to be inter*operable* in a functional sense, but the idea of being able to assemble diverse materials quickly for use in a specific education context does reflect at least one of the objectives of open architectures. In order to make materials interoperable in this sense, some of the development processes of the OpenGIS consortium may be useful. We need to have some understanding of the primary components or building blocks of educational events and we need to have some common ways of describing their properties.

Educational interoperability will not come easily. There are a number of problems with the concept of education interoperability which are common to education in general and others which are specific to GIS.

4.1 The Technological Basis of GIS

Since GIScience is based on a continually evolving technology, technical foundations and even some concepts change rapidly making it difficult to justify investing too many resources in the development of shareable education materials. How can we separate education about the technology from education about the concepts so that elements which do not change do not need to be constantly revised? Can we achieve this by breaking education materials into several smaller components? How small do these components need to be (i.e. what level of granularity is needed)? Additionally, since technology is central, open concepts are relevant for both our education materials and our technology and data. Is there some overlap between open education systems and open GIS technology of which we can take advantage?

4.2 Problems of Localization and Generalization in GIS

In the GIS education context, both concepts and data need to be *localized* in order to address:

- Differences imposed by local and federal institutions and regulations.
- Differing national data models, formats and standards, including semantic variations in classification systems.
- Cultural differences between both different geographic regions and different disciplines.
- Language variations both within and between language groups (i.e. South American Spanish versus European Spanish).

This leads to questions about what can and should be localized and what not. As well, many concepts and general topics such as geocoding and street networks are not easily *generalized* across various geographic regions. Is there some way to separate the general from the specific when preparing education materials for general use? Can this be achieved by separating concepts from context? Can we identify a level of granularity which achieves this?

4.3 The Multidisciplinary Nature of GIS

The way in which GIS is used differs considerably across disciplines. Thus, what needs to be learned also varies. As well, although GIS is assumed to have almost universal application, there is a general lack of spatial literacy. Is it possible to determine the fundamental core needed and to teach it broadly and generically across all disciplines?

4.4 The International Character of GIS

Given that there are only a handful of major GISystems used worldwide and that international standards for open systems and data exchange are currently being developed, the potential for materials developed by any instructor to be useful to colleagues around the world is quite high. As a result the sharing of education materials is already a well established state of affairs in the GIS education community [9]. This has moved the community to a critical stage at which it is now essential to identify and address education interoperability problems and issues.

4.5 Institutional Issues for Education Generally

While we would like to be able to share materials internationally, different incentive models for contribution create barriers to the type of international collaborative projects needed to make interoperable education work. At a minimum, the need for intellectual property protection and for financial return on investment of time vary considerably between the US and Europe. These models need to be clearly specified so that these differences can be accounted for when planning and conducting collaborative projects. In addition, as education materials are developed by various kinds of institutions, both private and public, mechanisms for promoting collaboration while providing for financial transactions between them are needed.

A further institutional issue relates to shifting education paradigms [2]. A large repository of on-line interoperable education materials provides an opportunity to move from "just-in-case" to "just-in-time" to "just-for-you" education, but not all educational institutions are prepared for these kinds of delivery mechanisms.

Finally, questions of granularity are not only technological, but they also need to be discussed at the institutional level. What is the appropriate level of interoperability from the institutional perspective? Should it be at the course level, the unit level, the exercise level or the component level? How can several different institutions benefit from shared instructional enterprises? Are different interoperable mechanisms needed for each level?

4.6 International Issues for Education Generally

Since GIS is international in character, it follows that any interoperable education activities need to account for international differences in education in general. These range from the obvious problem of language differences to more subtle issues of different education styles. Attention will need to be given to appropriate infrastructures and components for educational materials so that they will suit educational needs worldwide. Can materials prepared in English simply be translated to other languages? Are there vocabularies and/or dictionaries for GIS technical terms in all languages? Are there differences in how educational experiences should be structured in other regions? How might these differences be accounted for in the definition of education components?

5 Some Solutions - Materials Development

There is an urgent need for collaborative efforts and mechanisms which will assist in the discovery and use of the diverse but relevant educational materials already available on the web. Fortunately, the concept of interoperability for GIS education does not exist in a vacuum. A number of major and important projects are currently underway which provide significant foundations for GIS education interoperability. This and the next section provide brief descriptions of these projects within the context of interoperability for GIS education. The first section considers collaborative projects which are now developing shareable materials. The next section discusses some projects which are providing infrastructures for collaboration.

5.1 The Virtual Geography Department Project

The Virtual Geography Department (VGD) Project was begun in 1995 at the University of Texas Austin under the leadership of Professor Kenneth Foote with three years of funding from the National Science Foundation (NSF). It is an excellent example of collaboration in the development of education materials. While the topics range across the full spectrum of geographical inquiry, GIS materials are a major component of this resource. It is freely accessible via the web at `http://www.utexas.edu/depts/grg/virtdept/contents.html` [3].

The VGD provides a web-based clearinghouse of learning materials. Extended summer workshops have lead to the development of a common framework for the format and design of these materials. Stress is placed on the integration of curriculum through the creation and presentation of a range of teaching materials, including on-line course syllabi, texts, exercises, fieldwork activities and resource materials. While existing materials are linked through this framework, the development of new materials is encouraged. Materials are packaged using a standardized cover page including an abstract, table of contents, facts of publication and instructor's notes. This arrangement is particularly useful for the sharing of short, ephemeral materials which would not otherwise be published or distributed outside a single university department.

Within the spectrum of shareable materials development, the VGD sits at the extreme altruistic end. Contributors receive some limited recognition for their work and an opportunity to distribute useful materials more widely, but there is no monetary return for their effort. In fact, most of the contributors have been participants in the summer workshops where they learned how to put educational materials on the web. Their development of materials for the clearinghouse is thus an exercise in implementing that new knowledge. The ability of this project to survive past the end of the original project funding will demonstrate whether no-cost services such as this can be viable over the long-term.

5.2 The NCGIA Core Curricula

Like the original *NCGIA Core Curriculum in GIS*, the new on-line Core Curriculum in GIScience (GISCC - `http://www.ncgia.ucsb.edu/giscc`) and the

Core Curriculum for Technical Programs (CCTP - `http://www.ncgia.ucsb.edu/cctp`) currently under development will each be composed of over 50 units of materials organized as lecture notes, instructor's notes and supporting materials [7]. All materials are freely available on the web with development supported by base funding of the NCGIA in the case of the GISCC and with funding from NSF for the CCTP. In keeping with the spirit and success of the original Core Curriculum and to meet the same specific need in the GIS education materials market, the new Core Curricula concentrate solely on providing fundamental *course content assistance for educators* – formally as lecture materials, but adaptable for whatever instructional mode each course instructor wishes to use. Thus, as before, they are not comprehensive textbooks for students, nor are the materials designed to be used as distance learning materials. Instructors are encouraged to pick and choose amongst the materials on offer in order to develop courses suited specifically for their own students. Course design and materials presentation remain the responsibility of individual instructors.

Recognizing the need to reward contributors, the editorial procedure for the new GIS Core Curriculum was initially based on a journal metaphor. Each unit was to be overseen by a section editor, reviewed by peers and revised accordingly before being posted to the website. Authorship is clearly indicated and the format for citations given at the end of each unit. This procedure was put in place specifically to provide a strong academic incentive for contribution. Unfortunately, the incentives of citations and refereed publication have not proven strong enough to move commitments to prepare units to the top of most pledged authors' to-do lists. It was hoped that the GISCC would be fully populated within a year of its formal initiation, but as of June 1998, 2 years later, only 25 of the originally proposed 187 units were publicly posted. As a result, a new editorial procedure has now streamlined the process by removing the unrewarding section editor positions and offering additional rewards to authors in the form of NCGIA publications. By the end of 1998, over half of the units in a revised, shorter collection of units have been prepared.

At a minimum, the materials in the NCGIA's Core Curricula will be significant contributions to the global GIS education materials database. Since the materials are developed and distributed on the web, each unit can be easily tagged with appropriate metadata once specifications are complete. In terms of granularity, having units based on a single classroom session allows considerable flexibility in the organization of topics for a course.

5.3 The UNIPHORM Project

The UNIPHORM Project is funded under the EU PHARE Program in Multi Country Distance Education and has as its objective the development of course materials and of a service to support distance education in Open GIS for professionals. The partner institutions are the UNIGIS sites at Manchester/Huddersfield, Salzburg, Sopron, Bucharest and Debrecen and the PHARE Study Centres at Miskolc and CDOECS, Bucharest. The remit of the project is for the development of course materials at the UNIGIS sites and the subsequent delivery through PHARE study centers. It is yet another example of a collaborative materials development effort.

Since there are many partners involved in this project, materials development is organized around two fundamental components. First, topics are organized as a hierarchical tree. Each leaf or node on the tree contains materials which are developed as a set of PowerPoint slides. Thus flexibility exists in how individual slides can be organized both within a single leaf or node, or within the hierarchy of topics. The smallest interchangeable unit developed is a single slide and the related page of material provided to students.

This template is designed to allow a high level of control amongst several institutions and authors and yet retain a high level of flexibility for course engineering. It imposes some restrictions in the way material has to be structured and presented but the payoff is substantial in terms of managing complex and shifting resources and in providing cheap and effective creation and delivery of courses.

5.4 UNIGIS

UNIGIS is an international network of universities which together offer a postgraduate diploma and MSc in GIS by distance learning methods (`http://www.unigis.org`) Students complete ten modules each of which covers a substantive GIS topic and may elect to complete a research project to qualify for the MSc diploma [5]. The UNIGIS program has been taught from the UK since 1991 and thus its developers have already experienced many of the problems associated here with GIS education interoperability. Some of these include:

The need for a sustainable business model. Much of the material on the web is freely available and there has so far been a laudable ethos that the web is an arena for sharing knowledge. Visitors to the UNIGIS site, however, will find that most of the materials are behind a password which is available only to their students. The reason for this is, quite simply, that UNIGIS is a business. Within the UNIGIS network a system of royalty fees and concept payments allows those sites which originate materials to receive a return for their effort while giving other sites access to materials which they could not themselves have generated. Development of a generic "web market" would also allow UNIGIS to buy-in from other providers modules which cannot be generated internally.

The instability of the web for teaching purposes. While there is a huge amount of material already available for teaching purposes on the web, may problems exist. The quality of material is not guaranteed, there being no equivalent of peer review on the web. The continuing availability of material is not guaranteed – it may be that the great site upon which you've based your lecture may be taken 'off the air' by the site owner tomorrow. The legality of linking materials from web sites into ones own pages sometimes gives one pause for thought – do many academics actually understand the copyright implications of using web material?

The importance of cultural differences. Although GIS is often regarded as a technical subject, it is of course embedded in national and linguistic contexts. At present the UNIGIS materials have been authored primarily in the UK and so non-UK sites have the task of customizing these core materials to fit their local circumstances. This local customization process is not a trivial task.

6 Some Solutions - Infrastructures for Materials Development

6.1 The ESRI Virtual Campus and Knowledge Base

ESRI's Virtual Campus (`http://campus.esri.com`) is the first strong contender in private sector on-line GIS training. As might be expected the materials are very well designed, pleasant to use, very reasonably priced and extremely responsive to the market needs. The Campus was launched in 1997 and according to ESRI sources attracted over 1200 student in the first nine months. Currently the Campus offers several short, interactive on-line training courses in ArcView GIS. Each course is designed in a similar manner and contains well-structured content, examples, exercises and a short multiple-choice exam. The materials have been widely acclaimed and several universities have used or are considering using them as components in traditional campus-based courses (e.g., the Vrije Universiteit Amsterdam included ESRI's introduction to ArcView course in a recent campus course)

More important for the infrastructure of GIS education interoperability is ESRI's Knowledge Base. This is a database of GIS concepts, examples, exercises, and test questions which can be used to build learning situations within the context of their Virtual Campus. Materials within the database are structured in a uniform manner and adhere to a standardized set of component types. Courses can be quickly constructed out of these building blocks by determining what generic components and sequence is needed and then using the Knowledge Base to find and select a module for each specific element. The company plans to contract with third party authors who will create materials for the Knowledge Base. The authoring program will use a business model that will allow external authors to receive royalties when their materials are used in the Virtual Campus.

6.2 Instructional Management Systems (IMS)

The *Instructional Management Systems* Project (`http://www.imsproject.org`) represents a consortium of government, academic and commercial organizations who are developing a set of specifications and prototype software for facilitating the growth and viability of distributed learning on the Internet. Briefly, IMS seeks to provide a complete environment for the management of education materials, learning and administration. While it provides many facilities of general significance, the elements of the project of particular relevance to GIS education interoperability are their metadata specifications and mechanisms for authentication and commerce.

The IMS base metadata specifications have recently been approved by IEEE. IMS metadata properties that describe educational content include: Discipline, Concept, Coverage, Type, Approach, Granularity, Structure, Interaction Quality, Semantic Density, Presentation, Role, Prerequisites, Educational Objectives, Level, Difficulty, Duration. Recognizing that different disciplines will have different needs, the development of discipline-specific schemas and property definitions is encouraged. Given that the GIS education community is extremely international and active, the IMS project team has recognized that GIS provides an opportunity for the rapid, early development of some important demonstration activities.

Other aspects of the IMS project which are of interest to interoperable GIS education include:

- how IMS addresses the issue of intellectual property
- how lineage is recorded in the metadata descriptions
- authentication and how rights of access and use will be managed
- the commerce model being developed to handle financial transactions
- how IMS may provide assurance of quality through the use of review bodies (similar to the Michelan stars system), usage records and assessment of educational outcomes by external bodies.

From the perspective of the GIS education community, IMS does seem to provide some important immediate solutions for interoperability. In particular, the metadata schema provides a standardized means of describing and cataloguing the vast range of resources already available on the web. With the promised advent of metadata search engines on the web, properly tagged HTML documents could conceivably be discovered anywhere on the Internet. However, there are a number of issues of particular concern to GIS educators which may need special attention:

- There is a need to separate content from infrastructure.
- There is an acknowledged distinction between training and education. Can these share a common set of standards? Is all learning the same?
- There are several layers of interoperability needed: from technological (objects communicating) to semantics to institutional. Likewise, IMS is middleware in which technology is the foundation, policy and institutional matters are above this.
- Given the need for localization in geographic information science, how should learning profiles or educational settings be matched to metadata? Can we establish hierarchical schemas in metadata to address geographical or disciplinary foci?

- What is the appropriate level of granularity given the need for localization? Can we use nested hierarchies in metadata to address this? Can small objects be viable?
- IMS may provide the mechanisms needed to address the incentive problems.

6.3 The Open GIS Consortium (OGC)

Although OpenGIS specifications are not designed to meet instructional needs in particular, the inclusion of functional geoprocessing components in instructional materials points at the need to ensure that GIS educators who are developing interoperable education materials consider these new geoprocessing specifications during their materials development.

Like IMS, OGC provides a proven model for the development of community-wide specifications. Certainly an Education SIG in OGC would provide a vehicle for discussions of geoprocessing interoperability as it applies to education, however, the need for such domain specific geoprocessing specifications is not clear. On the other hand, there are some education needs which relate to OpenGIS. Merging interoperable educational services (from IMS) with interoperable GI services (from OGC) seems doable now. Interoperability in education is pretty much the same across domains, and GIS interoperability is not different for educational purposes. But in order to make products appear, both sides need to be aware of each other, providing input to IMS metadata definitions or OGC topics, and helping to explore and define business models.

7 What's missing? What do we need?

While each of these projects presents effective responses to various education problems, when considered across the spectrum of issues identified earlier, there remain unresolved problems. Still needed are:

- A clear picture of the various incentive models which can be used to encourage participation in collaborative projects supporting interoperability for education and which will lead to long-term sustainability of such activities.
- Identification of the appropriate granularity and specification of the range of educational component types (e.g. exams, units, concept modules, exercises, applets)
- Models for the development of shareable GIS educational materials which address the issues of generalization, localization and technological change.
- A fully functioning prototype of a database of education components which can be combined into various types of educational "events".
- Mechanisms for structuring shareable education components.
- The creation of a huge volume of relevant metadata associated with education materials already on-line.
- Attention to the critical but still unexplored issue of language differences in the context of shareable GIS education materials.

- Mechanisms for quality control.
- Digital "wizards" which would assist in the creation of metadata and the construction of education "events".
- A change in the education system which will support new education paradigms involving collaborative activities by educators.
- Dissemination of information about active projects and mechanisms which support collaborative and interoperable education.
- Information included in metadata which describe the context of shareable education objects, i.e. how is it used, what is the audience, what do people think who have used it?

8 Next Steps

The technology to allow the delivery of interoperable educational objects via the web will become available very soon. Furthermore the support for on-line learning and collaborative teaching from governments and significant higher education bodies is such that it is inevitable that such programs will expand dramatically in the short-term. Thus we should work to ensure that GIScience educators are as well placed as possible to take advantage of the opportunities, and avoid the pitfalls, which the shift towards on-line and collaborative teaching will generate. Three areas for concerted effort by the GI education community have been identified for immediate action.

8.1 Metadata for GIScience Education Materials

A major concern must be to ensure that the interests of GIScience are strongly represented in the super-disciplinary projects which are presently laying down the ground-rules for collaborative, on-line education interoperability. Just as the OGC are presently acting as a lobbying and technical development group to ensure that 'geography' is properly accommodated within emerging distributed computer environments, so too GIScience educators need to lobby to ensure that GIScience is properly represented in emerging on-line and collaborative educational initiatives. In particular, an international task force of GIS educators will soon begin work with the IMS project team to examine and refine where necessary the generic IMS metadata specifications. These extended specifications will be reviewed by the international community and, it is hoped, will lead to the preparation of structured metadata for large quantities of education materials already available on-line.

8.2 A Prototype Knowledge Base for GIScience Education Materials

Clearly, it is desirable to generate as quickly as possible a prototype GIScience knowledge base in order to learn what workloads are involved in creating such a structure and to create an exemplar which can be used at conferences and workshops to generate wider awareness. The potential of the work already done by the ESRI knowledge base project is impressive, thus, rather than create an entirely new structure, efforts are now underway to test the adaptability of this structure designed for proprietary training materials to its use with generic education materials "outside the ESRI firewall".

8.3 Incentive Models for Interoperable GIS Education

The move towards on-line and collaborative GIScience education should not be viewed purely, or even primarily, as a technical issue. For such GIScience education to be successful, academics will need to feel that it is worthwhile to spend their time creating shareable education objects. Higher Education institutions will need to be able to see how revenue might be generated from collaborative teaching initiatives. In other words, the *incentives* which might make the development and use of shareable GI education resources take-off need to be explored. A third task, therefore, is to research the motivations which lie behind current on-line resources initiatives and to try to anticipate what incentives might be necessary in future to encourage GIScience interoperable education projects to develop strongly.

9 Conclusion

The concept of interoperability as it pertains to making GIS education resources shareable and easily accessible is a goal worth pursing. Following the Soesterberg meeting, a small working group continues to advance work on these tasks. Progress on some, if not all, of these tasks will be observed through 1999. The involvement of the international GIS education community will be pursued. However, while a global effort might be possible, at a minimum, general understanding of a concept of interoperable GIS education and the specification of models for the development of appropriate education "objects" will make sharing our global education resources more feasible and productive.

10 Acknowledgements

The authors wish to acknowledge the contribution of all the participants at the Soesterberg meeting whose lively discussions and thoughtful responses provided the basis of what is presented in this paper. Partial funding for the meeting and subsequent development of this paper was provided by the National Science Foundation. Support for the meeting was also provided by the UNIGIS Consortium, Hewlett-Packard Netherlands and the Vrije Universiteit Amsterdam.

References

1. anonymous (1998). Industry Outlook '99; GIS melts into IT. *GEOWorld*. 11: 40–49.

2. Denning, P. (1996). Business Designs for the New University. *Educom Review* 31(6): 20–30.
3. Foote, K. E. (1997). The Geographer's Craft: Teaching GIS in the Web. *Transactions in GIS* 2(2): 137–150.
4. Goodchild, M. F., M. J. Egenhofer, R. Fegeas and C. A. Kottmann, eds. (1999). *Interoperating Geographic Information Systems.* New York, Kluwer.
5. Heywood, D. I., S. C. Cornelius and P. H. Cremers (1998). Developing a virtual campus for UNIGIS: an international distance learning programme for geographical information professional. In anonymous, ed. *Bringing Information Technology into Education.* Dortecht, Kluwer.
6. Heywood, D. I., K. K. Kemp and D. E. Reeve (1999). Interoperable education for interoperable GIS. In M. F. Goodchild, M. J. Egenhofer, R. Fegeas and C. A. Kottmann, eds. *Interoperating Geographic Information Systems.* New York, Kluwer: 443–458.
7. Kemp, K. K. (1997). The NCGIA Core Curricula in GIS and Remote Sensing. *Transactions in GIS* 2(2): 181–190.
8. Kemp, K. K., D. E. Reeve and D. I. Heywood (1998). *Report of the International Workshop on Interoperability for GIScience Education, IGE '98*, Soesterberg, The Netherlands, May 18–20, 1998. National Center for Geographic Information and Analysis, University of California Santa Barbara.
9. Kemp, K. K. and D. J. Unwin (1997). Guest Editorial. From geographic information systems to geographic information studies: An agenda for educators. *Transactions in GIS* 2(2): 103–109.
10. Miller, W. 1998. Personal communication, May 1998.

Adding an Interoperable Server Interface to a Spatial Database: Implementation Experiences with OpenMap™*

Charles B. Cranston[1], Frantisek Brabec[1], Gísli R. Hjaltason[1], Douglas Nebert[2], and Hanan Samet[1]

[1] University of Maryland, College Park, MD 20742, USA, {zben,brabec,grh,hjs}@cs.umd.edu
[2] U.S. Geological Survey, Reston, VA 22092, ddnebert@usgs.gov

Abstract. Many organizations require geographic data originating from diverse sources in their day-to-day operations. It is often impractical to maintain on-site a complete database, due to issues of ownership, the sheer size of the data, or its dynamic nature. OpenMap™ is a distributed mapping system that allows displaying together geographic data acquired from disparate data sources. In this paper, we report our experiences with building a "specialist" for OpenMap, allowing the OpenMap map browser access to data stored in SAND, a prototype spatial database system. DLG data from the U.S. Geological Survey were used to demonstrate the combined system. Key features of the OpenMap and SAND systems are described, as well as how they deal with the DLG data.

1 Introduction

Geographic data is being digitized at an ever increasing rate. In many cases this is driven by the day-to-day needs of the collecting entities: the public utilities whose paper maps are becoming frayed and unusable, various ecological and scientific entities who use the data in furtherance of their primary missions, or even mundane land-ownership tracking by local governments. In other cases new technologies, such as space shuttle imaging radar, are making digitized spatial information available almost faster than it can be recorded.

However, the Earth is very complex, far too complex to digitize in its entirety. Of necessity, it is *abstractions* of reality that are captured. In this process some details are inevitably lost. Moreover, each digitizing interest community has its own idea of which information it is important to retain. For example, given a river, the ecology community may be interested in the number of small frogs per kilometer of bank, a transportation agency may be more interested in the positions of current and future bridge crossing sites, while the Defense

* This work was supported in part by the U.S. Geological Survey under contract 1434HQ97SA00919, by the National Science Foundation under grant IRI-9712715, and the Department of Energy under Contract DEFG0295ER25237.

department may be more interested in knowing at what points the river can be forded by an M1 Abrams tank.

Because of these differing views, data digitized by one organization may not be easily shared with other dissimilar organizations, unless some method for interoperation can be devised. Yet such sharing is not only desirable but mandatory, as decision makers simply will not pay for essentially duplicative digitization efforts. Occasions arise when data simply *must* be shared. For example:

> An emergency response team requires the synthesis of a map that includes geological, soil properties, road network, water lines, demographic information, and public service facility locations such as hospitals and schools to be plotted for an urban area just impacted by an earthquake. No single agency is responsible for this variety of geospatial data yet "best-available" information must be assembled and printed for use by field personnel in paper and electronic form within 6 hours [9].

The OpenGIS (Open Geographical Information Systems) Consortium [4] is an open, industry-wide consortium of GIS vendors and users who are attempting to facilitate interoperability by proposing standards for GIS knowledge interchange. The goal of the consortium is to enable transparent interworking within any one community of interest, and to provide a framework for explicit conversion procedures when data is to be shared between differing interest communities. The consortium was founded relatively recently, so the standardization effort is still in its early stages. Nevertheless, a specification of an object model for GIS data has been approved by its members. The specification, called OpenGIS Simple Features, is in three variants, each tailored to a specific transport mechanism: ODBC/SQL92 (Open Database Connect/Structure Query Language) [12], Microsoft OLE/COM (Object Linking and Embedding/Component Object Model) [14], and CORBA (Common Object Request Broker Architecture) [13].

OpenMap™ is a product suite developed by BBN Technologies (now a division of GTE Internetworking) in a DARPA-sponsored project to demonstrate CORBA-based mapping. Whereas OpenGIS's Simple Features specification addresses the interface between a GIS database and a GIS application, OpenMap™ specifies an interface between a GIS application and its user interface (UI). OpenMap includes a user interface client, a client/server interface (implemented through CORBA), and a suite of *specialists* that implement the server side of the interface, making a particular kind of data source accessible to the user interface client. Thus, it provides a way to integrate geographical data from diverse data sources in a single map display. OpenMap has been used in technology demonstrations within the OpenGIS community, and its creators have initiated a dialogue concerning the need for an open application/UI interface.

SAND (Spatial And Nonspatial Data) [1] is a prototype spatial database system developed by our group. Its purpose is to be a research vehicle for work in spatial indexing, spatial algorithms, interactive spatial query interfaces, etc. The basic notion of SAND is to extend the traditional relational database paradigm by allowing row attributes to be *spatial* objects (e.g., line segments or polygons), and by allowing spatial indexes (like quad trees) to be built on such attributes, just as traditional indexes (like B-trees) are built on nonspatial attributes.

As part of a demonstration project involving OpenMap^{TM} we built a SAND specialist that makes data stored in SAND accessible to the OpenMap user interface client. In the demonstration, we used a SAND database populated with DLG (Digital Line Graph) data from USGS (U.S. Geological Survey). This paper discusses issues that arose in this work.

The rest of the paper is organized as follows. Section 2 discusses OpenMap^{TM} in more detail. Section 3 describes the SAND database system. Section 4 presents the SAND specialist for OpenMap, while conclusions are drawn in Sect. 5. For the interested reader, we have included in an Appendix a discussion of the issues arising in converting DLG data to SAND's native format.

2 OpenMap^{TM} and OpenGIS

A GIS system can be viewed as being divided into three tiers (see Fig. 1), the UI (User Interface), Application, and Database tiers. In the Database tier we have databases storing actual GIS data, in the Application tier we have applications that query the databases and process the result in some manner, and in the UI tier we have the graphical user interface where the query result is displayed to the user. The OpenGIS Simple Features specification addresses the interface between the Application and Database tiers (as well as between applications). OpenMap, on the other hand, specifies an interface between the UI and Application tiers. This interface is based on CORBA (Common Object Request Broker Architecture) [11], an industry standard middleware layer based on the remote-object-invocation paradigm. By *middleware* we mean shared software layers that support communication between applications, thereby hopefully achieving platform independence. Such a "layering" organizational paradigm has been extremely successful in networked computer communications (for example, FTP over TCP over IP over Ethernet). The recent adoption of the IIOP (Internet Inter-ORB Protocol) standardizes CORBA interoperation down to the TCP/IP protocol layer. Thus *any* two CORBA applications should be able to interwork.

The central component of OpenMap is the OpenMap Browser, its user interface client. It includes a map viewing area, navigation controls, and a layers palette, in addition to menus and a tool bar. A simplified version of the OpenMap Browser was implemented in Java (see Fig. 2), and a variant of it can be deployed on any Java enabled Web browser. The layer palette lists map layers available to the client. A map layer is a collection of related geographic objects, i.e., road network, railroad tracks or country boundaries. The layers come from data servers, termed specialists, that communicate with the OpenMap Browser using CORBA. The interface specification between specialists and the OpenMap Browser allows the Browser to request data objects intersecting a query rect-

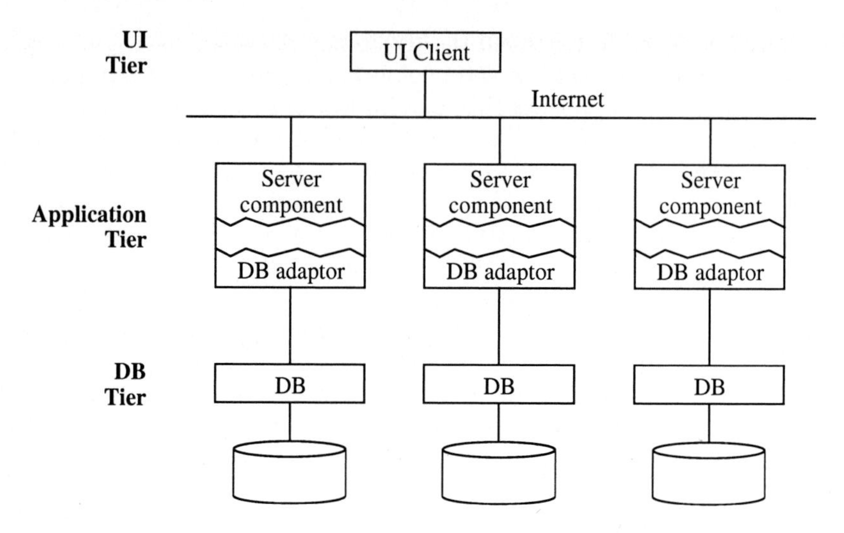

Fig. 1. A three tier view of a distributed GIS system

angle, where the data objects are graphical objects of various types, including line segments, circles, rectangles, polylines/polygons, raster images, and text. These can be specified either in lat/long coordinates or in screen coordinates. In addition, the interface provides support for custom palettes that allow the user to configure the specialist, and support for *gestures*, which allow the specialist to respond to mouse actions on the displayed graphics.

Specialists can be implemented either in C++ or Java. Among the components of OpenMap are classes for each language (called CorbaSpecialist and Specialist, respective) that encapsulate the common aspects of all specialists [2]. A custom data server can be created by extending these classes, adding only the specialized routines required to access a particular target database. Details of CORBA and session initialization, transfer of query rectangles from client to server, and transfer of GIS feature information from server to client are handled transparently.

3 SAND

SAND [6, 7], the spatial database system developed by our research group, is divided into two main layers, the SAND kernel, and the SAND interpreter. The SAND kernel was built in an object oriented fashion (using C++) and comprises a collection of classes (i.e., object types) and class hierarchies that encapsulate the various components. Since SAND adopts a data model inspired by the relational model, its core functionality is defined by the different types of tables and attributes it supports. Thus, the table and attribute class hierarchies are among the most important. The SAND interpreter provides a low-level procedural query interface to the functionality defined by the SAND kernel. Using

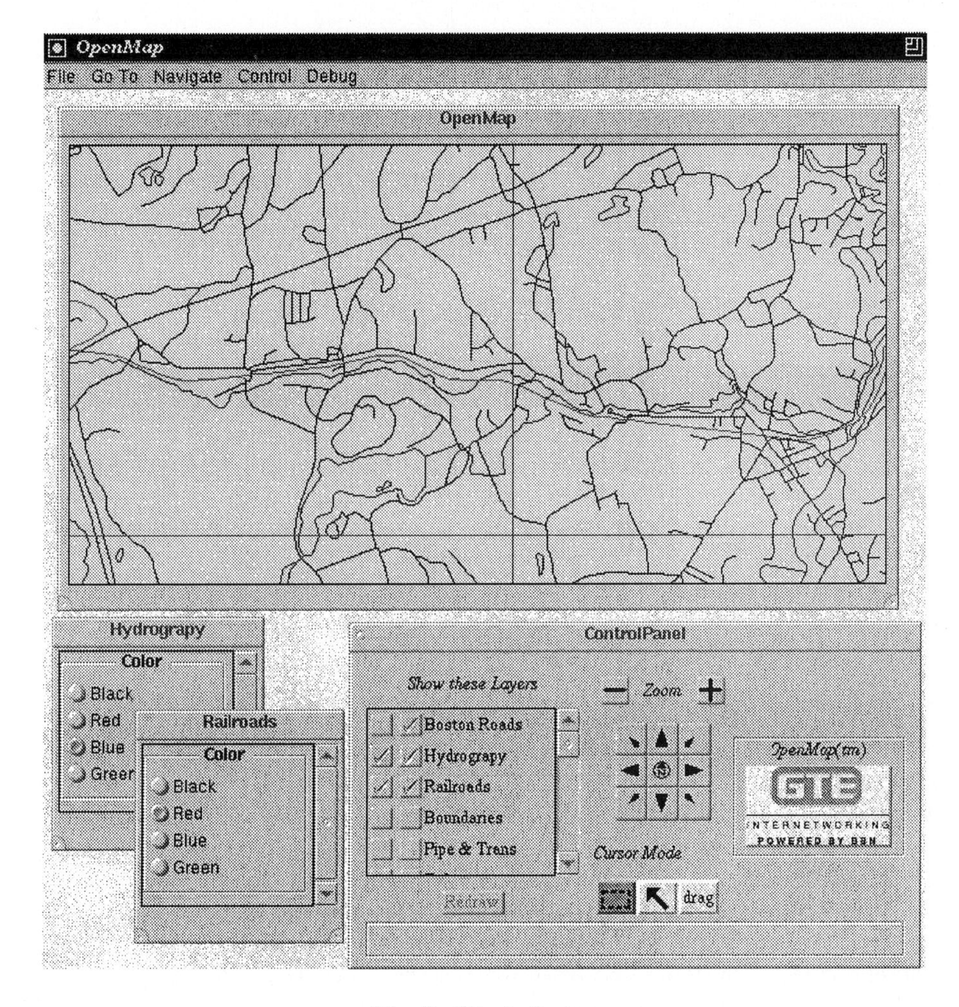

Fig. 2. Client display

the query interface provided by the SAND interpreter, we have built a number of useful tools. For example, the SAND Browser is an interactive spatial query browser, that allows the user to pose queries through graphical input. Also, we have built a prototype for a high-level declarative query interface to SAND, modeled on SQL.

3.1 Table Types

The table abstraction in SAND encapsulates what in conventional databases are known as relations and indexes. Tables are handled in much the same way as regular disk files, i.e., they have to be opened so that input and output to disk storage can take place. All open tables in SAND respond to a common set of operations, such as **first**, **next**, **insert**, and **delete**.

SAND currently defines three table types: relations, linear indexes and spatial indexes. Each table type supports an additional set of operations, specific to its functionality. The function of most of these operations is to alter the order in which tuples are retrieved, i.e., the behavior of **first** and **next**.

Relations in SAND are tables which support direct access by tuple identifier (*tid*). Ordinarily, tuples are retrieved in order of increasing *tid*, but the operation **goto** *tid* can be used to jump to the tuple associated with the given *tid* (if it exists).

Linear indexes for non-spatial attributes are implemented using B-trees [5]. Tuples in a linear index are always scanned in an order determined by a total ordering function. Linear indexes support the **find** operator, which retrieves from the disk storage the tuple that most closely matches a tuple value given as an argument. The **find** operator can also be used to perform range searches.

Spatial indexes are implemented using PMR-quadtrees [10, 16, 17]. They support a variety of spatial search operators, such as **intersect** for searching tuples that intersect a given feature, or **within** for retrieving tuples in the proximity of a given feature. Spatial indexes also support **ranking** [8], a special kind of search operator whereby tuples are retrieved in order of distance from a given feature.

3.2 Attribute Types

SAND implements attributes of common non-spatial types (integer and floating point numbers, fixed-length and variable-length strings) as well as two-dimensional and three-dimensional geometric types (points, line segments, axes-aligned rectangles, polygons and regions). All attribute types support a common set of operations to convert their values to and from text, to copy values between attributes of compatible types, as well as to compare values for equality. Non-spatial attribute types also support the **compare** operator, which is used to establish a total ordering between values of the same type. This is required so that non-spatial attributes can be used as keys in linear indexes. Spatial attribute types support a variety of geometric operations, including **intersect** which tests whether two features intersect, **distance** which returns the Euclidean distance between two features (used for the **ranking** operator), and **bbox** which returns the smallest axis-aligned rectangle that contains a given feature (i.e., its minimum bounding rectangle). Some spatial types support additional operations. For instance, the *region* type supports operations like **expand**, which can be used to perform morphological operations such as contraction and expansion, and **transform**, which can be used in the computation of set-theoretic operations.

3.3 The SAND Interpreter

The SAND kernel provides the basic functionality needed for storing and processing spatial and non-spatial data. In order to access the functionality of this kernel in a flexible way, we opted to provide an interface to it by means of an interpreted scripting language, *Tcl* [15]. Tcl offers the benefits of an interpreted language but still allows code written in a high-level compiled language (in our case, C++) to be incorporated via a very simple interface mechanism. Another advantage offered by Tcl is that it provides a seamless interface with *Tk* [15], a toolkit for developing graphical user interfaces.

The SAND interpreter provides commands that mirror all kernel operations mentioned in the previous section. In some cases, a single command may cause more than one kernel operation to be performed. In addition, the interpreter implements data definition facilities. The processing of spatial queries is supported by interpreter commands associated with spatial attributes, spatial indexes and bounding structures.

4 SAND Specialist for OpenMap^{TM}

In this section we describe a specialist for OpenMap that provides access to geographic data stored in SAND relations. We implemented this specialist in Java, and thus it is based on the Specialist class provided by BBN. Figure 3 shows the software components of an OpenMap session, where the structure of the SAND specialist is detailed. The user interface client uses CORBA middleware to communicate with various specialists, each of which provides access to a specific type of data source. The SAND specialist code communicates with the UI client with methods inherited from the Specialist class, and in turn invokes the SAND interpreter to perform the actual data access. The SAND specialist responds to requests for objects in a particular map layer intersecting a query rectangle. (In this case, each map layer corresponds to a SAND relation.) In addition, the SAND specialist directs the UI client to display a custom palette for each layer, where the color of the data objects in the layer can be set.

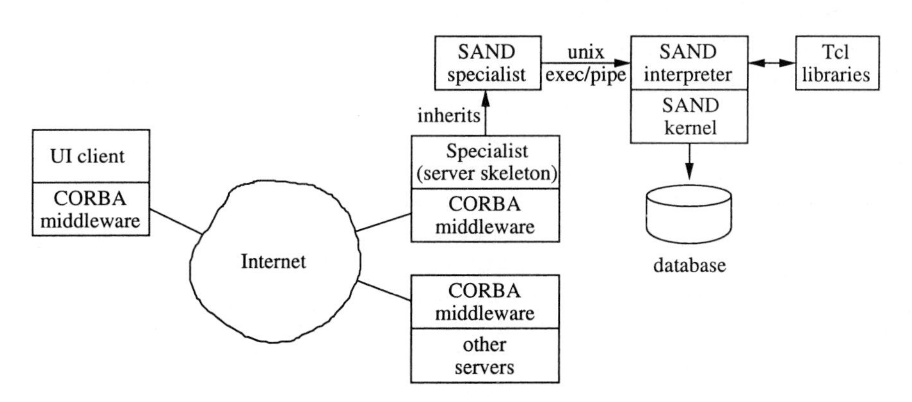

Fig. 3. Structure of the SAND specialist for OpenMap^{TM}

The data made available by the SAND specialist in our demonstration was obtained from USGS and is in the form of points, polylines and polygonal areas (see Sect. 5 for details of how this data was imported into SAND). Since the goal of the demonstration was to show how multiple maps from diverse sources could be overlayed on top of each other, it was undesirable to display filled polygons as they obliterate any map features in lower layers. Thus, we chose to focus on polylines and polygon boundaries, stored as line segments in the SAND relations representing each map layer. (An alternative approach would have been to represent polygonal areas with the SAND polygon attribute type, and convert from polygons to polylines or line segments in the server at run time.) A spatial index was built on the line segment attribute in the map layer relations in order to allow efficient spatial lookup.

4.1 Implementation of the SAND Specialist

The SAND specialist communicates with the SAND interpreter by passing it Tcl scripts that implement specific database queries using the low-level SAND query interface, and receiving back textual output through an I/O pipe. GIS features satisfying the query are translated into OpenMap Specialist objects, which are then passed on to inherited Specialist methods for transport to the client display. The Tcl script that the SAND specialist passes to the SAND interpreter selects a database, opens a spatial index, and then executes a query passing a rectangle as an argument (the only query currently specified in the OpenMap specialist interface is such a rectangle intersection query):

```
sand cd <directory>;
set index [sand open <indexname>];
$index first -intersect \
       {rectangle <left> <bottom> <width> <height>};

while {[$index status]} {
  puts [$index get];
  $index next;
}
$index close
```

The `<directory>` argument to the `sand cd` command specifies the file system directory that holds the database. The `sand open` command returns a handle to an open table (see Sect. 3.1), in this case an index on a spatial attribute. The handle, which corresponds to an underlying C++ spatial index object, can subsequently be used to perform actions on the index. The script initiates a spatial window query by invoking the table command `first` with a query rectangle, which loads the first tuple satisfying the query (if any exists), i.e., a tuple corresponding to a line segment that is intersected by the given rectangle. The table command `get` returns the contents of the current index tuple, in this case storing a line segment. The table command `next` loads the next tuple satisfying the query, or sets the table status to false if none exists. In fact, not only does the SAND kernel return the tuples satisfying the query one-by-one (through the SAND interpreter), but actually executes the query incrementally rather than batch style. The `while` loop outputs the line segment (which is received by the SAND specialist), and fetches another one until all line segments intersecting the query rectangle have been exhausted. At that point, the index file is closed.

This query plan, which makes use of a spatial index, is more efficient than the straightforward one of initiating a sequential scan of a relation, then testing each line segment for intersection with the query rectangle and only outputting those segments that actually intersect it. However, as it turned out in our experiment, the time saved was actually drowned out by communication costs. (We will discuss this at greater length in Sect. 5.)

The line segments as returned over the I/O pipe are of the form:

```
{line <x1> <y1> <x2> <y2>}
```

The SAND specialist parses this format, creates two OpenMap Specialist points, creates a Specialist line between the two points (using either black or a color determined by a control subpanel), and adds the line to the display list for return to the client.

Each coordinate value undergoes four data conversions. The data in the index file is in binary floating format. The SAND interpreter converts this to an ASCII string representation (conversion 1) to return it over the I/O pipe to the SAND specialist. The SAND specialist reads the ASCII string representation from the pipe and converts it back to binary floating (conversion 2). It must do this because Specialist's Point object creator takes binary floating arguments. Presumably Specialist must convert this machine-specific binary floating value into some machine independent wire format (conversion 3). Finally, the display client must convert from the wire format to some display-device specific format (conversion 4) for eventual display.

The Java source file for SAND specialist contains 275 lines of code. Of this, roughly 200 implement the basic server and another 50 implement the palette for control the color of the line segments in a SAND specialist map layer. The rest are comments and housekeeping "import" statements.

4.2 Sample OpenMap^{TM} Session with the SAND Specialist

Figure 2 shows the OpenMap UI client displaying three map layers obtained from the SAND specialist, Hydrography (i.e., rivers, lakes, etc.), Boston Roads, and Railroads. The window on the lower right is a control panel, where the user can select the layers to display as well as pan and zoom on the map display. The layer list on the left side of the control panel has two check boxes for each available layer. The right-hand one selects the layer for display, whereas the left-hand one causes the creation of a custom palette specific to the layer. The Hydrography and Railroads layers each have a palette that allows setting the color of their line segments (the two windows on the lower left). The color for the Hydrography layer is set to blue, and the color for the Railroads layer is set to red. The color of the Roads layer is set to the default color, black, since it does not have a palette associated with it. When the user clicks on the "Redraw" button, a query is sent to each selected server for any geographic objects visible in the display area.

5 Concluding Remarks

Our research agenda includes developing efficient ways to access spatial data using advanced indexing structures and thus one of our goals for the demonstration software was to showcase the speed advantage these make possible compared to the simple sequential scan paradigm commonly used in industry. This desire caused us to prematurely optimize our software, which became a problem during development. For example, we spent too much time worrying about cross-language linkage delay, the delay caused by using an I/O pipe to communicate with the SAND interpreter (as opposed to assembling all the code into one binary), the delay caused by our need to translate each data point between binary floating point and string representation four times, etc.

When the demonstration was finally assembled, we found that the speed advantage gained from our indexed access was completely swamped by the delays in other software components and by the communication cost. In addition, there is an inherent conflict between the incremental nature of our software and the batch nature of other software components in the demo. SAND was designed to be incremental; data are typically released from the query as soon as they become available. On the other hand, the Specialist class (which implements the CORBA communications) seemed to gather the complete data set before transmitting any data to the client, and in fact seemed to do significant processing before doing so. As an indication of this difference, for a typical query result size of a few thousand line segments, SAND would start returning data within five seconds and complete the query within 30 seconds. However, the Specialist class might then spend over two minutes marshalling the results before the client could show the user any feedback to the user's query.

The design decisions we faced in implementing the SAND Specialist were largely those common to most software development projects. We had to decide on an implementation language (Java or C++; we chose Java), and how our code would couple to existing systems (i.e., loosely or tightly coupled; we chose a loosely coupled approach). We needed to develop a data philosophy that would support the desired results, and to make that philosophy work with the data we could easily obtain. In retrospect, many of the problems we faced were due to premature optimization and a failure to test error paths early enough (i.e., errors occuring in various parts of the system weren't apparent, the only visible result being that no data was displayed).

The work described here can be extended in a number of ways. Some directions for future work include the following:

- Currently the SAND specialist invokes the SAND interpreter separately for each query, thereby incurring the overhead of opening and closing an I/O pipe. In order to allow the I/O pipe between the SAND specialist and the SAND interpreter to remain open between queries, we need to invent some explicit synchronization scheme.
- At this time, the non-spatial attributes of the SAND relations are not used by the SAND specialist. We plan to extend the SAND specialist to make

use of these attributes by modifying the display of the spatial features. This could be as simple as feature coloring, or as complicated as automatic legend placement.
- The OpenMap specialist interface supports gesturing thereby allowing the specialist to act on user interface actions. A simple example might be to display (in textual form) the non-spatial attributes of the database tuple closest to a mouse click.
- Extensions to deal with areal (i.e., polygon) data such as choropleths.
- Multi-resolution display: a road drawn as a line segment on a low-resolution display might be better represented at higher resolutions as a long thin rectangle or polygon. An intelligent specialist could make this translation.
- The addition of other interoperable interfaces to SAND, e.g., the OpenGIS SimpleFeatures interface. We expect that most of the lessons learned in implementing the OpenMap^{TM} server will be directly transferrable to work on implementing other interfaces. One such lesson is that one should take a careful look at network communication costs before expending too much effort in optimizing the local query execution of the database server.

References

1. W. G. Aref and H. Samet. Extending a DBMS with spatial operations. In O. Günther and H. J. Schek, editors, *Advances in Spatial Databases — Second Symposium, SSD'91*, pages 299–318, Zurich, Switzerland, August 1991. (Also Springer-Verlag Lecture Notes in Computer Science 525).
2. BBN Corporation. *Designing CORBA(Orbix/VisiBroker) Specialists for BBN's OpenMap*, 1997. Available as `http://javamap.bbn.com/projects/matt/development/specialist.html` on the web.
3. K. J. Boyko, M. A. Domaratz, R. G. Fegeas, H. J. Rossmeissl, and E. L. Usery. An enhanced digital line graph design. U. S. Geological Survey Circular 1048, 1990. (Also see `http://edcwww.cr.usgs.gov/glis/hyper/guide/usgs_dlg`).
4. K. Buehler and L McKee, editors. *The OpenGIS Guide — Introduction to Interoperable Geo-Processing*, Wayland, MA, 1996. OpenGIS Consortium. OGIS TC Document 96-001, available as `http://www.opengis.org/guide` on the web.
5. D. Comer. The ubiquitous B-tree. *ACM Computing Surveys*, 11(2):121–137, June 1979.
6. C. Esperança and H. Samet. Spatial database programming using SAND. In M. J. Kraak and M. Molenaar, editors, *Proceedings of the Seventh International Symposium on Spatial Data Handling*, volume 2, pages A29–A42, Delft, The Netherlands, August 1996. International Geographical Union Comission on Geographic Information Systems, Association for Geographical Information.
7. C. Esperança and H. Samet. An overview of the SAND spatial database system. *Communications of the ACM*, to appear.
8. G. R. Hjaltason and H. Samet. Ranking in spatial databases. In M. J. Egenhofer and J. R. Herring, editors, *Advances in Spatial Databases — Fourth International Symposium, SSD'95*, pages 83–95, Portland, ME, August 1995. (Also Springer-Verlag Lecture Notes in Computer Science 951).
9. Doug Nebert. WWW mapping in a distributed environment: Scenario of visualizing mixed remote data, 1997. Available as `http://www.fgdc.gov/publications/documents/clearinghouse/wwwmap_scenario.html` on the web.

10. R. C. Nelson and H. Samet. A consistent hierarchical representation for vector data. *Computer Graphics*, 20(4):197–206, August 1986. (Also *Proceedings of the SIGGRAPH'86 Conference*, Dallas, August 1986).
11. Object Management Group. *CORBA 2.0/IIOP Specification*, 1997. OMG formal document 97-09-01, available as `http://www.omg.org/corba/c2indx.htm` on the web.
12. Open GIS Consortium, Inc. *Open GIS Simple Features Specification for SQL Revision 1.0*, March 1998. Available as `http://www.opengis.org/public/sfr1/sfsql_rev_1_0.pdf` on the web.
13. Open GIS Consortium, Inc. *OpenGIS Simple Features Specification for CORBA Revision 1.0*, March 1998. Available as `http://www.opengis.org/public/sfr1/sfcorba_rev_1_0.pdf` on the web.
14. Open GIS Consortium, Inc. *OpenGIS Simple Features Specification for OLE/COM Revision 1.0*, March 1998. Available as `http://www.opengis.org/public/sfr1/sfcom_rev_1_0.pdf` on the web.
15. J. K. Ousterhout. *Tcl and the Tk Toolkit.* Addison-Wesley, 1994.
16. H. Samet. *Applications of Spatial Data Structures: Computer Graphics, Image Processing, and GIS.* Addison-Wesley, Reading, MA, 1990.
17. H. Samet. *The Design and Analysis of Spatial Data Structures.* Addison-Wesley, Reading, MA, 1990.

Appendix: Data Conversion – DLG to SAND

As the source for data in our demonstration we used data sets from USGS (U.S. Geological Survey), encoded in the DLG (Digital Line Graph) format [3]. In this section, we briefly describe the DLG format, and present issues that arose in the conversion of DLG data to a format readable by SAND.

Prior to describing the DLG format we point out that a DLG map of the same geographic area is divided into several layers. Each layer is represented in a separate DLG file:

- Hydrography: flowing water, standing water, and wetlands.
- Roads and trails.
- Railroads.
- Pipelines, transmission lines, and miscellaneous transportation.
- Hypsography: contours and supplementary spot elevations.
- Boundaries: state, county, city, and other national and state lands such as forests and parks.
- Public Land Survey System including town-ship, range, and section information.

Typically each of these layers is displayed in a different color in order for it to be easier to tell them apart. Since the DLG files are distributed in datasets covering a rather small region (or .5° square for 1:100,000-scale DLG maps), we merged several datasets together to represent a larger area.

The DLG format encodes information about geographic features, in the form of points, polylines and polygonal areas, together with associated non-spatial information. It has primarily been used to encode printed cartographic maps into digital form. As the name suggests, it is assumed that the map being represented forms a planar graph. Thus, DLG files are composed of node, line and area identifier elements. A single node is defined by its coordinates and may mark the start or the end of one or more lines or a point feature. Therefore nodes occur where lines intersect and at places on linear features where they are subdivided into separate line segments. A line (corresponding to an edge in the graph) is defined as sequences of line segments, with a node anchored at each end. Lines do not cross over themselves or any other lines in the map. An area is a contiguous region of a map bounded by lines. It is defined by a sequence of line references.

Along with their spatial information, nodes, lines and areas can carry feature codes[1]. These are numerical codes, used to describe the physical and cultural characteristics of the corresponding geographic features. For example, a feature code for an area might identify it to be a lake or swamp, and a feature code for a line might identify a road, railroad, stream, or shoreline. Features in DLG can have any number of feature codes.

In order to make the data stored in DLG files usable by our SAND system, we had to convert the DLG data to SAND's native format, which consists of a set of relations. Each relation in a SAND database contains an arbitrary, but fixed, number of attributes. Usually, only one of these attributes is of a spatial attribute type, but this is not a requirement. For our purposes we decided to focus on the line data provided by the DLG files. Nodes are mostly used to define line segments, and the spatial objects represented by a node type (e.g., wells, tunnel portals) were found to be of limited interest. The areas stored in DLG files are defined as a sequence of lines so each area was exported to the SAND database as such, but provisions were made to indicate that a set of lines originally defined a single area.

Although not currently used in the SAND specialist, we represented the feature codes of each feature in non-spatial attributes in the corresponding relation. We faced the dilemma that in DLG, a feature can have any number of feature codes, whereas the number of attributes in tuples of a given relation is fixed. To solve this, we divided the set of feature codes into three classes, primary, secondary, and the rest. Primary and secondary feature codes are meant to represent the most important characteristics of a feature. They are chosen in such a way as to make it very likely that there will be at most one primary and one secondary feature code for each feature. For example, if the feature is a road, the primary feature code would specify the type of the road (primary route, trail, footbridge, etc.), the secondary feature code would provide additional information (number of lanes, interstate route number, county route etc.) and all other feature codes would be stored together in a third attribute of the relation tuple storing the feature (in tunnel, on bridge, private etc.). Unfortunately, this strategy sometimes fails, i.e., a feature can have more than one primary or secondary feature code; for instance, a certain road segment could be part of several

[1] The usual term for the codes is "attributes". However, since they are a very different concept from "attributes" as used in SAND (to mean fields in tables), we use the alternative term "feature codes" to avoid confusion.

different interstate routes. In this case, only one of the feature codes is stored as the primary or secondary one.

Another issue that had to be resolved was the conversion of coordinates used to define the locations of spatial features. In our demonstration we used DLG files digitized from maps of scale 1:100,000; such DLG files define the location of objects with respect to the Universal Transverse Mercator (UTM) Projection, which is a map projection that preserves angular relationships and scale. However, OpenMap (and some of the spatial functions in SAND) assumes that spatial objects are specified using latitude and longitude coordinates. Thus we had to convert each DLG coordinate pair into latitude/longitude coordinates.

Interoperability in Practice: Problems in Semantic Conversion from Current Technology to OpenGIS

Gilberto Câmara[1], Rogério Thomé[2], Ubirajara Freitas[1], and Antônio Miguel Vieira Monteiro[1]

[1] Image Processing Division - DPI,
National Institute for Space Research - INPE,
P.O.Box 515,
12201-027 São José dos Campos, Brazil
{gilberto, bira, miguel}@dpi.inpe.br
http://www.dpi.inpe.br
[2] Geoambiente Sensoriamento Remoto,
Rua Anésia N. Matarazzo, 60, Vila Rubi,
12245-160 São José dos Campos, Brazil
geoamb@tecsat.com.br

Abstract. This work investigates the practical issue of mapping existing GIS to the OpenGIS standards [1]. We describe the data models used in three systems (MGE, ARC/INFO and SPRING) and analyse the problems involved when mapping them to OpenGIS. Our conclusion is that the OpenGIS standard has not been defined in a formal and unequivocal way, and therefore, there are indefinitions and competing alternatives for mapping existing GIS systems into the proposed standard.

1 Introduction

The issue of interoperability is currently subject to substantial efforts, both from an academical and an industrial perspective. In this issue, academia and industry have taken different, if complementary perspectives: whereas there is a major effort in the industry towards a consensus-based solution [1], researchers have concentrated efforts in theoretical issues, such as abstract models for semantic interoperability [2] [3].

This work tries to bridge the gap between the two approaches, aiming to describe and analyse the practical barriers to interoperability. We start from the pragmatic consideration that most end-users, which will eventually build interoperability frameworks, already have a large geographical database, organised around a commercial system and based on a proprietary data model. These users will probably soon face a decision as regards the introduction of technology which will support the OpenGIS standards, and will probably have a choice

[1] This work has been supported by FAPESP-Fundação de Amparo à Pesquisa no Estado de São Paulo

between different commercial implementations and migration paths from their existing environment. Therefore, in our assessment of interoperabilty in practice, we aim to understand issues such as:

- Is the OpenGIS proposal a truly generic model, which is able to provide semantic equivalents to concepts on existing proprietary data models ?
- What do real-world systems teach us about the problems of semantic interoperability and possible limitations of the OpenGIS approach?
- How effective and easy will be the migration from proprietary frameworks to environments such as OpenGIS?
- What sort of tools would simplify the migration from existing GIS to the OpenGIS framework?
- What lessons can be learned, from the academic perspective to interoperability, from considering the interoperability challenges to today's technology?

In order to address these questions, we have examined three existing GIS solutions: MGE [4], ARC/INFO [5] and SPRING [6]. We have chosen these systems because the first two are representative of existing technology and claim a significant proportion of GIS market share. The choice of SPRING is based on two reasons: this system has been developed by INPE, being therefore well known to the authors, and its data model explicitly supports the abstractions of *fields* and *objects*.

This work is divided in three parts. In Section 2, we briefly examine the semantic models used by MGE, ARC/INFO and SPRING. In Section 3, we describe a possible mapping between these systems and OpenGIS, which could be used in real-world migration to OpenGIS. We conclude with Section 4, where we consider the theoretical and practical consequences of our findings.

2 Semantic Models of Existing Systems

The semantic models of existing systems are a clear demonstration of the barriers faced by the interoperability issue in GIS. In the vast majority of cases, these semantic models have been derived based on practical considerations, mostly related to the data structures used for representing geographical data on a computer.

In order to present the data model of existing systems and of the OpenGIS model, we have used Rumbaugh's OMT diagrams[7], which capture the notions of *specialisation* ("is-a") and *aggregation*("has-a"). In what follows, semantic constructs of the different data models are marked in SMALLCAPS.

2.1 A Generic Reference Model for Geographical Data

Our working hypothesis for comparing the semantic models of different systems is that, given the great differences between them, a generic reference model is necessary to establish a common base into which each system's abstraction will be referred to. The conversion to OpenGIS, therefore, requires two steps: (a) mapping from the system's abstractions to the reference model; (b) conversion from the reference model to OpenGIS.

We will use an abstract formulation as a reference for comparing the semantic model of different systems: the concepts of *fields* and *objects* [8]. The *field model* views the geographical reality as a set of spatial distributions over the geographical space. Features such as topography, vegetation maps and LANDSAT images are modelled as fields. The *object model* represents the world as a surface occupied by discrete, identifiable entities, with a geographical location (with possible multiple geometric representation) and descriptive attributes. Human-built features, such as roads and buildings, are typically modelled as objects. For a more detailed discussion on these issues, the reader should refer to [9] [10] and [11].

In what follows, we will consider the following definitions:

- A *geographical field* (or *geo-field*) is defined by a relation $f = [R, V, \lambda]$, where R is a geographical region, V a set of attributes and $\lambda : R \rightarrow V$ is a mapping between points in R and values in V. Examples of geo-fields include: *thematic geo-fields*, (when V is a finite denumerable set), and *numerical geo-fields*, (when V is the set of real values), corresponding - respectively - to the intuitive notions of thematic maps and digital terrain models.
- Given a set of geographical regions $R_1, \ldots, R_n$ and a set of attributes A_1, $\ldots, A_n$ with domains $D(A_1, \ldots, D(A_n)$, a *geographical object* (or *geo-object*) is defined by a relation $[a_1, \ldots, a_n, S_1, \ldots, S_m]$, where a_i are its descriptive attributes ($a_i \in D(A_i)$) and S_i its geographical locations ($S_i \subseteq R_i$).

2.2 The MGE Data Model

The MGE ("Modular GIS Environment") data model uses three main concepts: CATEGORIES, FEATURE TYPES and FEATURES [4]. A geographic element is represented as a FEATURE. Features are instances of FEATURE TYPES, which may, in turn, be further grouped into CATEGORIES, as shown in Fig. 1. Each FEATURE TYPE is associated to an ATTRIBUTE TABLE.

Objects in MGE are modelled by a two-level hierarchy. An object (FEATURE) is an instance of a FEATURE TYPE, and FEATURE TYPEs can be aggregated into CATEGORIES.

MGE does not include an explicit notion of *fields*. Vector representations of thematic maps use the concepts of CATEGORIES and FEATURE TYPES, with two alternatives: (a) the thematic map may be considered as part of a single FEATURE TYPE (e.g. "Vegetation"), whose values stored as attributes in the ATTRIBUTE TABLE; (b) the thematic map may be considered as a higher-level entity (a CATEGORY such as "Land Cover") and each of its values (thematic classes) is modelled as a different FEATURE TYPE(e.g. "Urban Area", "Forest", "Agriculture"). Raster representations of thematic maps are modelled as REGULAR GRIDS and constitute a separate entity from their vector representation. Digital terrain models are stored separately as TIN or REGULAR GRID, depending on the chosen representation.

In resume,the MGE data model can be considered to be *object-based*, with a *two-level hierarchy*. Geo-fields (such as thematic maps or digital terrain models) are modelled directly through one of their representations.

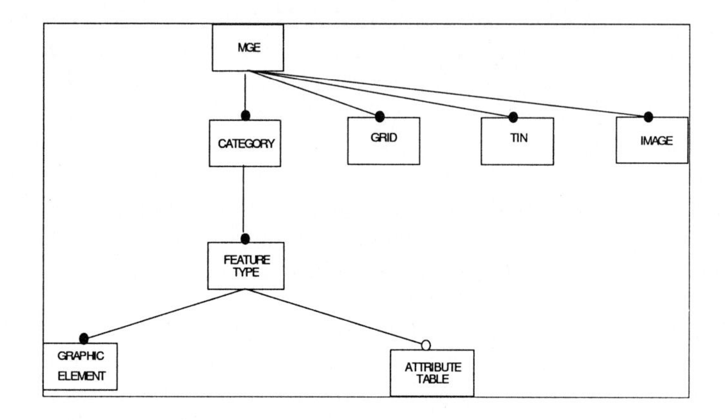

Fig. 1. MGE's Data Model

2.3 The ARC/INFO Data Model

The ARC/INFO data model [5] has four basic abstractions: COVERAGE, GRID, TIN and ATTRIBUTE TABLE. A COVERAGE is a vector representation of geographical data, associated to an ATTRIBUTE TABLE, which describes the map elements (points, arcs or polygons). In this model, the concept of *object* (or *feature*) does not exist explicitly; objects are implemented as rows of the ATTRIBUTE TABLE, which is required to maintain a unique index.

Thematic maps have two possible representations: their vector representation is mapped to a COVERAGE, where one or more fields in the ATTRIBUTE TABLE indicate the attributes associated to each geographical location. The raster representation of a thematic map uses an INTEGER GRID, associated to an ATTRIBUTE TABLE, which indicates, for each value in the grid, the corresponding attributes. Digital terrain models can be mapped either as an "FLOATING-POINT GRID" or as a triangular mesh (TIN).

In resume, the ARC/INFO data model is *representation-oriented*: instead of describing the world in terms of objects and fields, it allows the user to define and manipulate geometrical representations. He will therefore be responsible for externally defining the abstract entities and for mapping those entities to the most appropriate representation. The ARC/INFO data model is shown in Fig. 2.

2.4 The SPRING Data Model

SPRING is a public-domain GIS developed by INPE [6], available on the Internet (`http://www.dpi.inpe.br/spring`), whose data model (shown in Fig. 3) is

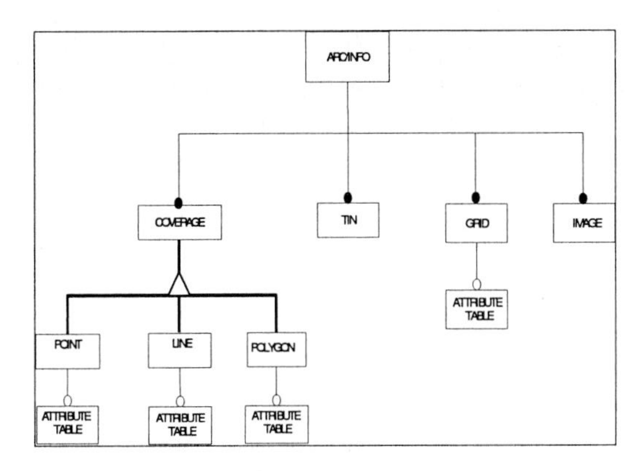

Fig. 2. ARC/INFO's Data Model

based on the abstractions of *fields* and *objects.* SPRING uses the definitions of GEO-FIELD and GEO-OBJECT presented in Sect. 2.1, and allows for two particular types of GEO-FIELDS:

- THEMATIC GEO-FIELDS are fields whose mapping function associates geographical locations to a finite denumerable set, and corresponds to the intuitive notion of a *thematic map.* Examples are soil, vegetation and land use maps.
- NUMERICAL GEO-FIELDS are fields whose mapping function associates geographical locations to the set of real values, corresponding to the intuitive notion of a digital terrain model.

Moreover, the model distinguishes between these abstract definitions and their geometrical representations, since:

- GEO-FIELDS can be associated simultaneously to vector and raster representations. THEMATIC GEO-FIELDS can be represented as a vector (polygon map) or as raster (integer grids). NUMERICAL GEO-FIELDS can be represented as vectors (contour maps, samples or TINs) or in raster format (floating-point grids). In other words, the relation between a GEO-FIELD and its representation is one of aggregation ("has-a") and not a specialisation ("is-a").
- GEO-OBJECTS can be mapped into different geometrical vector representations, with different topologies (polygon maps, networks and point maps). For this purpose, SPRING uses the auxiliary concept of GEO-OBJECT MAP, as explained below.

Since most applications in GIS do not deal with isolated elements in space, it is convenient to store the graphical representation of geo-objects together with its neighbours. For example, the parcels of the same city borough are stored and analysed together. These requirements lead SPRING to introduce the concept of a GEO-OBJECT MAP, which groups together geo-objects for a given cartographic projection and geographical region. Therefore, the geometric representations for *geo-objects* are maintained in instances of the class GEO-OBJECT MAP. The relation between GEO-OBJECTS and GEO-OBJECT MAP is one of "is-represented-by". Use of the model concepts has enabled the design of an user interface and a query and manipulation language for SPRING which allows manipulation of geographical data at an abstract level [12].

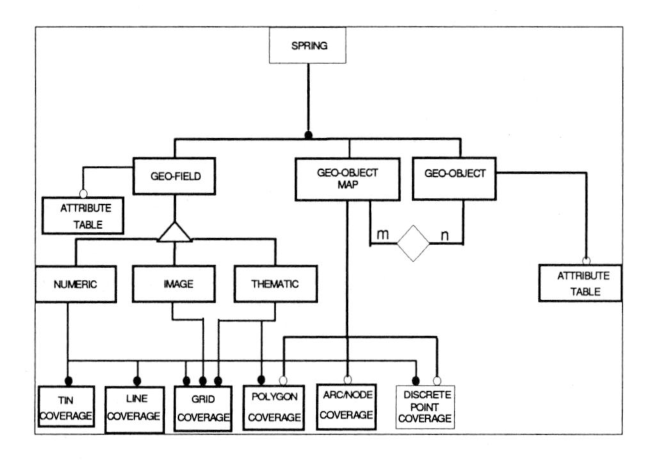

Fig. 3. SPRING's Data Model

3 Mapping into the OpenGIS Semantic Model

3.1 The OpenGIS Semantic Model

The OpenGIS model [1] is based on an abstract class (FEATURE) which has two specialisations: FEATURE WITH GEOMETRY and COVERAGE. The definition of FEATURE WITH GEOMETRY is associated with the idea of *geo-object* (as given in Sect. 2.1) and allows for complex geometrical representations to be associated to the same feature and for different features to share the same geometrical representation. The locational support for each feature is modelled by the concept of SPATIAL REFERENCE, whereby geometry structures (such as lines, points and polygons) which describe the geographical locations of the feature are related to a reference extent in a given projection.

In OpenGIS, COVERAGES are metaphors of phenomena over the Earth's surface, whose spatial domain is a *c-function*, which associates each location a set of attributes. A COVERAGE may be specialised into one of several geometrical representations, including: IMAGE, GRID, LINES and TIN. Fig. 4 illustrates the OpenGIS semantic model.

The OpenGIS specification does not provide a formal definition of FEATURE WITH GEOMETRY and COVERAGE, stating that these concepts represent "alternative ways of representing spatial information" [1]. Therefore, there is no exact equivalence in OpenGIS to the abstractions of *geo-objects and geo-fields*, presented in Sect. 2.1. This choice of industry-established terms, instead of simplifying the migration of existing systems, is likely to cause significant problems in the adoption of OpenGIS in real-life situations, as discussed in later sections.

OpenGIS is an evolving standard and, as of September 1998, the consortium had not published a conclusive definition for the idea of FEATURE COLLECTIONS, that would allow the expression of complex features, such as feature hierarchies.

It is important to note that there is a semantic mismatch between the concept of FEATURE WITH GEOMETRY and COVERAGE in OpenGIS. The definition of FEATURE WITH GEOMETRY is abstract and generic, whose relation to its geometry is one of aggregation ("a feature has many geometric representations"). Coverages are directly related to their geometric representations, by *specialisation* ("a grid coverage is a coverage").

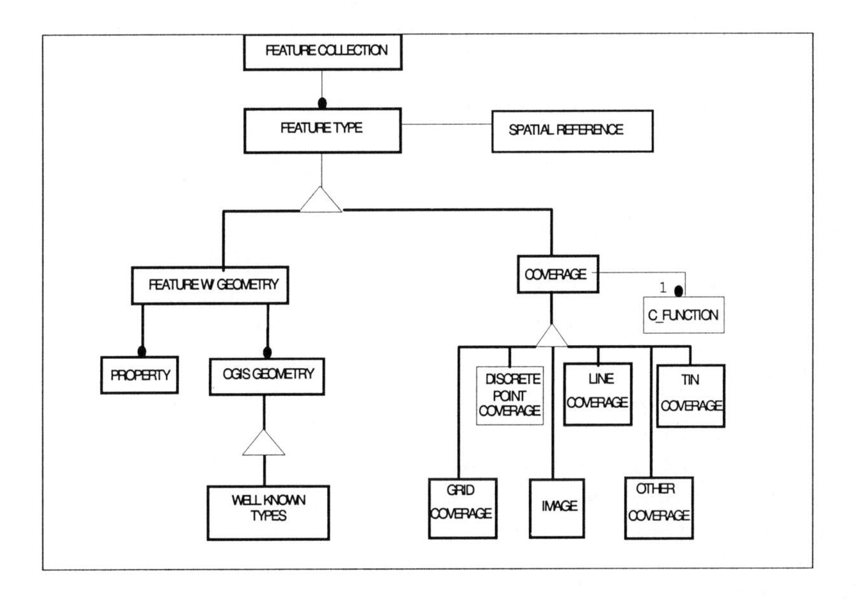

Fig. 4. OpenGIS Data Model

3.2 Mapping Existing Databases into OpenGIS

The scenario envisaged in this paper is a situation where an institution, which has an established geographical data base in a proprietary format, would like to migrate to the OpenGIS model. In this process, they will need to find approximate equivalents in the OpenGIS concept to their semantic models.

One important consideration here is that the mapping should be able to benefit, as much as possible, from the features and tools provided by OpenGIS. This will require an extra abstraction level which is not present in most semantic models of existing systems: that of establishing whether the data represent *objects* or *fields*.

3.3 Mapping MGE into OpenGIS

The mapping of MGE concepts into OpenGIS definitions faces a meaningful issue: the two-level feature hierarchy of MGE semantic model (CATEGORIES and FEATURE TYPES) requires the concept of FEATURE COLLECTIONS in OpenGIS to be fully defined; otherwise, a significant part of MGE's semantic richness will be lost in the translation.

In the case of *objects*, they are defined in MGE using the CATEGORY-FEATURE TYPE hierarchy, which would require an equivalent in OpenGIS, namely, the FEATURE COLLECTIONS-FEATURE WITH GEOMETRY hierarchy.

The issue is further complicated in the case of *thematic maps*, we have an ambigous situation. As discussed in Sect. 2.2, there are two possible ways of mapping vector representations of thematic maps in MGE: (a) using the CATEGORY-FEATURE TYPE hierarchy; (b) collapsing the CATEGORY-FEATURE TYPE notions into a single concept and using the ATTRIBUTE TABLE to store the values of the theme associated to each geographical area. When mapping to OpenGIS, situation (a) requires the abstraction of FEATURE COLLECTION to be defined in OpenGIS for a direct mapping to take place, and situation (b) is best handled by using the GEOMETRY COVERAGE definition in OpenGIS.

Digital terrain models in MGE (represented as TIN or GRID) are mapped in a straightforward fashion to OpenGIS GRID COVERAGE and TIN COVERAGE.

3.4 Mapping ARC/INFO to OpenGIS

When mapping ARC/INFO to OpenGIS, the user will first have to define whether the data being mapped refer to *fields* or to *objects*. Since ARC/INFO does not provide an explicit way of representing objects (they are implemented as unique indexes of the ATTRIBUTE TABLE), such a translation cannot be automatic but will require user intervention.

The issue is further complicated by OpenGIS's choice of the terminology COVERAGE. Some users will be tempted to automatically map an ARC/INFO COVERAGE into an OpenGIS COVERAGE, when in fact these are not exact equivalents. In the case of *objects*, ARC/INFO's COVERAGEs are best mapped into OpenGIS using the FEATURE WITH GEOMETRY concept, in order to benefit from the query functions defined by OpenGIS (which include topological operators).

Thematic maps in ARC/INFO are mapped directly to OpenGIS. Vector representations of such maps (which are ARC/INFO COVERAGEs) can be translated to OpenGIS LINE COVERAGEs, and raster representations (which are ARC/INFO GRIDs) are mapped into OpenGIS GRID COVERAGEs. Digital terrain models are mapped directly to their OpenGIS equivalents (ARC/INFO's GRID and TIN to OpenGIS's TIN COVERAGE and GRID COVERAGE).

3.5 Mapping SPRING to OpenGIS

Since the SPRING data model is based on the abstractions of fields and objects, its mapping into OpenGIS is somewhat simplified. OBJECTs in SPRING correspond directly to FEATURES WITH GEOMETRY in OpenGIS. However, the support for multiple representations in SPRING (with the idea of a GEO-OBJECT MAP) is not fully available in the current OpenGIS specification. In the special case when there is one GEO-OBJECT MAP asssociated to a set of OBJECTs, it can be mapped to the concepts of OPENGIS GEOMETRY and SPATIAL REFERENCE SYSTEM .

In the case of fields, the situation is more complicated. As discussed earlier, the relation of an OpenGIS COVERAGE to its subtypes is one of specialisation. In SPRING, THEMATIC and NUMERIC fields can have multiple representations (raster and vector), a notion which is consistent with the abstract definition of a field. Therefore, one SPRING THEMATIC field will be mapped to many OpenGIS COVERAGEs, depending on how many representations are associated with it. As a consequence, an important part of SPRING's semantic expressiveness will be lost in the process of translation to OpenGIS.

4 Conclusion: Practical Challenges to Interoperability

The main conclusions of the above discussion on the issues of mapping existing systems to OpenGIS are:

- Some systems have semantic models which are richer in content than the OpenGIS one (e.g. CATEGORY-FEATURE CLASS definition in MGE and GEO-FIELD definition in SPRING).
- The use of industry-established terminology in OpenGIS is a mixed blessing. Instead of simplifying the migration process, it may rather be a source of misunderstanding (e.g., the mapping of an ARC/INFO COVERAGE to OpenGIS FEATURES WITH GEOMETRY).

In each case examined, there were different mapping alternatives from the system to OpenGIS, which indicate that automatic migration to OpenGIS is not a recommended option and that a higher level of semantic modelling is needed *before* the actual mapping to OpenGIS takes place. This higher level of modelling would be an abstract description in terms of fields and objects (or a more sophisticated approach), along the lines of [2].

In our opinion, one of the main sources of the problems we have described is the choice of the OpenGIS consortium to use industry terminology, which is already content-rich and are associated by the users with existing semantic concepts (FEATURE from MGE and COVERAGE from ARC/INFO). Had OpenGIS chosen to describe its concepts in more abstract and formal terminology, some of these problems might have been avoided.

Another problem is the semantic mismatch between the notions of FEATURE WITH GEOMETRY and COVERAGE in OpenGIS, the first being an abstract definition and the second, directly linked to its geometric representations. This could be improved if the OpenGIS definition of COVERAGE be changed to an abstract one, where its relation to the representations is one of *aggregation*.

In conclusion, the analysis we have conducted has lead us to believe that there will be major challenges in practice to achieve interoperability, even with the adoption of the extensive work which is being pursued by the OpenGIS consortium. It also calls for theoretical work to be carried out regarding the issue of deriving rich semantic models, which could provide a general framework into which existing semantic models could be mapped, without loss of content and that would allow the later conversion to other models.

References

1. OGC - OpenGIS Consortium: Technical Specifications. [online] `http://www.OpenGIS.org/techno/specs.htm`. March 1998.
2. Yuan, M.: Development of a Global Conceptual Schema for Interoperable Geographic Information Interop'97. Santa Barbara, UCSB, 1997.
3. Gahegan, M.: Accounting for the Semantic Differences between Various Geographic Information Systems. Interop'97. Santa Barbara, UCSB, 1997.
4. INTERGRAPH Corporation: GIS-The MGE Way: An Intergraph White Paper. Huntsville, Ala, Intergraph Corporation, 1995. [online] `http://www.intergraph.com/mge/mgegis.htm`.
5. ESRI: Arc/Info Data Management Concepts, data models, database design, and storage. Redlands, CA, Environmental Systems Research Institute, 1994.
6. Câmara, G., Souza, R.C.M., Freitas, U. M.; Garrido, J.: Spring - Integrating Remote Sensing and GIS by Object-Oriented Data Modelling. Computers and Graphics, vol.20, n.3, 1996.
7. Rumbaugh, J., Blaha, M., Premerlani, W., Eddy, F., Lorensen, B., Lorenson, W.: Object-Oriented Modeling and Design. New York, Prentice-Hall, 1991.
8. Goodchild, M.: Geographical data modeling. Computers and Geosciences,vol.18, n.4, p.401-408, 1992.
9. Burrough,P.A., Frank, A.: Geographic Objects with Indeterminate Boundaries. London, Taylor and Francis, 1996.
10. Couclelis,H.: People manipulate objects (but cultivate fields): beyond the raster-vector debate in GIS. In: Frank,A.U.; Campari, I. and Formentini, U. (eds).: Theories and Methods of Spatial-Temporal Reasoning in Geographic Space. Berlin, Springer, Lecture Notes in Computer Science, n. 639, pp. 65-67, 1992.
11. Câmara, G., Freitas, U., Souza, R.C.M., Casanova, M.A., Hemerly, A., Medeiros, C.B.: A Model to Cultivate Objects and Manipulate Fields. Second ACM Workshop on Advances in Geographic Information Systems. Proceedings, ACM, Gaithersburg, MD., 1994., pp. 20-27.
12. Câmara, G., Freitas, U., Cordeiro, J.P.C., Barbosa, C.C.F., Casanova, M.A.: A Query and Manipulation Language for Integrating Fields And Objects in a GIS. Submitted to International Journal of Geographical Information Systems, 1999.

Building a Prototype OpenGIS® Demonstration from Interoperable GIS Components

Allan Doyle[1], Donald Dietrick[1], Jürgen Ebbinghaus[2], and Peter Ladstätter[2]

[1] GTE Internetworking, Cambridge MA 02138, USA,
[2] SICAD GEOMATICS, D-81739 Munich, Germany

Abstract. A software testbed was constructed by reusing components developed for a previous demonstration and by adding new components. These components were constructed using prototypes of software implementing early versions of the OpenGIS®[1] Simple Features Specifications[2]. The demonstration was presented at the Open GIS Consortium booth at GEObit in Leipzig, Germany in May 1998. The original demonstration from October, 1997 and the demonstration of May 1998 are described. Several issues regarding interoperability, data modelling, data scale, and usability are discussed. Data used in the demonstration include 1:1,000,000 scale Digital Chart of the World and 1:1,000 scale cadastral data for the city of Berlin. Technical issues are discussed and user feedback is presented.

1 Introduction

1.1 Background and Intention of the OGC/FGDC Demo

In August 1997, under the sponsorship of the Open GIS Consortium (OGC), a group of OGC members embarked on an ambitious demonstration [1][3] (which has subsequently become known as the FGDC demonstration, being named after its primary sponsor, the US Federal Geographic Data Committee) of how Open GIS would benefit the construction of distributed spatial applications. The decision to attempt the demonstration was made following the acceptance by the OGC Technical Committee of the OpenGIS®Simple Features Specifications for SQL, CORBA, and OLE/COM [5]. The specifications were accompanied by prototype implementations of the interfaces contained in the specifications. These

[1] OpenGIS is a trademark or service mark, or registered trademark or service mark, of the Open GIS Consortium, Inc. in the US and other countries.

[2] The software used in the demonstrations has not been tested for conformance to the Open GIS Consortium specifications, particularly given the fact that no such tests have been applied to any software at the date of the demonstrations. The specifications as they were available at the time were used in the construction of some of the software in order to provide a practical test of the specifications. Therefore the authors make no claims about actual conformance to current versions of the specifications.

implementations were to be used in the subsequent October FGDC demonstration. The driving force behind the FGDC demonstration was a scenario[3] written by Doug Nebert of the US Federal Geographic Data Committee (FGDC). That scenario called for use of a geospatial clearinghouse to discover spatial data and for the ability to instantly view the data sources without first having to download the data and load them into a local GIS. The demonstration was held in October, 1997 at the GIS/LIS exhibition in Cincinatti, Ohio in the United States.

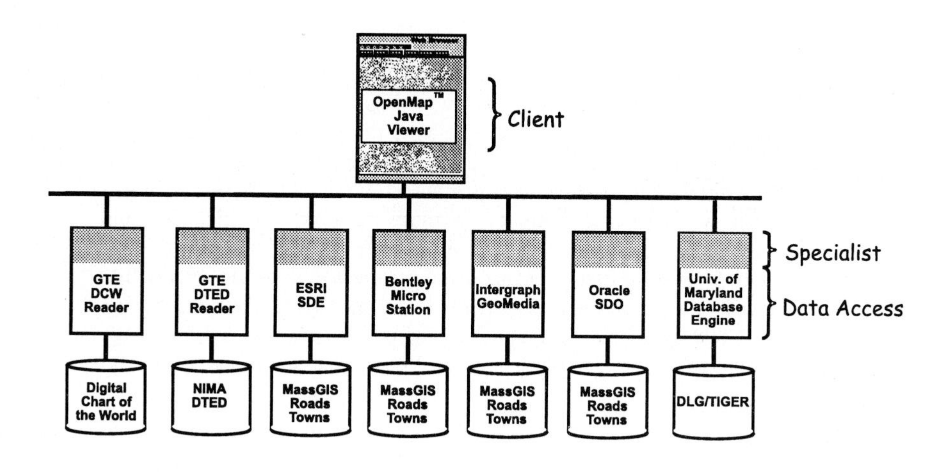

Fig. 1. Architecture and setup of the original FGDC demo.

1.2 Brief Description of the Original FGDC Demo Setup

Given the short amount of time and limited funding available to construct the demonstration, it was decided to use OpenMap^{TM}[4] which has its roots in US Defense Advanced Research Projects Agency (DARPA) research. OpenMap is a middleware system, which at the time provided most of the functionality needed to achieve the stated goals. What was missing were the various OGC vendors' implementations of the Open GIS Simple Features specifications. Four vendors (Bentley Systems, ESRI, Intergraph, and Oracle) each provided software which either implemented the nascent specification directly or which implemented portions of the specification. The vendors have subsequently upgraded much of the software that originally was used as they moved towards Open GIS conformance testing.

[3] `http://www.fgdc.gov/publications/documents/clearinghouse/wwwmap$\_$scenario.html`

[4] OpenMap is a United States trademark of BBN Corporation, a unit of GTE.

Figure 1 is a full component diagram of the original demonstration. Each grouping of modules represents an Internet-accessible service capable of providing graphical representations of the underlying database. OpenMap provided the integration architecture. Each service group is connected to an OpenMap software agent known as a Specialist. Using the OpenMap Specialist protocol[5] developed for DARPA, the OpenMap client is able to connect to any combination of the Specialist/Vendor "stacks". Further discussion of the architecture is provided below.

1.3 Intentions of the GEObit Demo

GEObit is an international trade fair for spatial information technologies and geo- informatics in Leipzig, which was organized for the first time in May 1998. The Open GIS Consortium wanted to use this trade fair as a platform to present Open GIS technology and to collect feedback from European GIS users, particularly from Germany and Eastern Europe. At the same time SICAD Geomatics had started some development efforts to implement the OpenGIS®Simple Features Specification for SQL. This presented an opportunity to extend the FGDC demo with a SICAD based server engine, the SICAD Geospatial Data Server, and data typically used in German GIS applications.

The choice of the data set was an important point. The FGDC demo employed small-scale data with little detail. For the GEObit demo it was decided to use large- scale (1:1000) cadastral data. This kind of data is used not only for cadastral purposes; it is also the base for most GIS applications in German municipalities. The particular dataset used for GEObit was supplied by the city of Berlin, and using this dataset had two advantages. First, the kind of data was well known for most of the GEObit visitors so they were able to view the demo and give qualified feedback. Figure 3 shows a picture of this data as shown in the demostration. Secondly, this was an ideal test case to find out how the OpenGIS®Specifications apply to these data and which difficulties would be encountered. During the preparation work for the demo, it would also be a good chance to gain experience in how to build applications from interoperable GIS-components in a joint project between two companies.

2 System Setup / Architecture

2.1 OpenMap

Rather than being a GIS, OpenMap is a geospatial middleware system, which allows rapid construction of information systems requiring a geospatial component. OpenMap consists of a set of Java components (since the GEObit demonstration, these components have been converted to Java Beans[6]) which assist in

[5] `http://javamap.bbn.com/projects/matt/development/specialist.html` and `http://javamap.bbn.com/projects/matt/development/source/Specialist.idl`

[6] `http://www.javasoft.com:80/beans/docs/spec.html`

the movement of geospatial information from a data source to the user's screen. Inherent in the architecture is the ability to simultaneously connect to multiple data sources, locally or via the Internet. For the FGDC demonstration and the GEObit demonstration, the OpenMap connection to the data sources was via IIOP (Internet Inter-ORB Protocol) [4]. Each OpenMap Specialist (a software agent running on the remote machine containing the data source) implements the same CORBA IDL (Common Object Request Broker Architecture Interface Definition Language). In order to bind a Specialist to a data source a programmer can use object inheritance to specialize the Specialist class. In other words, using either C++ or Java, a programmer can use the code supplied with OpenMap to build the data access agent for any given data source.

At its most basic level, a Specialist must be able to satisfy a single request from the client. This request is known as "GetRectangle." GetRectangle is invoked when the Java client's view of the world changes. When a user zooms or pans the map, or changes the projection, the action results in the client code invoking GetRectangle on the Specialists it is connected to. Paramters of the GetRectangle request include the current projection, the window parameters, and the latitudes and longitudes of the map's corners. Using this information, a Specialist can execute a spatial query into its database, find all the features that fall within the given area, and then use the geometries and attributes of the features to build a list of how the features should look on the map. For example, if a Specialist is connected to a database of roads when it receives a GetRectangle, it can find all the relevant roads, and then decide that the major roads are to be drawn as 3-pixel thick red lines and the minor roads are to be 1-pixel thick blue lines. Each feature's graphics are constructed by the Specialist and added to a list. When all the features have been found, the list is returned via the IIOP connection to the client code.

Then the client code simply looks at the list and, one by one, draws the appropriate graphics on the screen. This works for any kind of data that can be turned into a series of graphic objects. In practice, the GetRectangle call also includes information that allows a Specialist to receive parameters other than just the projection parameters. Query parameters can be handled so that a given Specialist can be directed to constrain the search along non-spatial dimensions.

2.2 SICAD Geospatial Data Server

The SICAD Geospatial Data Server[7] is a commercial software package, developed and distributed by SICAD Geomatics, Munich. Its main purpose is the management and storage of geodata. The SICAD Geospatial Data Server was first released in 1993. It is widely used, practically with every installation of SICAD/open, and now available in its third major release. All major UNIX operating systems as well as Windows NT are supported.

The SICAD Geospatial Data Server is middleware based on standard RDBMS software from Oracle and Informix. It extends these RDBMS by geometric and

[7] `http://www.sicad.com/technology/`

graphic datatypes. Spatial indexing is done by use of a quadtree. For efficient storage and retrieval the geometric data is stored in binary format, using either the BLOB datatype in Informix or RAW in Oracle. All geometric data belonging to one cell of the quadtree is stored within one datum of type BLOB or RAW. This results in a spatial clustering of the data.

The geographic objects in the SICAD Geospatial Data Server have a topologic structure in which common geometric primitives, e.g. a line shared by two adjacent areas, are stored only once. This means that geometries are stored without any redundancies. Together with the topologic relationships inherent to these data structures this data model is especially powerful for utility and cadastral applications. These applications are typically built to maintain large volumes of large-scale geographic data with frequent updates in a multi-user environment, and require high quality and accuracy of the data. Attribute information can be stored directly with the geometric data or in separate database tables. In the latter case, linkage information is maintained through the Geospatial Data Server.

Besides the basic query, update and retrieval functionality, the Geospatial Data Server provides functions to work in distributed environments typical of large utilities or municipalities. This includes accessing and overlaying the content of several databases, a central dictionary for distributed databases and advanced replication functions for differential update of remote databases.

2.3 Demonstration Architecture

The GEObit demonstration was implemented as a set of Specialists. These specialists, subclassing the Specialist CORBA IDL, were compiled into C++ classes for Orbix (a CORBA implementation from IONA Technologies, Ireland). Geodata management functions were provided using a prototype implementation of OpenGIS®Simple Features Specification for SQL based on SICAD Geospatial Data Server. Internet communication of the OpenMap client with Specialists was established by means of CORBA implementations deploying IIOP. On the Windows NT Client, Visibroker 3.2 for Java (CORBA implementation from Inprise, formerly Borland) was used to compile and integrate the Specialist IDL into the OpenMap client. On the Sun Solaris 2.6 server, Orbix 2.3 for C++ provided the ORB (Object Request Broker). With this system setup the suitability of CORBA technologies for inter-platform, inter-ORB and inter-language process communication could be tested. In the GEObit demonstration, access of an OpenMap client to a variety of geodata sets managed with different systems was presented. Geodata sets were

- of different types (e.g. raster data of elevation model, vector data of point, line and polygon features),
- of different scale (e.g. small scale political boundaries, large scale cadastral data) and
- of different information communities (e.g. cadastral surveying, utilities).

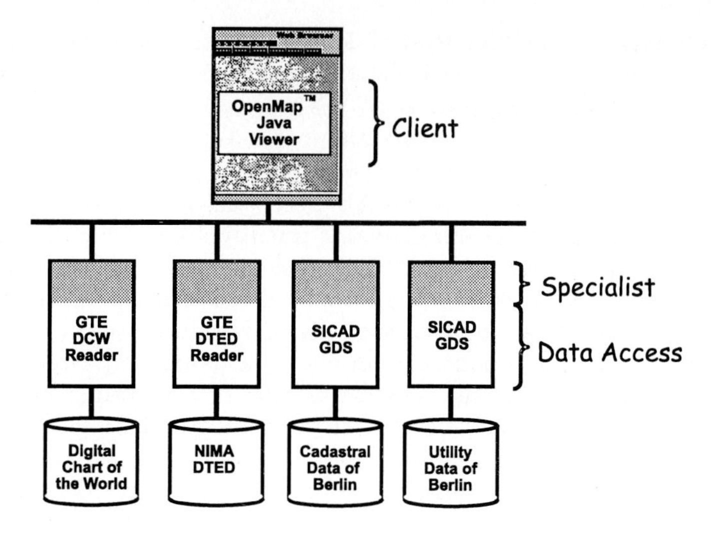

Fig. 2. Architecture and setup of the GEObit demo.

Therefore, besides the DTED and DCW Specialists which were already components of the FGDC Demo, SICAD Geospatial Data Server Specialists for accessing large scale cadastral and utility data for the city of Berlin were added (see figure 2). When OpenMap was set to a scale larger than an appropriate minimum scale, these new Specialists generated simplified graphics from existing geodatabases, which were used for all GEObit demonstrations of SICAD products.

The graphic objects, after being received and presented in the OpenMap client, allowed direct interactions with the respective Specialist objects via proxy objects in OpenMap. These interactions included inquiry and presentation of alphanumeric data for a selected object whose graphical appearance also was altered in OpenMap.

3 Semantic Issues

3.1 Geometry Model

The OpenGIS® geometry model, which is a basis for Simple Features for SQL implementations, doesn't define topological relationships between geometric primitives. This leads to redundancy of coordinates that exist in the geometry of more than one feature (e.g. border between two adjacent land parcels) since geometric primitives can't be shared by features.

As explained above, the SICAD Geospatial Data Server implements a topologic geometry model that controls redundancy. Because capturing, updating and storing topologic structures of geographic objects is efficiently supported by the SICAD geometry model, conversions were needed in the implementation of Simple Features for SQL.

Geometry conversions for generating Well Known Structures for OpenGIS®-Geometry could be achieved simply by navigating through topological structures gathering and ordering the coordinates of visited geometric primitives. But, since shared geometric primitives are visited multiple times in the course of generating Well Known Structures for the respective features, redundancy is introduced in the process of conversion. Therefore, it is very difficult to reproduce the topological structures by converting back from the generated Well Known Structures.

This is especially true for the SICAD Geospatial Data Server, which allows for controlled redundancy of geometric primitives. The Simple Features for SQL implementation is suitable to generate graphics for GIS viewers like OpenMap, but not for the exchange of geodata with GIS that are using topological geometry.

3.2 Definition of Geoobjects / Features

When working with geodata sets of different scales produced for different information communities, one finds that definitions of features as classes of geoobjects are hardly comparable. This affects not only definitions of feature attributes, but also their borders and geometry types, because they differ significantly. For example, roads in a cadastral application are defined by land ownership and are associated with Polygon geometries. This makes usage of existing cadastral road features (e.g. by transportation applications) difficult, since their road features are probably continuous between two nodes in a road network and associated with LineString geometries. Interoperability of these applications could be achieved, for example, by converting cadastral Polygon geometries to LineString geometries and assemble them in a MultiLineString to generate a transportation road feature from the cadastral road parcels. The prerequisite for such interoperability is a feature attribute in the cadastral application that can be used to select the road features which form a respective road feature in the transportation application.

In the GEObit demonstration the DCW (Digital Chart of the World, scale $1:1,000,000$) Specialist generated a small-scale OpenMap layer with the network of major roads, instead of using a SICAD Geospatial Data Server Specialist to generate a road network by converting cadastral geodata. Overlaying these OpenMap layers, the limitations of cross application analysis (e.g. finding the owners of land parcels adjacent to a major road) became obvious since the accuracy of the underlying geodata sets differed significantly.

3.3 ALK / ATKIS (German National Standards) vs. Simple Features

The Open GIS Consortium aims to provide a common object model that facilitates interoperability of GIS products and applications. On the German national level a standardization initiative was begun by the Association of Authoritative Surveying Administrations (AdV) many years ago. The latter initiative led to the production of cadastral (ALK) and topographic (ATKIS) base geodata of high accuracy covering most of Germany already. The underlying well-defined object model of hierarchical structured geoobjects associated with geometric primitives in a topologic geometry model was implemented by a variety of commercial GIS and database applications. Between them, geoobjects can be exchanged, and log files generated in the course of modifying geoobjects for correction with one GIS can be used to update replicated geodatabases managed with another GIS or database application. In the revision of the object model (ALKIS), functional interfaces are also introduced as methods of the geoobjects. In the GEObit demonstration, existing cadastral base data, managed with

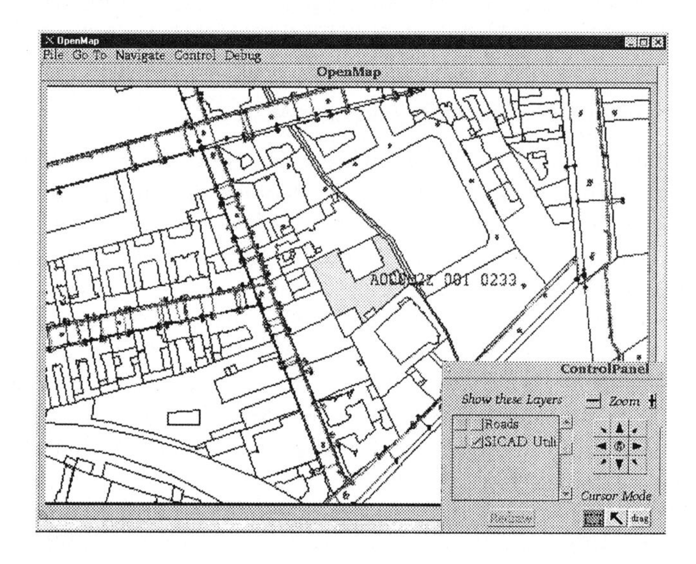

Fig. 3. Screen capture of demonstrated Berlin data.

SICAD Geospatial Data Server using the ALK object model, were converted to OpenMap graphic objects in a Specialist based on a prototype implementation of Simple Features for SQL. The conversion between the different geometry models was already explained above. Additionally existing arc and spline geometries had to be approximated by LineString geometries since they aren't defined in the OpenGIS® geometry model yet.

Conversion of object structures was only possible by omitting some of the associated data since complex textual annotations and geometric primitives used for portrayal still have to be defined by future OGC specifications extending the Simple Feature specifications. Therefore, graphical appearance of a cadastral layer in the OpenMap client didn't meet the requirements imposed on cadastral GIS applications in Germany.

3.4 Coordinate Transformation

Since the OpenMap client requires that coordinates of graphical objects created by a Specialist be provided in WGS84 coordinate system, the coordinate transformation implemented in SICAD Geospatial Data Server was applied to existing geodata stored in Soldner Berlin New coordinate system.

4 Technical Issues

4.1 Operating Systems

The Operating System differences were not an issue for this demonstration, because of the Java implementation of OpenMap and the use of CORBA implementations to provide IIOP-based object communication.

4.2 CORBA ORB

The CORBA IIOP connection between OpenMap and the Specialist was made up of products from two different vendors. The OpenMap specialist layer is based on the OpenMap Specialist IDL. Inprise Visibroker 3.2 for Java was used on the client side (OpenMap) of the interface, with the IDL compiler used to create the client interface code stubs as well as the Visibroker CORBA implementation jar (Java Archive) files used at runtime from the Specialist IDL. Iona's Orbix, version 3.2 was used to create the server-side (Specialist) C++ code stubs. The Orbix CORBA shared libraries, combined with OpenMap specialist support software, were used to create the C++ code for the specialist, running on a Sun Microsystems Solaris 2.6 machine.

The main technical issue with CORBA is, of course, to get the client and server communicating with each other. The Specialists wrote their Interoperable Object Reference (IOR) into a specific file whenever they were launched, and they wrote that file at a predetermined location accessible via an HTTP server. When the OpenMap client was started, it used the IOR information to contact the Orbix ORB. The Orbix ORB is a daemon process responsible for setting up communication between the client and the server. While the Specialists were set to use a dedicated port number for their communication, the ORB was still needed to handle the initial contact between the client and the Specialists graphical objects. The ORB, after receiving the request from the client, performed the handshaking required to get the client communicating with the respective Specialist server.

The only pressing technical problem was the repeated crashing of the Orbix ORB. For the GetRectangle calls to the Specialist, where the OpenMap client requested graphical objects to fill the screen for the geographic scale and location of the OpenMap window, the ORB was stable and performed as expected. The problems occurred when OpenMap was in the gesture mode. In gesture mode, the ORB was responsible for handling the message passing required to update the OpenMap display with identifying strings as the user moved the mouse over the objects on the screen. It was during this handshaking procedure that the Orbix ORB kept crashing, preventing the initiation of more message passing or GetRectangle responses. The ORB was crashing due to some internal memory management problems that could not be reliably reproduced at other physical locations.

The solution for this problem was to launch the ORB and have it use a log file to keep track of its state. The ORB has the capability to be started with a log file, causing it to be set to a certain state, aware of server objects that it can be responsible for. The ORB run command was inserted into an infinite loop, which restarted the ORB after every crash in its pre-crash state. The effect on the OpenMap performance was negligible. While the solution may appear to be distasteful, it was the path recommended by Iona on their website, to make their ORB more stable and persistent.

4.3 Internet Process Communication through Firewalls

In the case where OpenMap was being run as an applet in a web browser, the Visibroker Gatekeeper was required in order to bypass the inherent security restrictions imposed upon applets. Since an applet is only permitted to communicate with the computer it was loaded from, the Gatekeeper server was loaded on the applet serving computer and acted as a conduit for IIOP messages passing between the applet, ORB and Specialist.

The CORBA 3.0 Firewall specification addresses this problem by defining interfaces for passing IIOP through a firewall. It includes options for allowing the firewall to perform filtering and proxying on either side. This is very important for extending the secure use of CORBA to the Internet and across organizational boundaries.

5 User Feedback

Most interest and feedback came from representatives of larger German municipalities. These city administrations typically are using GIS software in a broad range of applications and departments. Many of them are using GIS software from different vendors and have difficulty sharing any geodata between departments due to the proprietary character of the software used. One representative of an East German city reported about six different GIS in use for the city administration.

The main question was if these governmental users would accept the simplification of official data, which was necessary in order to use the OpenGIS® Simple Feature Specification. This concerned mainly the graphical representation of data (which is not defined at all in the mentioned specification) and the missing map annotation (which is also not defined). It is to note here, that German authorities are using very sophisticated cartographic styles, especially in city planning applications. But without any exception everybody said that he would tolerate these deficiencies if he could gain direct access to heterogeneous sources of geodata.

The other very positive feedback came to the fact that the demonstration was allowed to query feature properties directly from the server, avoiding the overhead of transferring them at load time. It seems to be that more and more users consider a GIS also as just another user interface to access arbitrary data. Keeping database connections open and allowing the user to drill down by queries or navigation seems to be one of the major advantages of the OpenMap approach.

6 Summary

The importance of practical testing of specifications and ideas cannot be overemphasized. In the course of constructing the two demonstrations and in their operation at the respective locations in Cincinatti and Leipzig, the authors have learned many lessons. As discussed in the preceding sections, there were issues regarding software interoperability and quality, issues regarding data modelling differences, issues regarding data visualization, and issues regarding data scale differences. These kinds of issues were not necessarily unanticipated, but were nonetheless hard to conceptualize without actually trying things in an operational setting. Furthermore, it is much easier to revisit the interoperability specifications under development in the Open GIS Consortium once these lessons have been learned and recorded. An additional benefit of this kind of testbed is the opportunity to show previously uninitiated people what the benefits of the new technology can be. While the end-user community can read about Open GIS and interoperability, one cannot impart the full message about Open GIS without actually demonstrating it. The feedback from users can also influence the relative importance which is given to various parts of the specifications as revisions are made or new ones are developed.

References

1. Doyle, A.: FGDC Hosts Multivendor Interoperability Demo, OpenGIS®Newsletter, Vol. 2, No. 4, Open GIS Consortium, Wayland MA, December 1997.
2. Gouin, D., Morin, P., Clément, G., Larouche, C.: Solving the Geospatial Data Barrier, Published in: GEOMATICA, Vol. 51, No. 3, 1997, pp. 278-287. `http://www.globalgeo.com/products/wtepaper/default.asp?code=geo`
3. Nebert, D., Doyle, A.: Discovery and Viewing of Distributed Spatial Data: The OpenMap Testbed, Published in: J. Strobl and C. Best (Eds.), 1998: Proceedings of the Earth Observation & Geo-Spatial Web and Internet Workshop '98 - Salzburger Geographische Materialien, Vol. 27, Institut für Geographie der Universität Salzburg, ISBN: 3-85283-014-1.
4. Object Management Group (OMG): Internet Inter-ORB Protocol, Defined in: The Common Object Request Broker: Architecture and Specification, Revision 2.2, Release Date: February 1998, `ftp://ftp.omg.org/pub/docs/formal/98-07-03.pdf`
5. Open GIS Consortium (OGC), Inc.: OpenGIS®Simple Features Specification For SQL, Revision 1.0, Release Date: 17 March 1998, `http://www.opengis.org/public/sfr1/sfsql_rev_1_0.pdf`
6. Rose, T., Peinel, G., Scheer, T.: Deploying Cartography for Information Mediation in GEONET 4D, 3rd EC-GIS Workshop, Leuven, Belgium, June 1997 `http://ams.emap.sai.jrc.it/dg3gis/abstract97.htm`

Modeling and Sharing Graphic Presentations of Geospatial Data

Stefan F. Keller[1] and Hugo Thalmann[2]

[1] Federal Office of Topography,
Center of Competence INTERLIS,
CH-3084 Wabern, Switzerland,
stefan.keller@lt.admin.ch, http://www.swisstopo.ch
[2] a/m/t Software Service AG,
CH-8400 Winterthur, Switzerland,
mail@amt.ch

Abstract. Digitally capturing and maintaining geospatial data is expensive. The consequence is that there is a need for the integration of geoprocessing software into mainstream computer technology and a demand to share geospatial data. This demand includes consequently the sharing of specific graphic views on it. In this paper an implementable framework is presented for defining graphic views on geospatial data. This is based on an existing geospatial data definition language and a transfer format, called INTERLIS.

1 Introduction

People always looked at geographic or *geospatial* data through maps which are usually presented on paper media. With the fast progress of hardware technology and the advent of the World Wide Web the geospatial data more and more is also displayed on colored high-resolution screens. But digitally capturing and maintaining geospatial data is expensive which is also the case for the specialized computer systems that are still in use today. The consequence is that there is a need for the integration of geoprocessing software into mainstream computer technology and that there is a great demand to share geospatial data for the widespread use [4].

The same applies also to the production of paper and screen maps that are up-to-date and easily to reproduce, preferable with different geographic information systems (GIS). Therefore it is stated that the need for shared geospatial data consequently leads to deriving and sharing specific graphic views on it.

1.1 Federated and Interoperable Geographic Information Systems

Sharing geospatial data implies that heterogeneous GIS are "open" to a certain degree and are able to exchange data freely [8]. This can be called a data-centered approach because explicit description is especially important as well as rules that allow for the encoding of objects to a corresponding data access service

[15]. The access, query and manipulation of geospatial data is done through a data manipulation language or through an application programming interface (API). This is called a process-centered approach. In realistic environments *both* approaches are required for "interoperability" being "the capability for two or more computers to transmit data and to carry out processes as expected by the users" [16].

Because of the diversity of how GIS implemented geoprocessing [1] [18] and because data lives about ten times longer than software [20] we focus our standardization efforts on the data-centered approach.

It is assumed that the current geospatial infrastructure consists of federated networked systems which means that basically independent, heterogeneous systems are committed to follow common rules [15]. A user or an information community wants to become sometimes a data or service provider or a data or service consumer. Jones [14] wrote in a similar sense that "Multi-participant, cooperative GIS means that governments and agencies at the local, state, and federal levels and the private sector become partners in establishing and maintaining a GIS. Naturally, each 'partner' may participate in a different way".

1.2 Decomposing Geospatial Objects

Abstraction and decomposition are common techniques in information technology. This separation can also be applied to geographic information [18] and its consequent application is one of the prerequisites for the integration of geospatial data structures and geoprocesses into mainstream information systems.

A geospatial object can be decomposed in different aspects [15]:

1. Geometric, georeferenced attribute types (point, polyline, surface, area, solid)
2. Textual attribute types (string, number, date, time, etc.)
3. Other special types like a relationship attribute type
4. Possibly a reference to one or many graphic presentation definitions
5. Generic methods – like create, update or delete – and more complex ones

At this conceptual level geometric attribute types, for example, are treated like any other (complex) attribute type. This also means that one object can have zero, one or many geometric attributes. Note, that from a geospatial data modeling view there is no text label which denotes a certain attribute of a geospatial object. A text label belongs rather to the presentation model than to the geospatial data model.

1.3 Modeling Geospatial Objects

Openness of systems and common rules for data management are established through standardization. A well-defined geospatial data definition language – sometimes also referred to as a conceptual schema description language – that is based on predefined data types (as mentioned above) is like a "lingua franca" between different specialists and at the same time it remains computer readable.

With such a standardized language any user community has a common understandable and flexible tool for modeling geographic information. The result of this activity is a conceptual "user application schema". A conceptual schema is a crucial concept in database modeling which abstracts from system specific implementation issues and establishes a data structure common to the involved systems [6] [10] [4].

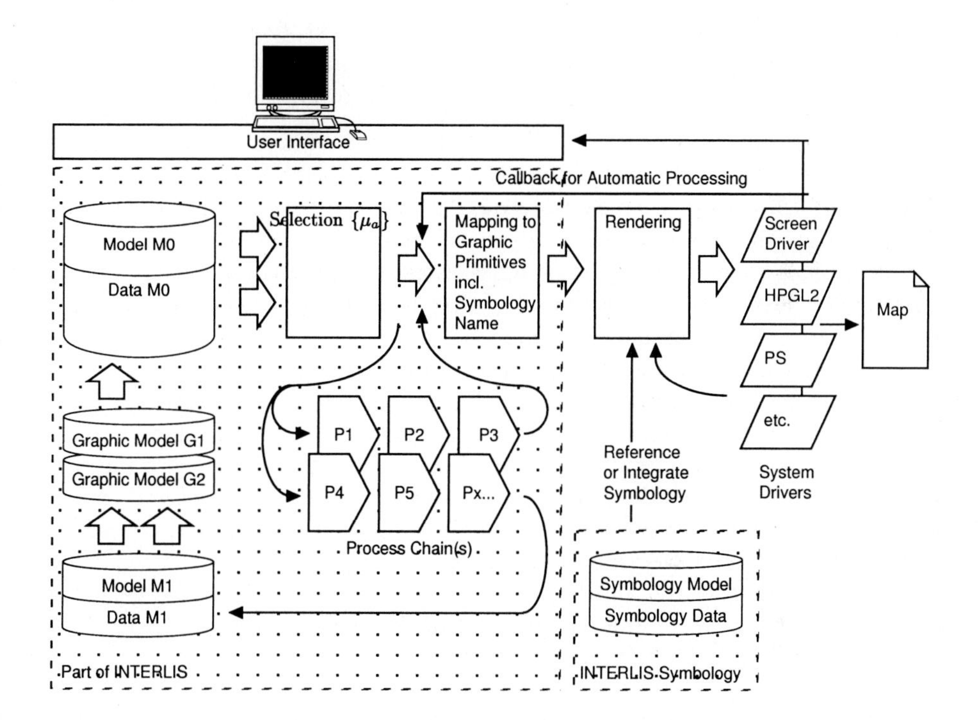

Fig. 1. The presentation model of INTERLIS: this describes graphic presentation as a multi-phase process which consists of a selection step and a mapping step which relates queried geospatial objects to graphic symbols.

1.4 The Graphic Presentation Process

Graphic Primitives Geospatial data is being visualized on paper maps or screens by use of elementary, generic graphic objects called *graphic primitives*. In the 2D case there are the following graphic primitives:

- *Text*: Text primitives have properties like fixed text, size, font, color etc.
- *Point*: Point (or icon) primitives consist of size and a line, surface combination

- *Line*: Line primitives have properties like width, dash style, patterning style, etc.
- *Surface*: Surface primitives may be filled, outlined, hatched, etc.

These graphic primitives are also objects that can be modeled with a geospatial data definition language.

Symbology Library A symbology library contains text, line, surface and point symbol object instances. Symbology library symbols that are defined and referred to by graphic primitives in a application specific graphic model (or schema) determine the final rendering on a graphic output device.

Selection, Mapping and Rendering The present research interest in the domain of graphic presentation of geospatial data seems to be rather low [13] [7] [9]. In [5] a language is proposed which consists of two components, a query language to describe what information to retrieve and a presentation language to specify how to display query results.

"The OpenGIS Abstract Specification" is a current extension of [4] which includes a topic volume called "Essential Model of Interactive Portrayal". The definition of portrayal is very similar to what is called here a "graphical presentation". In that specification the portrayal process is decomposed into the subprocesses *Selection* (including query constraints), *Generating Display Elements* (dealing with style), *Rendering* (dealing with image constraints) and *Displaying* (taking device characteristics into account).

Based on our research and implementation experiences it can be generally stated that the process of graphic presentation of geospatial data consists of the following steps [1]:

1. *What* is to be presented? The relevant object classes are selected. The information needs exactly to be communicated according to the intended message.
2. *How* are the resulting objects represented? The resulting objects of step one are mapped to graphic symbols (symbolized) such that the non-graphic information is "transcribed" to the graphic primitives in an unambiguous way [2].
3. *How* are the graphic symbols rendered and displayed on a graphic output device?

Step one can be resolved with a spatial and thematic query statement, possibly complemented with a filter (c.f. [13]). Step two defines how graphic symbols are generated and implies all problems of visualization, generalization and graphic layout. Step three is a combination of rendering and displaying. Step one and two are enough to explicitly describe any symbol; step three is device dependent and therefore not subject of data-centered standardization.

This approach is the key for sharing not only the geospatial core data and its related schema but also the conceptual definition of specific graphic views among different systems. Therefore, one can define the graphic presentation process of geospatial data as follows:

Graphic presentation is a multi-phase process that consists of a selec-tion step and a mapping step which relates queried geospatial objects to graphic symbols; they in turn are rendered and displayed on system specific graphic output devices

In the next section it is shown that abstraction and decomposition can also be applied to the graphic presentation of geospatial data and how it was implemented in INTERLIS.

2 The Approach of INTERLIS

Faced with federated GIS in the Swiss government the Directorate of Cadastral Surveying initiated a standardization process in the 1980s with a small project team. This team consists of technical representatives from industry, universities and government. The state-of-the-art in information technology – specifically database theory – has been analyzed. Then, instead of compiling an ultimate all-purpose feature or object catalogue (see for example [17]), the project team first specified a textual, geospatial data definition language, called INTERLIS [11].

2.1 A Data Definition Language and a Transfer Mechanism

In INTERLIS encoding rules are attached to the geospatial data definition language, which specify for example the encoding into a file format. This enables to communicate not only the description of the data structure, as a mandatory part of meta-data but also the object instances without worrying e.g. about object codes. In other words the georeferenced objects are self-describing their structure. This information can be accessed before the data is actually transferred. Ultimately this enables so called "plug-and-play" data sets. See [11] for a complete implementation specification of such a mechanism.

2.2 The Mapping Language and the Presentation Model

To describe and share graphic primitives with their properties a standard presentation model is introduced, which is defined itself in an INTERLIS schema.

Specific graphic primitives of a graphic presentation have a well-known relation to geospatial objects. Therefore a graphic presentation can be seen as functional mapping of data model objects to presentation model objects:

$$GraphicPresentation = Function(GeospatialDataModel). \tag{1}$$

The graphic primitives generated by the graphic presentation function according to Equation 1 exist only temporary. There is no need to store them in a database because they can always be generated from geospatial data objects.

In typical computer aided drawing systems it could be argued, for example, that there exists no explicit data model; it seems that there are only graphic primitives stored directly in the system. But in fact there is an internal fixed data model from which graphic objects are generated by a system dependent mapping function at run time.

Finally every graphic presentation has to be created on some existing output device. Here the established graphic standards come into play, therefore this is considered as being outside the scope of this new standardized framework.

2.3 Summary of the Process

This is a summary of the whole process (see Fig. 1 and [12]):

- First a selection of relevant object classes is performed.
- Through the mapping function (1) the geospatial objects are mapped to graphic primitives. For the specification of this a selection part of the geospatial data definition language is needed as well as an interface to symbols being application specific instances of graphic primitives.
- The graphic interface definition (consisting of specific procedure calls) describes the graphic primitives text, line, surface and point symbol. The graphic primitives exist only temporary whereas the referenced symbols are persistently stored and maintained in a system specific manner.
- The presentation model consists of a graphic interface (i.e. an abstract interface definition) and a symbology library. The presentation model is defined in INTERLIS and will be called `ILI_PRESENTATION` or `SYMBOLOGY`.
- The symbology library contains text, line, surface and point symbol definitions. The symbology library determines the rendering of a graphic primitive to an output device.
- Graphic primitives are rendered and displayed by a driver program to a graphic output format or output device.

2.4 An Example

This example is taken from an existing cadastral surveying application schema and shows geodetic points that have to be presented in scale range lower or equal than the sclae number 500 (1/500) and above 500 (e.g. 1/10,000). In scale range below 1/500 a text and a point symbol for each geodetic point are displayed. Different symbols should be used for different kinds of geodetic points (i.e. `GP1` or `GP2`). In scale 1/10,000 only the GP1 symbol is specified, there will be no GP2 symbols displayed and there is no accompanying text specification.

```
MODEL BaseModelMO =

  TOPIC Geodetic_Points =

    TABLE Geodetic_Point =
      Number: TEXT*12;
      NumPos: Point2D;
```

```
      NumOri: OPTIONAL LabelOri; !! Default: 100.0
      NumHAli: OPTIONAL HALIGNMENT; !! Default: Center
      NumVAli: OPTIONAL VALIGNMENT; !! Default: Half
      SymbolOri: OPTIONAL LabelOri; !! Default: 0.0
      Geometry: Point3D;
      !! etc.
      Type: (GP1, GP2);
      Lineage: OPTIONAL TEXT*30;
    CONSTRAINT
      UNIQUE Number;
      UNIQUE Geometry;
    END Geodetic_Point;

  END Geodetic_Points.

END BaseModelM0.


GRAPHIC MODEL GraphicModelG1
  DEPENDS ON BaseModelM0
  BASED ON SYMBOLOGY =

  PRESENTATION
    Resolution; !! User defined parameters

  TOPIC Geodetic_Points =

    VIEW Geodetic_Point_View =
      ALL OF BaseModelM0.Geodetic_Points.Geodetic_Point;
    END Geodetic_Point_View;

    GRAPHIC Geodetic_Point_Symbology
      BASED ON Geodetic_Point_View =

      MyGP1_Symbol: WHERE Type == "GP1"
        SIGNATURE SYMBOL
          Symbology: "GP500_GP1";
          Geometry: Geometry;
          Orientation: SymbolOri;
          Priority: 100;
        END;
      MyGP2_Symbol: WHERE Type == "GP2" && Resolution <= 500
        SIGNATURE SYMBOL
          Symbology: "GP500_GP2";
          Geometry: Geometry;
```

```
          Orientation: SymbolOri;
          Priority: 100;
        END;
      MyGP_Text1: WHERE Resolution <= 500
        SIGNATURE TEXT
          Symbology: "Helvetica-10-normal";
          Text: Number;
          Geometry: NumPos;
          Orientation: NumOri
          HAlignment: NumHAli;
          VAlignment: NumVAli;
          Priority: 110;
        END;
      MyGP_Text2: WHERE Resolution > 500
        SIGNATURE SYMBOL
          Name: "GP10000_GP";
          Geometry: Geometry;
          Orientation: SymbolOri;
          Priority: 100;
        END;
    END Geodetic_Point_Symbology;

  END Geodetic_Points.

END GraphicModelG1.
```

The first model (BaseModelM0) defines the structure and the constraints of the base data. A separated graphic model, called GraphicModelG1, refers to BaseModelM0. Then the INTERLIS standard presentation model, ILI_PRESENTATION or SYMBOLOGY, is referred to because in the graphic presentation of the objects are going to be defined. With the keywords BASED ON SYMBOLOGY the graphic interface and the symbology library are inherited. The following PRESENTATION clause allows to define parameters which can be used later on to control certain things. In this case the parameter Resolution is being declared. After the opening of the topic Geodetic_Points, all objects from table Geodetic_Point are selected in a database view (Geodetic_Point_View). The keyword GRAPHIC introduces the graphic definitions and refers to the before established view. Then, the named graphic definitions follow. A WHERE filter can optionally inserted before the SIGNATURE declaration introduces the definition of a graphic primitive. This is the second process step which maps objects to predefined symbols. For each presentation definition the graphic interface is called with the arguments, like:

```
MyGP1_Symbol:
  SIGNATURE SYMBOL
```

```
  Symbology: "GP500_GP1";
  Geometry: Geometry;
  Orientation: SymbolOri;
  Priority: 100;
END;
```

The argument `Priority` of each symbol definition, for example, controls the display priority (e.g. with value 100). The graphic interface of the `SYMBOL` signature was inherited from `SYMBOLOGY` together with the standard presentation model. The symbols – or display elements – are assigned by reference with a unique name (e.g. `MyGP1_Symbol`). The management of these graphic primitives is system and device dependent but they are neutrally encoded in a symbology library which is structured according to the standard graphic presentation model.

2.5 Discussion

The proposed framework has been implemented using commercial GIS and draft INTERLIS Version 2. Based on these experiences we can discuss the gained insights.

One Way Process Prototype implementation of the proposed approach showed that graphic presentation can be seen as a one way process, without any manual interaction. The base data and the related processes are processed from the left of Fig. 1 to the right.

Derived Geospatial Models By collecting the graphic presentation definitions in a separate model the flexibility of different views on the same spatial object is also structurally properly expressed. This is like encapsulating the presentation model definition from the geospatial data model. Manual editing of base data is still possible within this framework given that any change in the structure is being modeled and stored as a separate object. The table defining these objects can either inherit the structure of the original base table or make a relationship to it, so that only these attributes need to be stored that really have changed. The consequent implementation of this framework enables multi-scale databases where changes in the base model are propagated to the derived models (see `Model M1` in Fig. 1).

Process Chains In Fig. 1 there are process chains indicated. These are subprocesses consisting of procedures, which can be automated given the information is there from the geospatial objects or the presentation function calls. Examples are calculating the offset of a text label up to automated name placement algorithms.

Generalization Model and cartographic generalization processes are hard to automate [19]. Those who are possible to handle by the computer can be integrated as a subprocess mentioned above. For those generalization problems, which require manual editing derived geospatial models are proposed.

2.6 Compatibility With Other Standards

There are not many standards in this domain. The international standardization body, ISO [13], rather provides a general framework for interoperability where implementation standards have to comply.

Currently, the Open GIS Consortium [4] is an emerging standardization initiative from industry which has a rather process-centered technical approach. It tries to standardize e.g. simple feature (or object) types and feature catalogue services. With this service the properties of an object can be accessed through a API call. This is very similar to a specific schema. While this is useful for interactively exploring geospatial data sets this is probably not adequate for getting an overview of the overall structure and the relationships. Certainly it is not meant as a standard tool to model a user application schema of a user or information community. As mentioned before a working group for the "Interactive Portrayal" of geospatial data in the World Wide Web has recently been constituted.

The approach presented here could complement these efforts in many respects: INTERLIS delivers a compact, unifying geospatial data definition language including a presentation part and a file format. Therefore it fits well into the international standards so far and will continue to enhance the means for information communities to document their needs and share their geospatial data.

3 Conclusions and Future Work

The framework presented here has been tested in a prototype and the benefits can already be identified: graphic presentation definitions and even symbology libraries can be shared and distributed in a federated systems environment. The vision is to enable distributed providers who are specialized on graphic presentation services.

Taking the diversity of present GIS into account it is almost impossible to agree on a common set of attributes for defining graphic symbols, so we had to choose a balance between completeness and implementation costs. Therefore further work is required for assessing the proposed standard presentation model. Although this framework is based on 2D it can be extended to 3D in next versions by introducing solids. But due to unresolved geoprocessing and geospatial modeling issues this remains to be done.

From a process-centered view identifying appropriate process chains taking into account the properties of the chosen output device is very important. Very important is also the investigation of the human part in the process for allocating interactive processes like continuous feedback while editing.

Additional experiments need to be made in order to exploit the full range of cartographic problems in GIS, especially the interoperability of graphic presentation of geospatial data.

Acknowledgements

We would like to thank all the people who contributed to this paper, especially to Michael Germann, Sepp Dorfschmid, Hans Ruedi Gnägi and Rene L'Eplattenier.

References

1. Bartelme, N.: Geoinformatik – Modelle – Strukturen – Funktionen. Springer Verlag, Berlin Heidelberg, (1995) 414p
2. Bertin, J. (1983), Semiology of Graphics: Diagrams, Networks, Maps. University of Wisconsin Press
3. Booch, G.; Rumbaugh, J.; Jacobson, I.: Unified Modeling Language Semantics and Notation Guide 1.0. San Jose, California USA, Rational Software Corporation (1997) (http://www.rational.com)
4. Buehler, K. and McKee, L.: The OpenGIS Guide 2nd edition. Open GIS Consortium Inc., May 22 (1996) 103p ISBN pending (http://www.opengis.org)
5. Camara, G.; Casanova, M.A.; De Freitas, U.M.; Cordeiro J.P.C.; Hara, L.: A Presentation Language for GIS Cadastral Data. In: Proceedings of ACMGIS'96 (ed. R. Laurini) (1996)
6. Codd, E.F.: A relational model of data for large shared data banks. Communications of the ACM, Vol. 13 **(6)**, (1970) 377–387
7. Egenhofer, M.: Spatial SQL: A Query and Presentation Language. IEEE Transactions on Knowledge and Data Engineering **6** (1) (1994) 86–95
8. Goodchild, M.F.; Egenhofer, M.J.; Fegeas, R.: Interoperating GISs – Report of a Specialist Meeting Held under the Auspices of the Varenius Project Panel on Computational Implementations of Geographic Concepts. December 5–6, 1997, Santa Barbara, California, (1997) 64p. (http://ncgia.ucsb.edu/conf/interop97/)
9. Goh, P-C.: A Graphic Query Language for Cartographic and Land Information Systems. International Journal on Geographical Information Systems, **1**(4) (1989) 327–334
10. Hull, R. and King, R.: Semantic Database Modeling: Survey, Applications and Research Issues. ACM Computing Surveys. Sept. (1987) 201–260
11. INTERLIS: INTERLIS – A Data Exchange Mechanism for Land Information Systems, Version 1 Revision 1, Federal Department of Cadastral Surveying, Berne, (1997) 32p (http://www.gis.ethz.ch/)
12. INTERLIS: Draft INTERLIS Version 2, Federal Department of Cadastral Surveying, Berne, (1998) 51p (http://www.gis.ethz.ch/)
13. ISO/TC211: ISO/TC211 Geographic information / Geomatics, Family of standards numbered ISO 15046 – part nn, (1994) (http://www.statkart.no/isotc211/welcome.html)
14. Jones, J.T. and Slutzah, R.P.: A Framework for Multi-Participant/Cooperative GIS: Process, Public Records and Data Products. In: URISA 1994 Annual Conference Proceedings. Washington, D.C.: Urban and Regional Information Systems Association, Vol. 1, (1994) 92–101

15. Keller, S.F.: Modeling and Sharing Geographic Data with INTERLIS. In: Computers & Geosciences, Special Issue on Systems Integration Within The Geosciences, (in press)
16. Markley, R.W.: Data Communications and Interoperability. Prentice–Hall Intl. Inc. (1990), ISBN 01-31-993402
17. SDTS: Spatial Data Transfer Standard (SDTS). Federal information processing standards 173. National Institute of Standards and Technology (NIST), MD (1992)
18. Voisard, A. and Schweppe, H.: Abstraction and Decomposition in Interoperable GIS. ICSI Technical Report No. 97006, March 1997 (1997) (ftp.ICSI.Berkeley.EDU)
19. Weibel, R.; Keller, S.F.; Reichenbacher, T.: Overcoming the Knowledge Acquisition Bottleneck in Map Generalization: The Role of Interactive Systems and Computational Intelligence. In: Frank, A.U.; Kuhn, W. (ed.): Spatial Information Theory, A Theoretical Basis for GIS. Lecture Notes in Computer Science, 988, Berlin et al., (1995) 139–156
20. Zehnder, C.A.: Informationssysteme und Datenbanken. Verlag der Fachvereine, Zürich, (1985) 253p

Towards OpenGIS Systems
The Vector Data Storage Component Evaluated

Frits C. Cattenstart[1] and Henk J. Scholten[2]

[1] Geo Information division of the Dienst Landelijk Gebied for sustainable land and water management, P.O. box 20021, 3502 LA Utrecht, the Netherlands
cattenstart@dlg.agro.nl

[2] Department of Regional Economics, Free University of Amsterdam, De Boelelaan 1105, 1081 HV Amsterdam, the Netherlands
henk@geodan.nl

Abstract. The Open GIS Consortium has defined and classified two dimensional geometries for the SQL92 Data Model Architecture in order to improve interoperability of spatial information systems. This specification was used as a reference to evaluate the implemented data models of two commercial GIS packages. It is concluded that the data models of both commercial packages are not OpenGIS compliant. They generally follow the rules as set by the Open GIS Consortium, but there are still deflections. These deflections are related to the concepts of the package developers and the way geometric entities are handled in their software. The fact that deflections exist should encourage a more in depth study of the OpenGIS Data Model Architecture in order to capture all occurrences of geometric phenomena. A first contribution was made by examining geometries in the Dutch Topographic Map scale 1 : 10.000.

1 The Challenge

In 1994 the Open GIS Consortium (OGC) was launched with the mission of improving the interoperability of GIS. At the heart of this mission was a need to find a common standard for the structuring of topological entities, among which the structuring in a SQL environment. This common standard has evolved into OGC "Simple Features Specification For SQL" (NN, 1998). However, from a practical point of view this specification is only of value once it is embodied within systems that are used in practice. In 1997 two major commercial packages came on the market. Environmental Systems Research Institute Spatial Data Engine (SDE) 3.0.2 for ORACLE version 7 and ORACLE Corporation Spatial Data Option (SDO) 7.0 for ORACLE version 7.3.3 (and later, with the same functionality Spatial Data Cartridge (SDC) 8.0 for ORACLE 8.0). Both with the intention of using the OGC standards as the basic building block for structuring vector databases in Relational Database Management Systems (RDBMS).

It was a challenge to evaluate how well SDO and SDE apply the OGC specification and furthermore establish whether or not the practical application of these tools and standards meets the needs of users for the topological structuring of geographic data. To do this a case study data set from the Dienst Landelijk Gebied in the Netherlands was implemented in both SDO and SDE (Cattenstart, 1998).

The concepts and implementations of spatial data in relational databases cover many aspects like storage structure, data format, indexing and search algorithms. This article concentrates on the OGC spatial Data Model Architecture for two dimensional SQL92 oriented environments and its practical impact.

2 The OGC Vector Data Representation

"OpenGIS" can be defined as "the interoperability of geospatial data and geoprocessing resources" (NN, 1996). In practise it means that data resources and the calling of process functionality can be shared between different (parts) of organisations and software packages that process spatial data. To the user data storage should be transparent between systems. This wish calls for a uniform all inclusive data model in order to communicate geometric data correctly.

A spatial object consists of one or more geometries which consist of data types. The following geodata types for two dimensional space are specified in the Open Geodata Model:

1. **Point** - a zero dimensional topology. This type specifies a geometric location,
2. **Curve** - a one dimensional topology. This type specifies a family of geometric entities including line segments, line strings, arcs, b-spline curves, and so on,
3. **Surface** - a two-dimensional topology. This type specifies a family of geometric entities including areas and surfaces that are defined in other ways.

In the Open Geodata Model it is possible to combine elements of the same data type to a geometry collection. In Fig. 1 the building blocks for the geometry specification and their hierarchy are shown. In the vector representation there is an other kind of element that is of great importance. It can be named Node which has the spatial characteristic that it is an intermediate location on a curve or surface boundary. A Node can have attribute data. A Node might for example be a dam in a river, a crossing of roads, or the junction of three state borders. The most significant difference with Point is that Node is directly coupled to geodata type Curve or Surface boundary and that it is not necessarily located on an explicit co-ordinate. Fig. 1 could be extended as shown in Fig. 2.

It is recommend that OGC clarify the notion of Node.

The SDO data model also identifies the geometric entities Point, Curve and Surface as well as Multi element geometries (NNa, 1997). SDO has no explicit implementation of the entity Node. However, the user can define a co-ordinate to act as a Node.

The SDE data model identifies the geometric entities Nil, Point, Curve and Surface and also Multi element geometries (NNb, 1997). The Nil entity is introduced as an object without contents as the physical outcome from a spatial function with no geometric interaction. SDE has no explicit implementation of

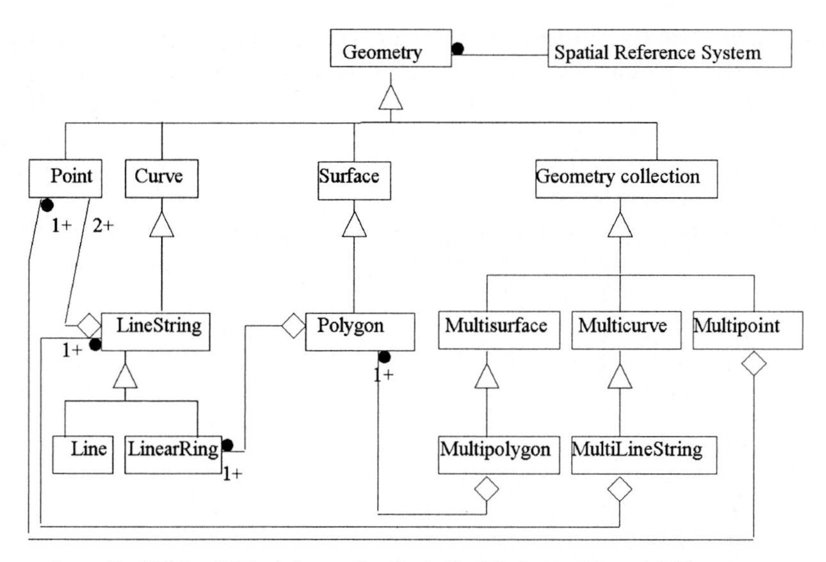

Fig. 1. Entity building block hierarchy

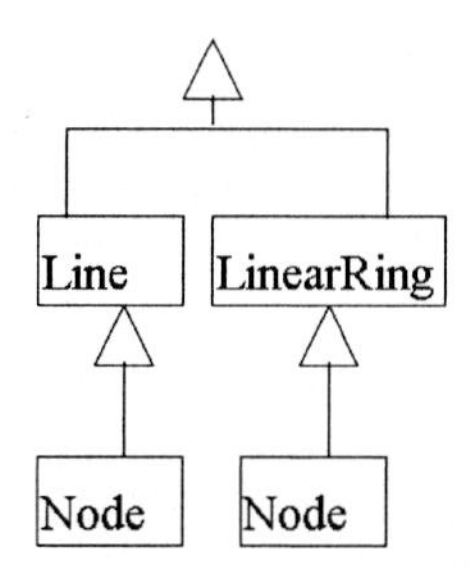

Fig. 2. Proposed extension of the entity building block hierarchy with Node

the entity Node. SDE uses an attribute called "Measure" in order to attach attribute values to a specific co-ordinate.

3 The SQL92 Implementation Model Data Architecture

The OGC Model Storage Architecture has on object based approach. This means that every geometry is self supporting, not dependent of any other entity. Related to administrative systems this is an advantage since the geometric entity can be fully incorporated within the rules of set theory mathematics. For example, a parcel may have several contracts, in which case the geometry might occur in the database as many times as there are contracts. Discarding a contract will invoke discarding the associated geometry. This will not effect the other contracts since they have their own geometry specification of the same parcel. On the other hand the geometric data structure lacks any kind of normalisation. Co-ordinate duplication occurs in any instance where geometries interact. For example, when roads are presented as lines and a junction is considered, the co-ordinate of the junction is stored as many times as there are lines ending/starting at the junction. When polygon entities are considered, two polygons that are neighbours have a common boundary. The co-ordinates of this boundary are stored separately for each polygon and are thereby present in the database in duplicate. OGC embraces actual developments in the GIS world as they are. However, with this trend OGC moves away from an essential property of spatial data: *Intrinsic topological relationship.* Shared co-ordinates only have to be stored once and when correctly referenced reveal spatial relations by simple searching in tables in stead of computer intensive relationship calculations (Cattenstart, 1998). Since this is not the case determining topological relationships is in the hands of software developers who have their own concepts and implementations in this respect. It is recommended that OGC investigate this perception.

4 The OGC, SDO and SDE Data Models Compared

In this chapter the OGC data model is used as a reference to evaluate the SDO and SDE data model. For each geometric data type the OGC assertions and if applicable special occurrences are presented. For each assertion and occurrence presented it is denoted if it is supported by the SDO and SDE data model. In some cases support is unkown due to imperfections of the checking routines in the packages.

In the OGC concept geometric elements are categorised into simple and non simple geometries. OGC has defined extensive (set theory based) rules to distinguish simple from non simple geometries (NN, 1998). This distinction is made in order to correctly communicate geometries and their properties between systems.

4.1 The Point Data Type

The OGC definition of a Point is that it is a zero dimensional geometry and represents a single location in co-ordinate space.

OGC has not determined special occurrences of Point.

4.2 The MultiPoint Data Type

The OGC definition of a MultiPoint is that it is a zero dimensional geometric collection. A MultiPoint is called simple in the OGC data model if no two Points in the MultiPoint are equal (have identical co-ordinate values). MultiPoint as defined by OGC is supported by both SDO and SDE. SDO and SDE support simple and non simple Multipoint.

Table 1. Assertions for Point data type

OGC Assertion	SDO	SDE
Has an ordinate value for each dimension	Yes	Yes

Table 2. Assertions for MultiPoint data type

OGC Assertion	SDO	SDE
The elements of a MultiPoint are restricted to Points	Yes	Yes
The Points are not connected or ordered.	Yes	Yes

4.3 The Curve Data Type

A curve is a one-dimensional geometric object stored as a sequence of points, with the subclass of curve specifying the form of the interpolation between points. The OGC specification defines only one subclass of curve, LineString, which uses linear interpolation between points. This is also the case in SDO and SDE. The set of OGC assertions covers the assertions in SDO and SDE. A Curve is called

Table 3. Assertions for Curve data type (subclass LineString)

OGC Assertion	SDO	SDE
Each Curve has a start point and an end point	Yes	Yes
The boundary of a Curve consists of its start point and end points	Yes	Yes
The sequence of points determine the global morphology	Yes	Yes
A Curve is defined as topologically closed (= no gaps)	Yes	Yes
Each consecutive pair of points defines a Line segment	Yes	Yes
Only one algorithm determining the form of all Line segments	Yes*	Yes

* *In ORACLE 8i-Spatial extended. In 8i-Spatial a curve may consist out of sections. Each section can have a different algorithm that determines its shape.*

simple by OGC if it does not pass through the same point twice. In the SDO data model there is no distinction between simple and non simple curves. A curve may cross itself. However, a self-crossing LineString does not have an implied interior. In SDE the user can determine if curves may self cross.

The distinction between simple and non simple curves has great impact on geometry classification. The next set of special occurrences are considered:

Table 4. Occurrences of Curve data type (subclass LineString

Occurrence	OGC	SDO	SDE
Start point equals end point	Closed	Yes	Yes
Simple closed	LinearRing	Yes	Yes
Non simple closed	Non simple	Yes	Yes
Non linear interpolation	Undefined	No*	No
Zero length Line	Simple ?	Yes	No
Zero length Line segment	Simple ?	Yes	No
Self crossing	Non Simple	Yes	No
Non sequential segments overlay	Non Simple	Yes	No
Non sequential points overlay	Non simple	Yes	No

* *In ORACLE 8i-Spatial supported. In 8i-Spatial a curve may consist out of sections. Each section can have a different algorithm that determines its shape.*

4.4 The MultiCurve Data Type

A MultiCurve is a one-dimensional Geometry Collection whose elements are Curves. MultiCurve is a non-instantiable class in the OGC specification, it defines a set of methods for its subclasses and is included for reasons of extensibility. SDO and SDE support subclass MultiLineString. A MultiCurve is simple if and

Table 5. Assertions for MultiCurve data type (subclass LineString)

OGC Assertion	SDO	SDE
All elements of the MultiCurve are Curves	Yes	Yes
A MultiCurve is topologically closed	Yes	Yes

only if all of its elements are simple and the only intersections between any two elements occur at points that are on the boundaries of both elements.

4.5 The Surface Data Type

In planar two dimensional space a Surface is a planar two dimensional geometric object.

A Surface is called *simple* by OGC when it consists out of a single 'patch' that is associated with one 'exterior boundary' and zero or more 'interior' boundaries that do not cross.

Polyhedral surfaces are formed by 'stitching' together simple surfaces along their boundaries. SDO and SDE do not support polyhedral surfaces.

The only instantiable subclass of Surface defined in the OGC specification is the Polygon. A Polygon is a planar surface, defined by one exterior boundary and zero or more interior boundaries. Each interior boundary defines a hole in the polygon.

Table 6. Assertions for Surface data type (subclass Polygon)

OGC Assertion	SDO	SDE
The boundary of a polygon consists out of LinearRings	Yes	Yes
A polygon may have zero or more interior boundaries	Yes	Yes
Polygons are topologically closed	Yes	Yes
No two Rings in the boundary cross	Yes	Yes
Exterior and interior boundaries are not connected	?	Yes
Boundaries of polygons do not cross	Yes	Yes

Due to these assertions Polygons are considered simple.

In SDO an additional assertion is made:

1. A Polygon must at least be determined by three co-ordinates.

In SDE the next additional assertions are made:

1. Co-ordinates of the exterior boundary must be in counter clockwise sequence,
2. Co-ordinates of the interior boudaries must be in clockwise sequence,
3. Multiple interior boundaries that touch are combined into one interior boundary,
4. An interior boundary that touches the outer boundary is converted into an inversion of the outer boundary.

The fact that polygons are composed out of LinearRings has impact on the geometry classification. The next set of special occurrences are considered:

Table 7. Occurrences of Surface data type (subclass Polygon)

Occurrence	OGC	SDO	SDE
Zero length boundary segment	Simple ?	Yes	No
Zero perimeter	Simple ?	?	No
Zero area	Simple ?	?	No
Boundary cross	Non simple	No	No
Boundary segment overlap	Non simple	No	No
Boundary segments touch	Non simple	Yes	No
Touching islands	Simple	Yes	No
Islands overlap	Non simple	No	No

4.6 The MultiSurface Data Type

A MultiSurface is a geometry whose elements are Surfaces. They are made up from Polygons. A MultiPolygon is a MultiSurface whose elements are Polygons.

Table 8. Assertions for MultiSurface data type (subclass Polygon)

OGC Assertion	SDO	SDE
The boundary of a MultiPolygon is a set of closed curves corresponding to the boundaries of its element Polygons	Yes	Yes
A MultiPolygon is defined as topologically closed	Yes	Yes
Each curve in the boundary of the MultiPolygon is in the boundary of exactly one element Polygon, and every curve in the boundary of an element Polygon is in the boundary of the MultiPolygon.	Yes	Yes
The interiors of two Polygons that are elements of a MultiPolygon do not intersect	Yes	Yes
The Boundaries of any two Polygons that are elements of a MultiPolygon may not 'cross' and may touch at only a finite number of points	Yes	Yes

The assertions for MultiPolygon prevent topological overlap of polygons that are elements of the MultiPolygon. The SDO and SDE data model are identical in that respect.

5 Discussion

The user is not interested in the way data are organised in the database. She/he only wants to store and retrieve. What is important to the user is that the system is capable of accommodating the users view of the world. In that respect it is important that the system is capable to accept the geometries as they are fed into the system. Unfortunately this is not always the case.

From the previous paragraph it can be derived that basic differences in the data models exist due to the acceptance of sequential co-ordinates at the same location and the acceptance of Line and boundary segments to cross or overlap within the geometry. Additionally SDE modifies interior boundaries of a Polygon when they interact. Within the OGC simple feature class no overlap or cross is allowed. As will be shown in the next paragraph it will place several types of occurrences in the non simple feature class where exchange of data between systems is difficult.

Regarding Point geometries there are no differences between OGC, SDO and SDE.

Regarding Line type geometries the most important differences are:

1. Zero length Line,
2. Zero length Line segments,
3. Self crossing,
4. Line segment overlap.

In the case of zero length Lines and zero length Line segments it should be noticed they these deflections from a simple geometry do not change the morphology of the geometry. From a data model view or storage volume perspective it may seem to be unnecessary to have two or more co- ordinates at the same location, but the system is not capable of determining why the co-ordinates are there. For example the co-ordinates or zero Line segment might be there as placeholders or foreign keys.

Self crossing and segment overlap does not change the properties of the geometry in its self. Self crossing becomes a problem when determining what to do with it (whether it is a true intersection or not). There are no uniform rules on how to deal with it in a certain context. To avoid such problems self crossing and segment overlap is defined not to be allowed.

Polygon boundaries are constructed using Lines and LineStrings. In all models the restrictions on these Lines and LineStrings also apply to Polygon. Specifically related to Polygon the most important differences are:

1. Zero area,
2. Zero perimeter,
3. Crossing and overlap of the internal and external boundaries,
4. Interactions between internal and external boundary.

"Zero area" and "zero perimeter" are related to the acceptance of co-ordinates at the same location and was discussed in the previous section on linear types.

"Crossing" and "overlap" are also related to the validation rules for Lines, but for Polygon there is the principle issue of validating boundary segments that do not determine the polygon characteristics. For example the common boundary segment of two internal boundaries. The common boundary has no relationship with the Polygon morphology or characteristics like perimeter and area. The question is whether this segment is a essential part of the geometry. In the next paragraph some examples are given.

OGC has made effort to exactly as possible define simple geometries. All other instances of geometries are considered non simple. Rules on converting non simple to simple geometries are absent. This leaves room for software developers to implement their own package specific rules on converting non simple to simple geometries. Developers are free to implement their definition of "simple and non-simple" as a proprietary superset of what OGC as defined.

6 Effects of Data Storage of Geometries

In order to determine the morphology and characteristics of geometries, as they emerge in every day practice, a representative geodata set of large volume is visited. In this case a data set which is intensively used by the Dutch GIS community is the Dutch National Topographic Map (TOP10vector) scale 1 : 10,000. The data set contains over 15 million polygons, 75 million separate lines and 1.5 million points (Cattenstart, 1992).

A program filter in conjunction with visual inspection was used to find specific occurrences of geometry morphologies. A set of morphologies was derived from the dataset and some are shown in Table 9. Based on the method of "strong inference" (Platt, 1964) these morphologies were inserted and tested in SDO and SDE whether they are valid or not.

In Table 9 the morphologies of geometries that are tested are presented. In the SDO and SDE column the package reaction is denoted as:

1. **Accepted:** The morphology is accepted as presented,
2. **Modified:** The morphology receives a co-ordinate modification. This can be discarding, addition or movement,
3. **Converted:** The morphology is split, joined or reclassified into an other (set of) geometry type(s),
4. **Rejected:** The morphology was not accepted,
5. **Unknown:** Due to imperfection of the checking routines the acceptance is unknown.

It is remarked here that in the case of modification not only the morphologies shape is changed but also related information might be lost. In SDE this might be a "Measure", in SDO a foreign key to other data which is coupled to the co-ordinate column. In Table 9 in the validation columns a code is provided referring to the next set of remarks:

0 = Zero length segment,
1 = LineString or polygon boundary is self-intersecting,

2 = Polygon does not close properly,
3 = The number of points is less than required for geometry,
4 = Polygon patch has no area,
5 = Co-ordinate discarded or added without loss of morphology.

From Table 9 it is clear that there is a variety of occurrences of geometries. Related to the OGC data model both simple and non simple geometries were found. This leaves room for interpretation and geometric modification of inserted entities. The table shows evidence of this manipulation.

Occurrence 1 shows overlap of Points and an SDO and SDE identical validation.

Occurrences 2 through 5 show variations of crossing, overlap and co-ordinates at the same location of lines. It shows how SDO and SDE deal with these geometries differently.

Occurence 6 is a more principle problem. Line segments overlap here and therefor the line is classified non simple. However, although crossing and overlap are not allowed in SDE the line is accepted as is.

Occurrence 7 shows how topological overlap is identically approached in the OGC, SDE and SDO data model.

Occurrences 8 through 13 show variations of crossing and overlap of polygons and are all validated identically by SDO and SDE in accordance with the OGC data model.

Occurrence 14 shows SDE merging both islands into one due to its data model rules.

Occurrence 15 shows a principle difference between SDO and SDE. In SDO the common internal boundary is preserved, in SDE discarded. The question is if the common boundary is an essential part of the geometry. The OGC data model does not provide rules in this case.

Occurrence 16 and 17 show although the data models are equal there are different interpretations leading to different results.

Occurrence 18 shows that in all data models dangling Line segments in polygon boundaries are not accepted.

Occurrence 19 shows an anomaly that is treated differently by SDO and SDE.

Table 9 is evidence of the fact that the data models of OGC, SDE and SDO differ from each other with the result that the same dataset appears differently in SDO and SDE. To the user it is important to understand geometry modifications and decomposition necessities since a phenomenon or entity in the users view is recomposed. On improper retrieval the user may think that the stored entities are corrupted, or even worse, that they do not longer exist in the database.

7 Conclusions and Future Developments

The comparison between the OGC Data Model Architecture and the SDO and SDE reveals differences. The SDO and SDE data model are therefor not fully OpenGIS simple feature compliant.

Table 9. Special geometry morphologies and their validation in SDO and SDE.

#	Geometry type	Morphology	Origin	OGC type	SDO valid	SDE valid
1	Point	1 x 2	2 geometries at the same location. Overlapping geometries.	Multi point	Accepted	Accepted
2	Line	1 x 2	Zero length line. Digitising error? Resolution error?	Simple	Accepted	Rejected code: 0
3	Line	2 3 1 4 5	Closed line.	Simple	Accepted	Accepted
4	Line	3 2 1 4 5	Crossing line with interior.	Non simple	Accepted	Accepted
5	Line	1 2 3 4 5	Zero length segment Digitising error ?	Simple	Accepted	Modified code: 0
6	Line	1 2 4 3 5	Reverse direction Resolution error?	Non simple	Accepted	Accepted
7	Polygon	1 4 1' 4' 2' 3' 2 3	Topological Overlap	Multi polygon	Rejected code: 1	Rejected code: 1
8	Polygon	5 4 1 3 2	Polygon with zero length segment.	Simple	Accepted	Modified code: 0
9	Polygon	6 5 1 2 3 4	Overlapping segments	Non simple	Rejected code: 1	Rejected code: 2
10	Polygon	8 7 4 3 5 6 1 2	Boundary inward loop with no co-ordinate at the crossing	Non simple	Rejected code: 1	Rejected code: 1
11	Polygon	9 8 4 3 5 7 6 1 2	Touching inside island.	Non simple	Rejected code: 1	Rejected code: 1
12	Polygon	8 7 3 4 6 5 1 2	Boundary outward loop with no co-ordinate at the crossing	Non simple	Rejected code: 1	Rejected code: 1
13	Polygon	8 7 4 3 5 6 1 2	Inward overlapping segment.	Non simple	Rejected code: 1	Rejected code: 1

14	Polygon	4' 3' 4 3 1' 2' 1 2	Touching islands	Simple	Accepted	Converted Modified code: 5
15	Polygon	4' 3' 1' 2' 4 3 1 2	Islands sharing a segment boundary.	Simple	Accepted	Converted Modified code: 5
16	Polygon	4' 3' 4 3 1' 2' 1 2	Islands overlapping.	Non simple	Rejected code: 1	Rejected code: 1
17	Polygon	6 5 8 7 3 4 1 2	Crossing outer boundary with co-ordinates at the crossing.	Multi polygon	Accepted	Modified code: 5
18	Polygon	6 5 3 4 1 2	Outer boundary with overlapping segments	Non simple	Rejected code: 1	Rejected code: 2
19	Polygon	8 7 5 6 4 3 1 2	Outer boundary segments split polygon.	Multi polygon	Rejected code: 1	Rejected code: 2

SDO and SDE differ in their set of geometry validation rules and the way they interpret the OGC specification. Both systems will react differently when the same data are inserted. This will lead to problems when data from these packages are integrated. To the user this is an most unwanted situation and is a rationale for the need of OpenGIS.

OGC has classified geometries into simple and non simple features. Rules for converting non simple to simple geometries are absent. This leaves room for software developers to implement their own concepts in dealing with certain types of geometries. This situation calls for a further extension of geometry definitions and rules to convert from non simple to simple entities in the OpenGIS Data Model Architecture.

OGC has distinguished data types Point, Curve and Surface. There is the notion of Node, defined as an intermediate location on a curve or surface boundary. A Node can have attributes. The OpenGIS Data Model Architecture is object based. Geometries are stored as independed entities. Normalised storage of geometries is thereby absent as well as intrinsic topological relationship. OGC is recommended to perform research on the notion of Node and intrinsic topological relationship.

From a users point the positive message is that with proper care, spatial data, can be incorporated, managed and set available in both packages, following the rules of OpenGIS. Still there is much work to be done before the OGC specification and its implementation in products like SDO and SDE will provide GIS products which are truly interoperable in character.

Future development of OpenGIS might concentrate around more in depth definitions of simple and non simple geometries, rules upon conversion from non simple to simple geometries, research and rules upon extension of the data model with entity types like Node or user compiled combinations and standards on determining topological relationships.

The link between the planar two dimensional data model and the three dimensional data model will become stronger and non linear interpolation between co-ordinates in planar 2D might become of more importance in this respect.

In all cases the user should be helped in her/his task to operate on data where the physical storage and representation schema of these data is totally transparent to her/him. However, the bumpy road to this goal leads through practical, scientific, technical and commercial interests.

References

1. Cattenstart, G.C., 1992, Van IGDS naar ARC/INFO. Projectgroep Digitaal Topografisch Basisbestand. Ministerie van Landbouw en Visserij, Nederland.
2. Cattenstart, G.C., 1998, Open Boundaries in GIS; OpenGIS and the SDO and SDE implementation. MSc dissertation at the Manchester Metropolitan University in conjunction with the Free University of Amsterdam.
3. NN, 1996, OpenGIS. Revision 0 of the Open GIS Consortium. Inc. Internet website www.OpenGIS.org.
4. NN, 1998, OpenGIS Simple Features Specification For SQL. Revision 1.0 of the Open GIS Consortium. Inc. Internet website www.OpenGIS.org.
5. NNa, 1997, Oracle Spatial Data Option User's Guide reference release 7.3.3. Part A53264-02.
6. NNb, 1997, SDE version 3.0 Administrators Guide. ESRI Redlands.
7. Platt, J.R., 1964, Strong Inference. Science 146, pp 347-352

Semantic and Geometric Aspects of Integrating Road Networks

Harry Uitermark, Anton Vogels, and Peter van Oosterom

Dutch Cadastre, P.O. Box 9046, 7300 GH Apeldoorn, The Netherlands
Phone: +31 55 528 5163, Fax: +31 55 355 7931.
Email: {uitermark,vogels,oosterom}@kadaster.nl

Abstract. A prototype of a Geographic Database Integrator is under investigation and development. One of the long term goals of the Geographic Database Integrator is to reduce the need for operator intervention in update operations between objects in different databases. This paper focuses on the research related to road network elements from two independently surveyed and maintained topographic databases, one at large scale and one at mid-scale. Central to the issue of update propagation is certainty of equivalence of different road object representations. Therefore, precise definitions of road segment and road junction are important. In both the large and mid-scale geographic data sets, the roads are area features, although the whole may be considered as one linear road network. In order to find the junctions and road segments, the constrained Delaunay triangulation is applied. Using these well defined elements, a strategy for finding equivalent or corresponding road network elements has been developed. This forms the basis for processing and propagating updates in the road network from one database to the other database.

1 Introduction: Scope, Context and Related Work

Geographic Database Integration is the process of establishing links between corresponding objects in different, heterogeneous and autonomously produced databases of a certain region [15]. The purpose of geographic database integration, in general, is to share geo-information between different sources. Sharing geo-information is a communication process. In communication the semantics or meaning of data is important and touches at the heart of interoperability. In this paper geographic database integration is being studied in the context of *update propagation*, that is the reuse of updates from one geographic database to another geographic database [20]. Geographic database integration gets more and more attention nowadays since the digitizing of traditional map series has ended. In these map series corresponding objects were only linked implicitly by a common reference system, the national grid [21]. In order to make these links more explicit, geo-science researchers and computer scientists have developed various strategies. In computer science *schema integration* has been the dominant methodology for database integration; see for example [11]. This approach has been extended for geographic databases; see [6] for an overview and see [2] for a fine example. Geo-scientists on the other hand have adopted methods from communication theory like *relational matching* [21] and, in our case, from the field of AI [14]. In our approach the construction and use of an *ontology* for geographic databases [9, 18] makes it possible to inspect the result of the geographic database integration process for *inconsistencies* [13].

But despite these advances in geographic database integration methodologies, there is still a problem that can not be solved by these methodologies alone, that is the demarcation of *homologous* entities, suitable for update propagation especially in the case of road networks; see Section 2. The remainder of this paper is organized as follows. Section 3 introduces the semantics of road networks, specifically the definitions of road segments and road junctions. In Section 4 it is demonstrated how these road segments and road junctions are demarcated by a constrained Delaunay triangulation algorithm. Also the relationship with the road center lines (skeleton) is discussed in this section. A six step update propagation method for road elements is given in Section 5. Note that a road element is a road segment or a road junction. Finally, descriptions of the conclusions and future work end this paper.

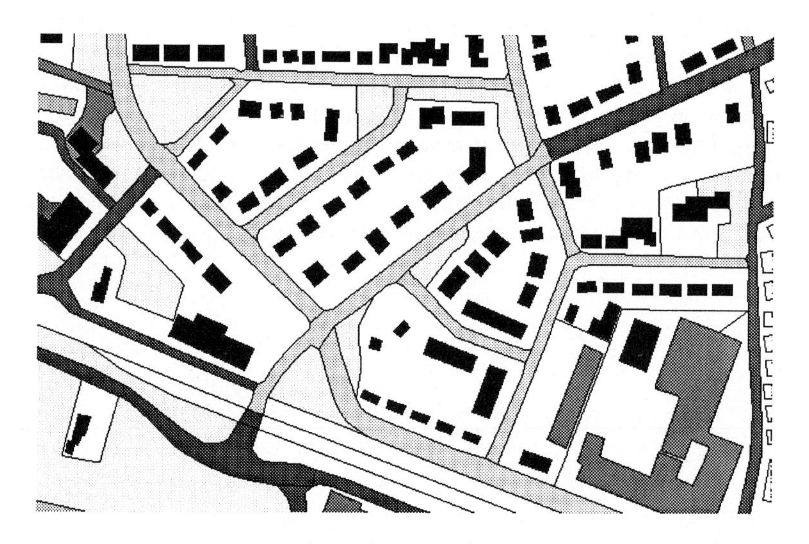

Fig. 1. The mid-scale topographic database TOP10vector.

2 The Demarcation of Homologous Entities

In previous research the propagation of updates in building objects (e.g. houses, building blocks, garages, annexes, etc.) has been studied [20]. Building objects share the property that they are demarcated quite naturally. In contrast, road elements are sometimes demarcated in a haphazard way; see Fig. 1 and Fig. 2.

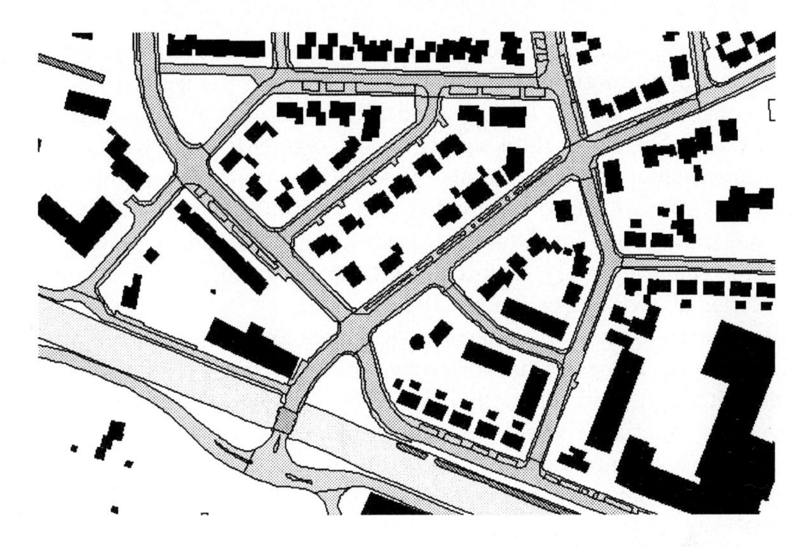

Fig. 2. The large-scale topographic database GBKN.

Here road elements from the mid-scale topographic database correspond to several different road elements from the large-scale topographic database. This n:m correspondence relationship is not suitable for pin-pointing an update from one database to another. So it is necessary to demarcate entities in such a way that 1:1 (or 1:n or n:1) correspondence relationships can be established; that means the demarcation of homologous entities.

3 Semantic Aspects of Road Networks

In general people might observe road networks, as a collection of line segments and nodes (junctions at point locations). In small scale geographic data sets roads are also represented as line features, but in large and mid-scale geographic data sets, the roads and junctions are represented as area features. In this research the GBKN (large scale base map, scale 1:1,000; see Fig. 2) and the TOP10vector (mid-scale map, scale 1:10,000; see Fig. 1) are used. In both geographic data sets roads are represented as area features. Although the roads are area features, it is also useful to think about the linear network topology, because the road network is a complex whole. It is not possible to look at one piece of the road network and forget about the other parts. A single change could affect multiple related road segments and road junctions.

As explained in the previous section, it is difficult to use complex road polygons for update propagation. Therefore, the road network is split in multiple elements. Before this can be done it is important to define these road elements in an unambiguous manner. The road element definition of the *Nationaal Wegenbestand* (NWB) [4] is used, which adheres to the European CEN standard Geographic Data Files (GDF) [5]. The definitions cover road segments and road junctions. A road segment is the reference unit on which users can put attribute information, road junctions are nodes which connect the segments. The NWB sees road segments as line features and road junctions as points. In the large and mid-scale geographic data sets the road segments and road junctions are area features, still it is tried to adhere the NWB as much as possible. The NWB declares that roads have to be divided into segments and junctions if three or more roads come together. Roads also have to be divided at places where the street name changes, or the maintenance is done by a different organization, or at the border of a village or a municipality.

This research focuses on the geometric aspect and the roads are just divided at the junctions. The road network now consist of two types of area features: road segments and road junctions. The NWB has detailed criteria for defining the type of a junction. When two T-junctions are close together, you have to treat them as one junction when one of the extended boundaries of a road lies between the boundaries of the road on the other side of the junction; see Fig. 3. Otherwise you get two independent T-junctions; see Fig. 4.

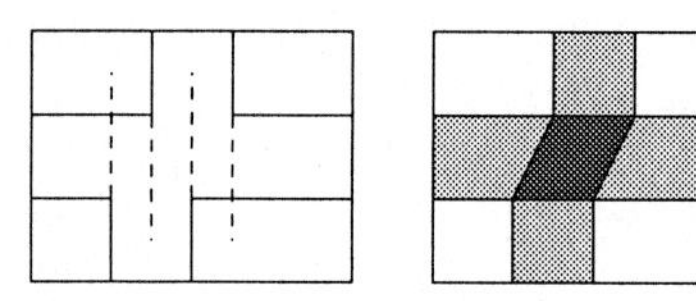

Fig. 3. These 'T-junctions' result in one road junction area and four road segments.

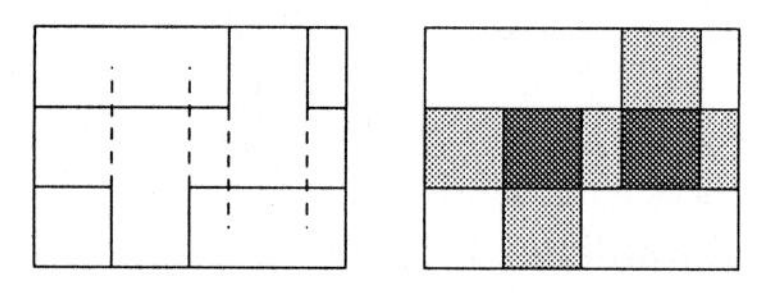

Fig. 4. These T-junctions result in two road junction areas and five road segments.

It is not possible to use GBKN information directly to update the TOP10-vector. Generalization and aggregation play a role in converting GBKN updates into TOP10vector updates; see Figs 1 and 2. In the GBKN speed ramps and small roundabouts are represented as small area features. The same is true for sidewalks and parking strips. None of these objects are represented in the TOP10vector. Updates in these small objects may not have any influence alone, but several small updates together could create an update which might have enough relevance to be propagated.

4 Geometric Aspects of Road Networks

The method used to find the junctions in the road network is based on triangulating the road area. This triangulation is used to compute a skeleton of the road. The nodes in the road skeleton define the location of the junctions and the edges of the surrounding triangle are used to separate the road network in road segment areas and road junction areas. The *constrained Delaunay triangulation* (CDT) algorithm described in [17] is used to compute the triangulation. A CDT over a planar set of n vertices together with a set of non-intersecting edges has the properties that all specified vertices and edges can also be found in the output and the result is as close as possible to the unconstrained Delaunay triangulation [8]. That is the circumcircle of any triangle has no vertex inside unless the vertex is at the other side of a constraining edge. In the case of the road network the set of separate input vertices is empty and the set of edges consists of the road boundary edges.

4.1 Constrained Delaunay Triangulation

The applied algorithm runs in $O(n \log n)$ time, which is asymptotically optimal [10]. The algorithm is based on the concept of two other algorithms. The first algorithm is the unconstrained Delaunay triangulation (UDT) algorithm of Lee and Schachter [7] and the second algorithm is the CDT algorithm of Chew [1]. More details about the algorithm and its implementation can be found in [16].

In general the input of the CDT algorithm is a graph $G = (V, E)$ in which V is a set of n vertices (separate points and the end points of the input edges) and E is a set of edges, the so called *G-edges*. Two different kinds of edges appear in a CDT: G-edges, already present in the graph, and *D-edges*, created by the CDT algorithm. If the graph has no G-edges then the CDT and the UDT (unconstrained Delaunay triangulation) are the same. The applied algorithm is based on the *divide-and-conquer* paradigm. The graph can be thought of to be contained in an enclosing rectangle (the domain). This rectangle is subdivided into n separate vertical strips in such a way that each strip contains exactly one region (a part of the strip) which in turn contains exactly one vertex. After dividing the graph into n initial strips, adjacent strips are pasted together in pairs to form new strips. During this pasting new regions are formed of existing regions for which the combined CDTs are calculated. This pasting of adjacent strips is repeated following the divide-and-conquer paradigm until eventually exactly one big strip, consisting of exactly one big region, is left for which the CDT is calculated.

4.2 Interpreting Triangles of the Road Network

The method used to derive the skeleton from the CDT is based on Wilschut *et al.* [22]. In the triangulation four different types of triangles cover the road area based on the number of G-edges in the boundary of the triangle: 0-triangle, 1-triangle, 2-triangle, and 3-triangle. The 3-triangle is an exception and does only occur when there is a non-connected road area with triangular shape in the input data set. A junction can be found by a triangle which has only D-edges and no G-edges, that is a 0-triangle; see the light triangles in Figs 5 and 6. A T-junction is a single 0-triangle with no neighbor 0-triangles. A normal crossing (4-way junction) is defined by 2 adjacent 0-triangles; e.g. the junctions at the bottom center in Fig. 6. In general a n-way junction is defined by (n-2) adjacent 0-triangles.

The 1-triangles form the building blocks of connecting road segments. Finally, the 2-triangles define the end points of the road network, that is, the dead-end

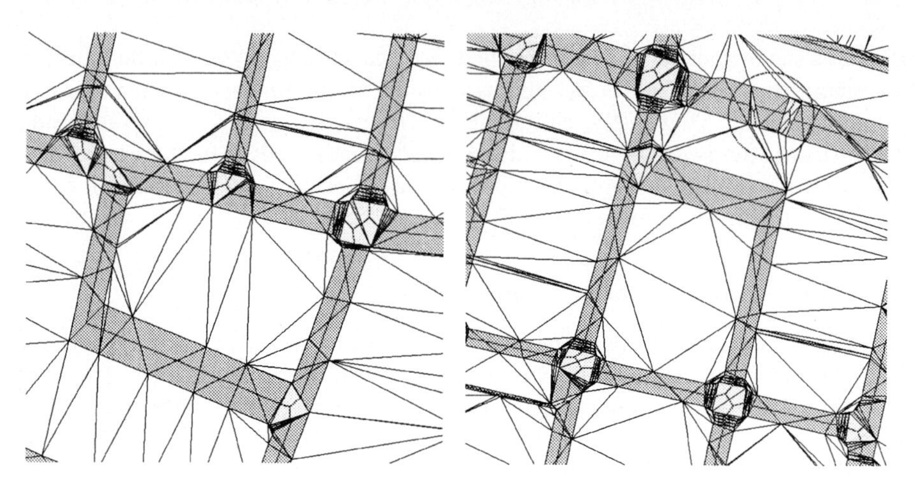

Fig. 5. Triangulated road network

Fig. 6. Another road network

streets. Also, a small lump in the boundary of a road segment may result in 2-triangle and therefore in a small dead-end street; see the top right road segment in Fig. 6. A solution for this 'problem' is to (virtually) remove the 2-triangle smaller than a certain threshold area adjacent to a 0-triangle. In such a case, also the original 0-triangle becomes a (virtual) 1-triangle, that is, part of a connecting road segment and not a junction. Further, due to the fine distribution of vertices on the road boundary, the two 0-triangles defining a 4-way junction may not be topologically adjacent. In this situation another approach is needed to couple the two 0-triangles for one 4-way junction: if the distance between the center points of the two 0-triangles is less than a certain threshold, then the 0-triangles can be coupled. Note that this is also true for n-way junctions in general.

The separation (demarcation) between road segments and road junctions is defined by the edges of 0-triangles and leaving out the shared edges of 0-triangles in case of n-way junction with $n > 3$. A last point of attention is the location of the junction (point) and the separation edges. In case of the T-junction, there may not be a close vertex at the road boundary at the other side of the road; see Fig. 7: in the top horizontal road, the junction is about 30 meters to the east of the actual location of the junction. This may be solved by adding additional (intermediate) vertices within long road boundary edges; whenever they are longer than a certain maximum length (e.g. the average width of a road, about 15 meters); see Fig. 8. Note that adding additional vertices will increase the computing resources (memory and time) during the triangulation and subsequent processes. First applying line generalization may reduce the required computing resources during the triangulation. It has also the advantages that it removes some virtual 2-triangles and that close, but no direct neighbor 0-triangles, may become direct neighbor 0-triangles (beneficial for finding 4- and higher way road junctions); see Fig. 9.

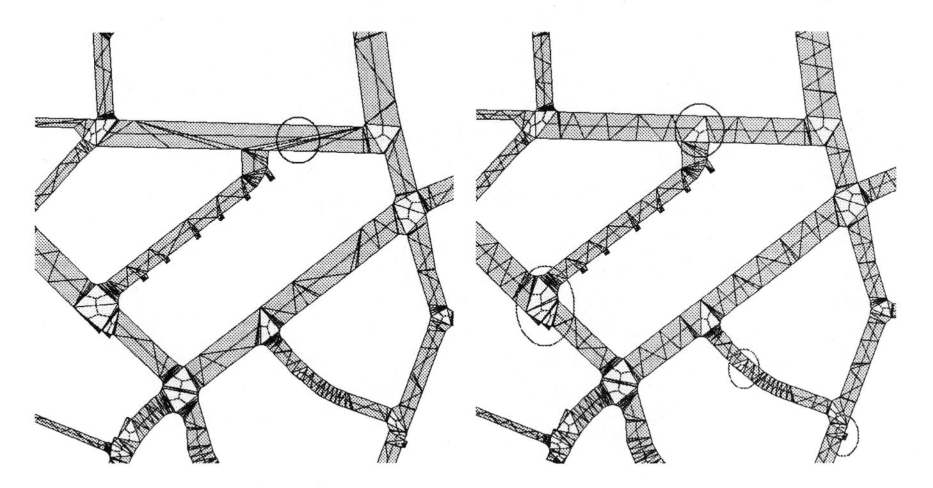

Fig. 7. Road junction with displaced 0-triangle (center node).

Fig. 8. Add more intermediate points: better 0-triangles.

4.3 Road Center Lines

Once the 0-, 1-, and 2-triangles are obtained it is not only possible to find the road junctions and dead-end roads, but it is also relatively easy to find a skeleton of the road. That is, the corresponding linear network based on the road center lines. The construction of the skeleton is based on following the middle of the internal edges of the 1-triangles. In a 0-triangle the center of the triangle is connected to the middle of all three edges. In a 2-triangle the middle of the D-edge is connected to the common point between the two G-edges, that is, the end-point of the road. Finally, this method needs some post processing in order to remove the 'dip' in a straight road in T-junction and also to make one center point of 4-way junction instead of two connected center points of T-junctions.

A general method for solving these problems is described in detail by Gao and Minami [3]. Their method is based on looking at the *trend-lines* of the parts of the center lines within a certain radius around the node (or average of a group of nodes within radius distance of each other). A pair of trend-lines with nearly the same angle is replaced by a straight line connecting the two corresponding center lines. In case there are more pairs of trend-lines with nearly the same angle, then the intersection(s) of the corresponding straight line connections is (are) computed. The other center lines are connected through their trend-line to the straight line. The location of the junction node is the (average) intersection point, to which all center lines are connected.

The method, described in this paper, to obtain center lines is a vector based approach. It assumes a topologically correct input of the road boundaries. In case the vector data is inaccurate, a raster based approach may be more appropriate. A description of this method is given by Thomas [12], in which the raster is represented by a compact run-length encoded binary image.

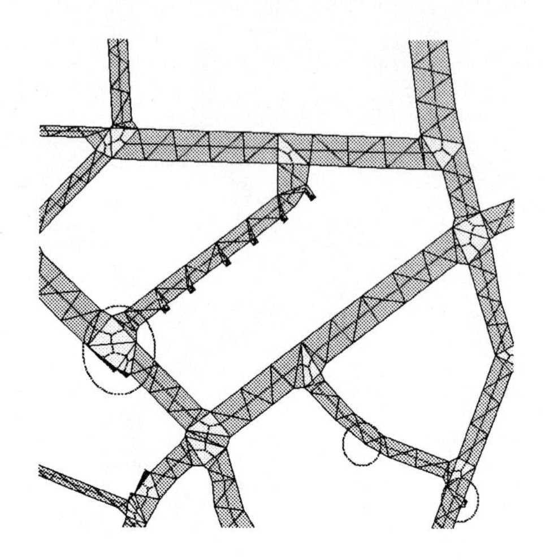

Fig. 9. Apply line generalization: less fake 0-triangles

5 Update Propagation

As explained before road update propagation is different from building update propagation. The most important difference is that buildings are usually unattached to other objects, whereas road segments are connected to road junctions and other road segments; it is a complex whole even after the road polygons are divided into road segments and road junctions. To reduce the complexity and find the 1:1 (or 1:n or n:1) correspondences in order to propagate the relevant updates we propose the following six steps:

- *Step 1: Synchronization of GBKN and TOP10vector to the same moment in time.* In general every 6 to 12 months updates in the field are measured and used to update the GBKN. On the other hand the TOP10vector is updated every four years. Before updates can be propagated to the other database, the two databases have to be synchronized. Synchronization means rolling back the GBKN in time until its date is the same as the TOP10vector date. This is possible because every object in the GBKN database has two time stamps: Tmin and Tmax[19]. Tmin is the date an object has been added to the database. Tmax is the date an object has been replaced by one or more other objects. These 'old' objects remain in the database, but are not 'valid'. If you bring back the database in time, the old objects become valid and represent the desired moment in the past.
- *Step 2: Create road segments and road junctions areas in the GBKN and the TOP10vector.* This step is based on the constrained Delaunay triangulation and is explained in the previous section.
- *Step 3: Find corresponding road segments and the road junctions areas between GBKN and TOP10vector.* The correspondences between road elements is found by computing the overlap between the elements from both the

GBKN and TOP10vector databases; the method is explained in [20] and is similar to finding correspondences between building updates.

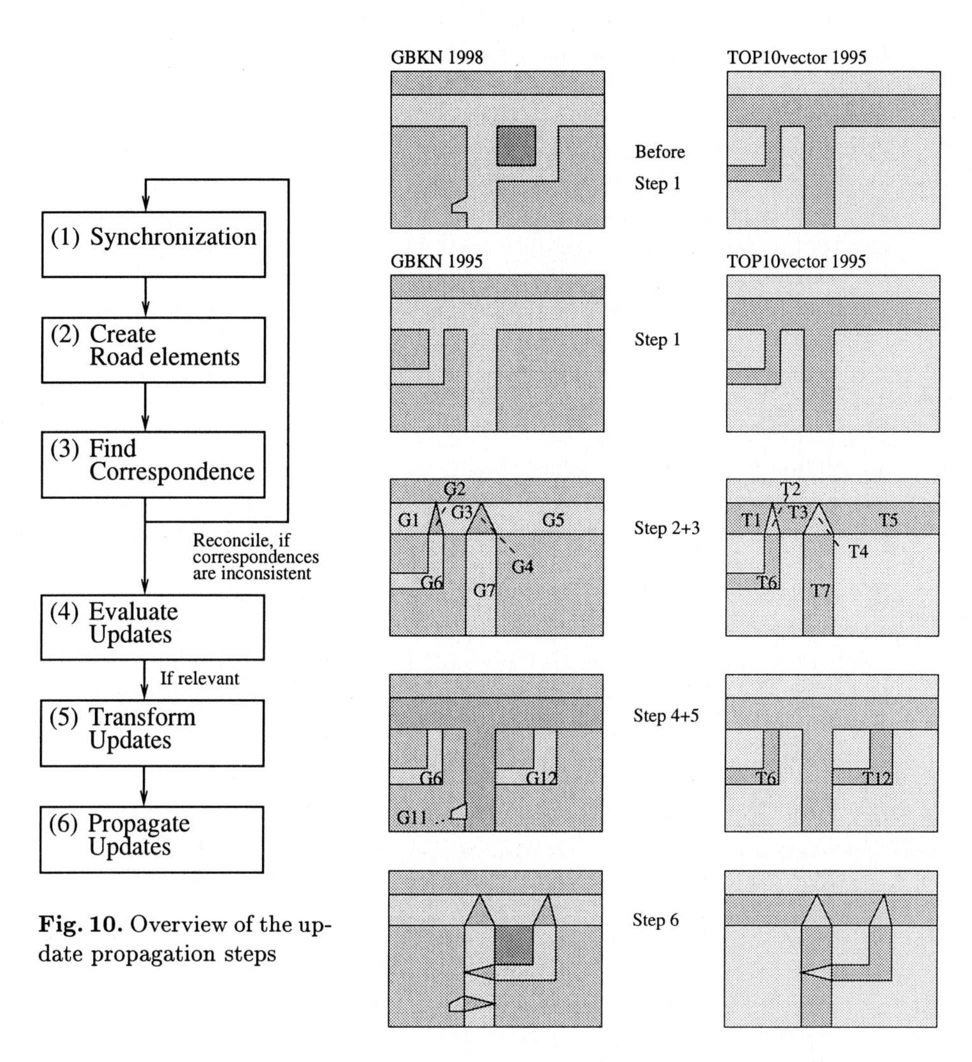

Fig. 10. Overview of the update propagation steps

Fig. 11. Changes in the road network.

Before it is possible to propagate updates, line feature updates (as a result of a change in the road boundary edges) have to be transformed into area feature updates. This has to be done because surveyors in the field will always collect (point and) line measurements. Attention has to be paid to which lines have been used for creating road objects, which areas are effected by deleted, changed or new lines. We also have to find an answer to the questions when and how GBKN road updates should and could be propagated into the TOP10vector?

- *Step 4: Decide if an update is important enough to be propagated to TOP10-vector.* Determine whether the update affects the TOP10vector objects while respecting generalization and aggregation rules and relevance of an object for the TOP10vector. For example, in Fig.11 the new road element G12 is relevant, but the new road element G11 is not relevant because it is too small.
- *Step 5: Transform GBKN updates into TOP10vectors updates.* The object definitions are not the same. For example, in the TOP10vector a ditch belongs to the road, but this is not the case in the GBKN. Before changing an update it is helpful to analyze the relationship between the objects in both geographic databases. We analyze correspondences at geometric, attribute and semantic levels to understand if and how database integration could be possible. Making it possible to find a mechanism to change GBKN updates into TOP10vector updates. For example, in Fig.11 the GBKN road element G12 is first generalized and its classification is adjusted, and it then becomes TOP10vector road element T12.
- *Step 6: Propagate the updates into the TOP10vector.* An update has changed an original object. If that update has to be propagated to TOP10vector, it has to be propagated to the TOP10vector object, which corresponds with the GBKN object, which has been updated. It is not just possible to propagate the new object and remove the old version just like propagation buildings. If there is a new road, it is also necessary to connect the new road to the existing road segments, road junctions and to other nearby areas. That is, it has to be fitted into the TOP10vector topology structure. For example, in Fig.11 the GBKN road element G12 is connected to road elements G5 and G7, in the same manner T12 has to be connected to T5 and T7 in the TOP10vector.

Fig.11 shows an example area with changes in the road network. In this figure only the roads are shown. One old road disappears, one new road appears.

6 Conclusion and Future Work

In this paper the importance of well defined road network elements is argued for the purpose of geographic database integration in general and update propagation in specific. It was shown that the constrained Delaunay triangulation gives a good basis for the demarcation of road segments and road junctions. However, a few refinements are required such as removing small lump (or very small dead end street) and grouping neighbor or very close T-junctions to one n-way junction.

Future work consists of experimenting with the update propagation steps described in the previous section with real data sets. Further, investigation into other 'linear' feature types, such as railroads and water ways, is planned. The question is whether the same method for defining objects is valid for these feature types. Finally, we have to consider update propagation, and geographic database integration in general, not only on a 'feature by feature' basis, but in a more integrated way. Very often the different feature types are embedded in the same planar topology structure and do heavily influence each other.

References

1. L. Paul Chew. Constrained delaunay triangulations. *ACM*, ?:215–222, 1987.
2. T. Devogele, C. Parent, and S. Spaccapietra. On spatial database integration. *International Journal of Geographical Information Systems*, 12:335–352, 1998.
3. Peng Gao and Michael M. Minami. Raster-to-vector conversion: A trend line intersection approach to junction enhancement. In *Auto-Carto 13*, pages 297–303, April 1997.
4. L. Heres and P. Plomp J. den Hartog-Sprockel. NWB-kernmodel en extenties. Technical Report LAV2_007, RWS/AVV, 1997.
5. L. Heres et al. Geographic data files (GDF). Technical report, CEN/TC 287, Brussels, 1995.
6. R. Laurini. Spatial multi-database topological continuity and indexing: a step towards seamless gis data interoperability. *International Journal of Geographical Information Systems*, 12:373–402, 1998.
7. D. T. Lee and B. J. Schachter. Two algorithms for constructing a delaunay triangulation. *International Journal of Computer and Information Sciences*, 9(3):219–242, 1980.
8. Franco P. Preparata and Michael Ian Shamos. *Computational Geometry*. Springer-Verlag, New York, 1985.
9. Ravi. Geo-information terrain model. A Dutch standard for: Terms, definitions and general rules for the classification and coding of objects related to the earths surface (NEN-3610). Technical report, Ravi Netherlands Council for Geographic Information, Amersfoort, 1995.
10. M. I. Shamos and D. Hoey. Closest-point problems. In *Proceedings of the 16th Annual Symposium on the Foundation of Computer Science*, pages 151–162, October 1975.
11. S. Spaccapietra, C. Parent, and Y. Dupont. Model independent assertions for integration of heterogeneous schemas. *VLDB Journal*, 1:81–126, 1992.
12. Federico Thomas. Generating street center-lines from inaccurate vector city maps. *Cartography and Geographic Information Systems*, 25(4):221–230, 1998.
13. H. T. Uitermark. *Ontology-based map integration*. PhD thesis, Wageningen Agricultural University, 1999. In preparation.
14. H. T. Uitermark, P. J. M. van Oosterom, N. J. I. Mars, and M. Molenaar. Propagating updates: finding corresponding objects in a multi-source environment. In T. K. Poiker and N. Chrisman, editors, *Proceedings of the 8th International Symposium on Spatial Data Handling, Vancouver, Canada, 1998*, pages 580–591, 1998.
15. Harry T. Uitermark. The integration of geographic databases. Realising geodata interoperability through the hypermap metaphor and a mediator architecture. In *Second Joint European Conference & Exhibition on Geographical Information (JEC-GI'96), Barcelona, Spain*, pages 92–95, 1996.
16. Joost van Bemmelen and Wilko Quak. Master's thesis: Cross country movement planning. Technical Report FEL-93-S205, TNO Physics and Electronics Laboratory, August 1993.

17. Joost van Bemmelen, Wilko Quak, Marcel van Hekken, and Peter van Oosterom. Vector vs. raster-based algorithms for cross country movement planning. In *Auto-Carto 11*, pages 304–317, October 1993.
18. P. E. van der Vet and N. J. I. Mars. Bottom-up construction of ontologies. *IEEE Transactions on Knowledge and Data Engineering*, 10:513–526, 1998.
19. Peter van Oosterom. Maintaining consistent topology including historical data in a large spatial database. In *Auto-Carto 13*, pages 327–336, April 1997.
20. F. A. van Wijngaarden, J. D. van Putten, P. J. M. van Oosterom, and H. T. Uitermark. Map integration. Update propagation in a multi-source environment. In R. Laurini, editor, *5th ACM Workshop on Advances in Geographic Information Systems ACM-GIS'97, Las Vegas, Nevada, USA*, pages 71–76, 1997.
21. V. Walter and D. Fritsch. Linking objects of different spatial data sets by integration and aggregation. *GeoInformatica*, 5:1–23, 1997.
22. Annita N. Wilschut, Roelof van Zwol, Niek Brasa, and Jan Flokstra. Road collapse in magnum. In R. Laurini, editor, *6th ACM Workshop on Advances in Geographic Information Systems ACM-GIS'98, Washington DC, USA, 6-7 november 1998*, 1998.

Assessing Semantic Similarities among Geospatial Feature Class Definitions

M. Andrea Rodríguez[1,2], Max J. Egenhofer[1,2,3], and Robert D. Rugg[1,4]

[1] National Center for Geographic Information and Analysis, University of Maine, Orono, ME 04469-5711, USA
[2] Department of Spatial Information Science and Engineering, University of Maine, Orono, ME 04469-5711, USA
[3] Department of Computer Science, University of Maine, Orono, ME 04469-5752, USA
[4] Department of Urban Studies and Planning, Virginia Commonwealth University, Richmond VA 23284-2008, USA
{andrea,max}@spatial.maine.edu, rugg@vcu.edu

Abstract. The assessment of semantic similarity among objects is a basic requirement for semantic interoperability. This paper presents an innovative approach to semantic similarity assessment by combining the advantages of two different strategies: feature-matching process and semantic distance calculation. The model involves a knowledge base of spatial concepts that consists of semantic relations (is-a and part-whole) and distinguishing features (functions, parts, and attributes). By taking into consideration cognitive properties of similarity assessments, this model represents a cognitively plausible and computationally achievable method for measuring the degree of interoperability.

1 Introduction

Since the first studies on interoperability, progress has been made concerning syntactic interoperability, i.e., data types and formats, and structural interoperability, i.e., schematic integration, query languages, and interfaces (Sheth 1998). As current information systems increasingly confront information and knowledge issues, semantic interoperability becomes a major challenge for the next generation of interoperating information systems.

In information systems, semantics relates the content and representation of information to the entities or concepts in the world (Meersman 1997). The problem of semantic interoperability is the identification of semantically similar objects that belong to different databases and the resolution of their schematic differences (Kashyap and Sheth 1996). Schematic heterogeneity can only exist, and therefore be solved, for semantically similar objects (Bishr 1997). Studies have suggested the use of an ontology (Guarino and Giaretta 1995) as a framework for semantic similarity detection (Bishr 1997, Kashyap and Sheth 1998). On the one hand, a possible approach is to create a knowledge base in terms of a common ontology, upon which it is possible to detect semantic similarities and to define a mapping process between concepts (Lenat and Guha 1990, Kahng and McLeod 1998). On the other hand, we can expect that in a realistic scenario new concepts will be added to or eliminated from the ontology. There may be different ways to classify a concept based on the specific application and the degree of detail of the concept's definition. Hence, the reuse and integration of existing domain specific ontologies becomes necessary (Kashyap and Sheth 1998, Mena *et al.* 1998).

This paper presents a computational model for similarity assessment among entity classes. We use the term entity classes to describe concepts in the real world and to distinguish their semantics from the semantics of data modeled and represented in a database. By concepts in the real world, we mean the cognitive representation which a person uses to recognize and categorize objects or events (Dahlgren 1988). Naturally, achieving semantic representation of objects in a database implies a good understanding of the semantics of the corresponding concepts in the real world. Consequently, our work considers studies done by cognitive scientists in the area of knowledge and behavior as well as by computer scientists in the domain of artificial intelligence.

The similarity model assumes a common ontology that includes the real world concepts' distinguishing features and interrelationships. A feature-matching process, together with a semantic distance computation, provides a strategy to create a model that satisfies cognitive properties of similarity assessment. In particular, we capture the idea that similarity assessment is not always a symmetric evaluation: similarity is a result of the commonalities and differences between two concepts, and the relevance of the distinguishing features (functions, parts, and attributes) may differ from one to another. In addition, is-a relations are complemented with part-whole relations to create an ontology that better reflects the interrelationships between concepts.

We focus on the domain of spatial information and we combine two existing sources of information, WordNet (Miller 1995) and the Spatial Data Transfer Standard (USGS 1998), to create a common ontology that is used for the development of a prototype. The scope of this study includes only the evaluation of similarity within this common ontology. The analysis of how to integrate two domain specific ontologies is left for future work.

The remainder of the paper is organized as follows. Section 2 reviews different approaches to the evaluation of semantic similarity. Section 3 describes the components of the definition of entity classes. In Section 4 we present our similarity model, and we illustrate its use with an example in Section 5. Finally, conclusions and future work are presented in Section 6.

2 Methods for Comparing Semantics

Most of the models proposed by psychologists are feature-based approaches, which use features that characterize entities or concepts (for example, properties and role). Using set theory, Tverski (1977) defined a similarity measure as a feature-matching process. It produces a similarity value that is not only the result of common features, but also the result of the differences between two entity classes. A different strategy for feature-based models is to determine a semantic distance between concepts as their Euclidean distance in a semantic, multidimensional space (Rips *et al.* 1973). This approach describes similarity by a monotonic function of the interpoint distance within a multidimensional space, where the axes in this space describe features of concepts. Krumhansl (1978) introduced the distance-density model based on a distance function for similarity assessment that complements the interpoint distance with the density of the space. This model assumes that, within dense regions of a given stimulus range, finer discriminations are made than within relatively less dense subregions.

A shared disadvantage of feature-based models is that two entities are seen to be similar if they have common features; however, it may be argued that the extent to which a concept possesses or is associated with a feature may be a matter of a degree (Krumhansl 1978). Consequently, a specific feature can be more important to the meaning of an entity class than another. On the other hand, the consideration of common features between entity classes seems to match the way people assess similarity.

With a different approach, computer scientists have defined similarity measures whose basic strategies make use of the semantic relations between concepts. These semantic relations are typically organized in a semantic network (Collins and Quillian 1969) according to which the links between nodes denote concepts. The semantic distance results in an intuitive and direct way of evaluating similarity in a hierarchical semantic network. For a semantic network with only is-a relations, Rada *et al.* (1989) pointed out that the semantic relatedness and semantic distance are equivalent and we can use the latter as a measure of the former. They defined conceptual distance as the length of the shortest path between two nodes in the semantic network. This distance function satisfies metric properties of minimality, symmetry, and triangle inequality.

Although the semantic distance models have been supported by a number of experiments and have shown to be well suited for a specific domain, they have the disadvantage of being highly sensitive to the predefined semantic-network architecture. In a realistic scenario, adjacent nodes are not necessarily equidistant. Irregular density often results in unexpected conceptual distance measures. Most concepts in the middle to high sections of the hierarchical network, being spatially close to each other, would therefore be deemed to be conceptually similar to each other. In order to account for the underlying architecture of the semantic network, Lee *et al.* (1993) argued that the semantic distance model should allow for weighted indexing schema and variable edge weights. To determine weights the structural characteristics of the semantic network are typically considered, such as the local density network, the depth of a node in a hierarchy, the type of link, and the strength of an edge link.

Some studies have considered weighted distance in a semantic network. Richardson and Smeaton (1996) used a hierarchical concept graph (HCG) derived from WordNet (Miller 1995) to determine similarity. They defined weights of links in a semantic network by the density of the HCG, estimated as the number of links, and by the link strength, estimated as a function of a node's information content. Likewise Jiang and Conrath (1997) proposed the use of information content to determine the link strength of an edge. The information content of a node is obtained from the statistical analysis of word frequency occurrences in a corpus. The general idea of the information content is that, as the probability of occurrence of a concept in a corpus increases, informativeness decreases, such that the more abstract a concept, the lower its information content.

Richardson and Smeaton (1996) and Richardson *et al.* (1994) used a hierarchical network and information theory to propose an information-based model of similarity. Their approach to modeling semantic similarity makes use of the information content as described above, but it does not include distance as a basic strategy for similarity assessment. Conceptual similarity is considered in terms of class similarity. The similarity between two classes is approximated by the information content of the first superclass in the hierarchy that subsumes both classes. In the case of multiple inheritance (Cardelli 1984), similarity can be determined by the best similarity value among all various senses the classes belong to. The information-content model requires less information on the detailed structure of the network. On the other hand, many polysemous words and multi-worded classes will have an exaggerated information content value. The information-content model can generate a coarse result for the comparison of concepts, because it does not differentiate the similarity values of any pair of concepts in a sub-hierarchy as long as their "smallest common denominator is the same (Jiang and Conrath 1997).

In the cognitive-linguistics domain, Miller and Charles (1991) discussed a contextual approach to semantic similarity. They developed a measure for similarity that is defined in terms of the degree of substitutability of words in sentences. For words from the same syntactic category and the same domain, the more often it is possible to substitute one word by another within the same context, the more similar the words are. The problem with this similarity measure is that it is difficult to define a systematic way to calculate it.

Based on our analysis of current models for semantic similarity, we propose a combination of the features-matching process and the evaluation of semantic distance. We expect that this interpreted model will provide a similarity measure that is not only cognitively plausible, but also computationally achievable.

3 Components of Entity Class Definitions

Important components of the entity class definitions are the semantic relations among classes. We select a specific domain, spatial information systems, and describe the set of entity classes and their semantic relations as an ontology. In artificial intelligence, the term ontology has been used in different ways. Ontology has been defined as a "specification of a conceptualization" (Gruber 1995) and as a "logical theory which gives an explicit, partial account of a conceptualization" (Guarino and Giaretta 1995). Thus, an ontology is a kind of knowledge base that has an underlying conceptualization. For our purpose, an ontology will be used as a body of knowledge that defines (1) primitive symbols used in the representation of meaning and (2) a rich system of semantic relations connecting those symbols.

The most common semantic relation used in an ontology is the is-a relation, also called the hypernymic or superordinate relation. This relation goes from a specific to a more general concept that resembles the generalization mechanism of object-oriented theory (Dittrich 1986). The is-a relation is a transitive and asymmetric relation that defines a hierarchical structure, where terms inherit all the characteristics of their superordinate terms.

Mereology, the study of part-whole relations (Guarino 1995), plays another important role for ontology. Studies have usually assumed that part-whole relations are transitive such that if a is part of b and$b$ is part of c, then a is part of c as well. Linguists, however, have expressed concerns about this assumption (Cruse 1979, Iris *et al.* 1988). Explanations of the transitive problem rely on the idea that part-whole relations are not one type of relation, but a family of relations. Winston *et al.* (1987) defined six types of part-whole relations: component-object (e.g., pedal-bike), member-collection (e.g., tree- forest), portion-mass (e.g., slice-cake), stuff-object (e.g., steel-bicycle), feature-activity (e.g., paying-shopping), and place-area (e.g., oasis-desert). Chaffin and Herrmann (1988) extended the previous classification with a seventh meronymic relation, phase-process (e.g., adolescence-growing up). For this work, we only consider the component-object relation with the properties of asymmetry and (with some reservations) transitivity.

When defining entity classes, the part-whole converse relations do not always hold. For example, we can say that a building complex has buildings, i.e., building complex is the whole for a set of buildings; however, buildings are not always part of a building complex. Thus, we distinguish the two relations, "part-of" and "whole-of," to be able to account for such cases.

Although the general organization of the entity classes is given by their semantic relations, this information is not enough to distinguish one class from another. For example, a hospital and an apartment building have a common superclass building; however, this information is insufficient to differentiate a hospital from an apartment building. Considering that entity classes correspond to nouns in linguistic terms, we borrow Miller's (1990) description of nouns and propose to assign what he called *distinguishing features* to each class. Distinguishing features include parts, functions, and attributes.

Parts are structural elements of a class, such as roof and floor of a building. We could make a further distinction between "things" that a class must have ("mandatory") or can have ("optional"). Note that parts are related to the relation part-whole previously discussed. While the relation part-whole works at the level of entity-class definitions and forces us to define all the entity classes involved, part features can have items that are not always defined as entity classes in our model. Function features are intended to represent what is done to or with a class. For example, the function of a college is to educate. Thus, function features can be related to other terms such as *affordances* (Gibson 1979) and *behavior* (Khoshafian and Abnous 1990). Attributes correspond to additional characteristics of a class that are not considered by either the set of parts or the set of functions. For example, some of the attributes of a building are age, user type, owner type, and architectural properties. Using a lexical categorization, parts are given by nouns, functions by verbs, and attributes by nouns whose associated values are given by adjectives or other nouns.

In addition to semantic relations and distinguishing features, two more linguistic concepts are taken into consideration for the definition of entity classes. Entity classes are associated with concepts represented in natural language by words. Natural language understanding distinguishes two properties of the mapping between words and meanings, polysemy and synonymy. Polysemy arises when the same word may have more than one meaning, different *senses*. Synonymy corresponds to the case where two different words have the same meaning (Miller *et al.* 1990). Our entity-class definition incorporates synonyms, such as *parking lot* and *parking area*, and different senses of entity classes, such as the case when a *bank* could be an elevation of the seafloor, a sloping margin of a river, an institution, or a building.

4 A Computational Method for Assessing Similarities of Entity Classes

We introduce a computational model that assesses similarity by combining a feature-matching process with a semantic distance measurement. While our model uses the number of common and different features between two entity classes, it defines the relevance of the different features in terms of distance in a semantic network.

For each type of distinguishing features (i.e., parts, functions, and attributes) we propose to use a similarity function $S_t(c_1, c_2)$ (Equation 1) that is based on the *ratio model* of a feature-matching process (Tversky 1977). In $S_t(c_1, c_2)$, c_1 and c_2 are two entities classes, t symbolizes the type of features, and C_1 and C_2 are the respective sets of features of type t for c_1 and c_2. The matching process determines the cardinality (#) of the set intersection ($C_1 \cap C_2$) and the set difference ($C_1 - C_2$), defined as the set of all elements that belong to C_1 but not to C_2.

$$S_t(c_1, c_2) = \frac{\{C_1 \cap C_2\}_\#}{\delta_t(C_1, C_2)} \qquad \text{with} \qquad (1)$$

$$\delta_t(C_1, C_2) = \{C_1 \cap C_2\}_\# + \alpha\{C_1 - C_2\}_\# + (1 - \alpha)\{C_2 - C_1\}_\#$$

This similarity function yields values between 0 and 1. The extreme value 1 represents the case when everything is common between two entity classes, or when the non-common features between two entity classes do not affect the similarity value (i.e., the coefficient of the non-common features is zero). The value 0, in constrast, occurs when everything is different between two entity classes. The weight α is determined as a function of the distance between the entity classes (c_1 and c_2) and the immediate superclass that subsumes both classes. This corresponds to the least upper bound (l.u.b.) between two entity classes in partially ordered sets (Birkhoff 1967). When one of the concepts is the superclass of the other, the former is also considered the immediate superclass (l.u.b.) between them. The distance of each entity class to the l.u.b. is normalized by the total distance between the two classes, such that we obtain values in the range of 0 to 1. Then, to obtain the final values of α, we define an asymmetric function (Equation 2).

$$\alpha(c_1, c_2) = \begin{cases} \frac{d(c_1, l.u.b.)}{d(c_1, c_2)} & d(c_1, l.u.b.) \leq d(c_2, l.u.b.) \\ 1 - \frac{d(c_1, l.u.b.)}{d(c_1, c_2)} & d(c_1, l.u.b.) > d(c_2, l.u.b.) \end{cases} \tag{2}$$

The assumption behind the determination of α is that similarity is not necessarily a symmetric relation (Tversky 1977). For example, "a hospital is similar to a building" is a more general agreement than "a building is similar to a hospital." It has been suggested that the perceived distance from the prototype to the variant is greater than the perceived distance from the variant to the prototype, and that the prototype is commonly used as a second argument of the evaluation of similarity (Rosch and Mervis 1975, Krumhansl 1978). Hence, we assume that the non-common features of the concept used as a reference (the second argument) should be more relevant in the evaluation.

An interesting case occurs when comparing a class with its superclass or vice versa. Since subclasses inherit all features of their superclasses, only subclasses may have non-common features. It can be easily seen that when comparing a class with its superclass or vice versa, the weight associated with the non-common features of the first argument is 0 (α) and the weight for the non-common features of the second argument is 1 ($1-\alpha$). By considering the direction of the similarity evaluation, a class will be more similar to its superclass than the same superclass to the class. Currently and for the purpose of calculating the weight α, the part-of relation is treated like the is- as relation. The difference of these two relations depends upon the inheritance property of the is-a relation. The effect of the part-of relation can be illustrated when comparing a building with a building complex or vice versa. With our model, a stronger similarity is found between the building and the building complex than between the building complex and the building. Note, however, that the similarity between the whole and its parts could also be higher, since there is not an inheritance property for this semantic relations that forces the parts to have all the features of the whole.

When searching for an entity class, synonyms are incorporated at the beginning of an evaluation of similarity. In addition, synonyms are also taken into account in the matching process of parts, functions, and attributes. Each term (entity class, part, function, or attribute) is treated in the same way as its synonyms. Words with different semantics or senses (polysemy) are also included. We handle different senses of entity class as independent entity classes with a common name. For parts, functions, and attributes, we first match the senses of the terms and then we evaluate the set-intersection or set-difference operation among the set of features. A term in one sense might have a set of synonyms, therefore, we match terms or their synonyms that belong to the same sense. For example, the verb "to play" has two different senses in our database, play for recreation and play for competition. For any entity class that has the function "to play," the knowledge base also includes the sense of the word such that the system can find the synonyms of "to play" for the respective sense.

The global similarity function $S(c_1, c_2)$ is a weighted sum of the similarity values for parts, functions, and attributes (Equation 3), where ω_p, ω_f, and ω_a are weights of the similarity values for parts, functions, and attributes, respectively. These weights define the importance of parts, functions, and attributes that might vary among different contexts. The weights all together must add up to 1.

$$S(c_1, c_2) = \omega_p \cdot S_p(C_1, c_2) + \omega_t \cdot S_t(c_1, c_2) + \omega_a \cdot S_a(c_1, c_2) \tag{3}$$

5 An Example

We have implemented a software prototype for the similarity assessment. It used WordNet (Miller 1995) and the Spatial Data Transfer Standard (SDTS) (USGS 1998) to derive a knowledge base. From SDTS we extracted the entity classes to be defined, their partial definition of is-a relations, and the attributes for entity types. By using WordNet we complemented the is-a relations with the part- whole relations and we obtained the structural elements (parts) of entity types. Finally, functions were derived from verbs explicitly used in the description of entity classes, augmented by common sense.

To illustrate the use of our model for interoperability, consider an urban-planning application that deals with the rehabilitation of the downtown of a city. To accomplish the goal, planners have decided to analyze and compare the downtowns of cities of similar sizes that are considered high quality examples of urban life. In the first instance, planners are concerned about the functional components of the downtown, i.e., entity classes, and they have left for *a posteriori* analysis the geometric distribution of these components.

Maps of each downtown are obtained from different spatial databases and we face the problem of comparing the semantics of entity classes. For the time being, we assume that maps are based on a common ontology because they were created by using the same conceptualization. Although the assumption of a unique ontology simplifies the problem of interoperability, different classifications within the same ontology remain possible. For example, what was identified as a sidewalk on one map could be identified as a path in another one using different criteria. This type of problem resembles the abstract level incompatibility discussed by Kashyap and Sheth (1996) when describing the schematic heterogeneities in multidatabases.

Our approach to accomplish the planners' objective is to evaluate the semantic similarity by searching for the best match, entity-to-entity, between two downtown maps. A portion of the knowledge base used for this application, representing an ontology with only is-a relations, is shown in Figure 1. Entities that represent cases of polysemy (i.e., belonging to different entity classes with same name but multiple meanings) and entities that belong to classes with multiple superclasses (i.e., an entity class with multiple inheritance) are highlighted. Figure 2 shows the complete description of an entity class, i.e., its distinguishing features and its semantic relations.

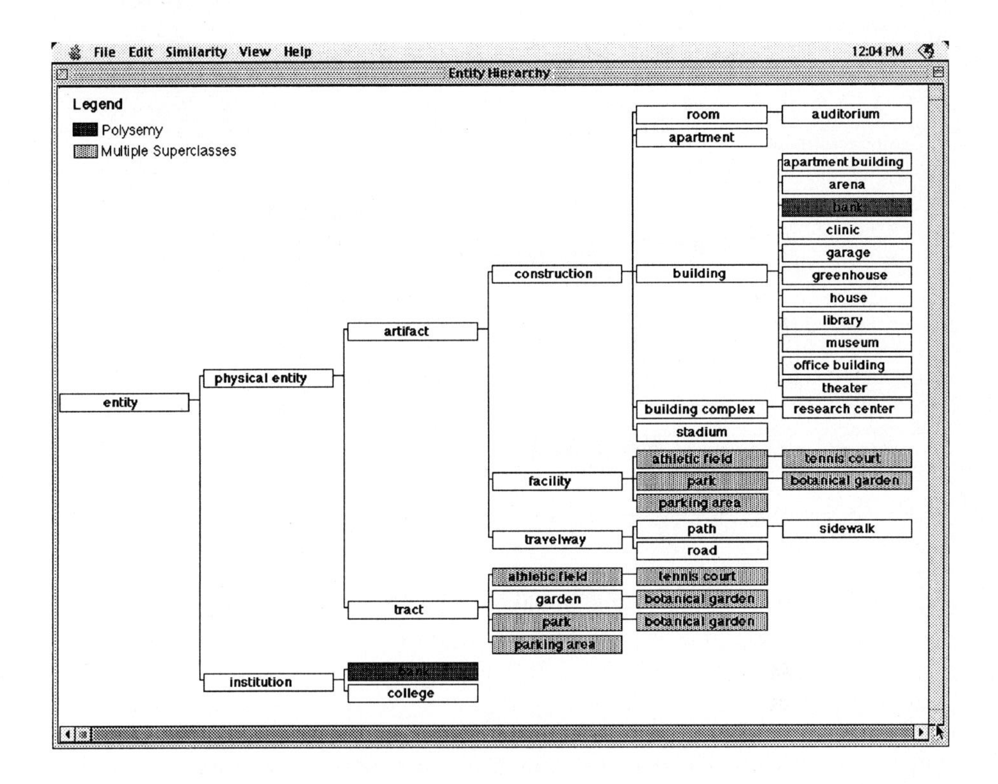

Fig. 1. Entity class hierarchy (is-a relations).

Since the planners in our example are mostly concerned with the functional components of the downtowns, they may assign a higher weight to the function features: for example, 50for attribute features. For this application, the direction of the evaluation is determined by the target downtown (the downtown to be redesigned) against which the ideal downtowns are compared.

Figure 3 shows a similarity assessment between a *stadium* and all other possible entity classes in the knowledge base. The similarity evaluation (Equation 1) is performed in two steps. Firstly, the set-union and set- difference operations are determined between the set of features of *stadium* and the set of features of each entity class in the knowledge base. For example, *stadium* has five parts, four functions, and six attributes. Taking *arena* as an example of one of the entity classes to compare against *stadium*, *arena* has eight parts, four functions,

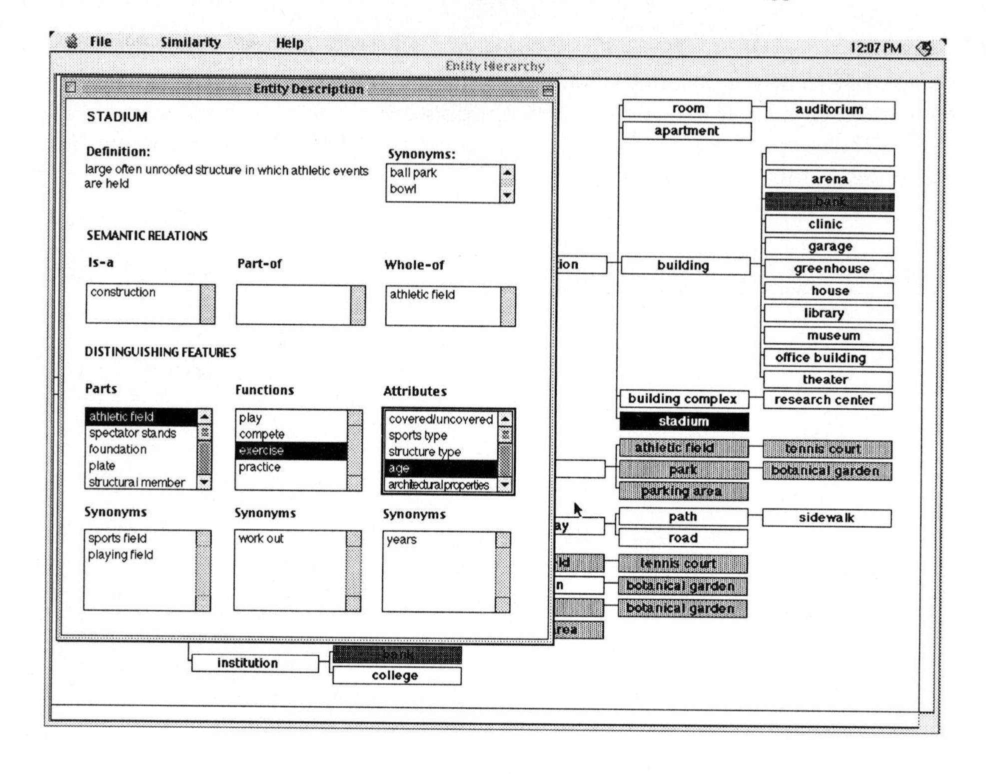

Fig. 2. Distinguishing features and semantic relations of an entity class.

and ten attributes. The set of common features between *stadium* and *arena* includes four parts, four functions, and four attributes. Secondly, the weight ((Equation 2) is determined based on the is-a relation and part-whole relations between *stadium* and the rest of the entity classes. Considering the comparison between *stadium* and *arena*, the common superclass between these two entities is *construction*, leading to a weight (equal to 0.33 when the direction of the evaluation goes from *stadium* to *arena*. This value of (reflects the fact that for our knowledge base *stadium* is a more general concept than the concept of *arena*. Numerically, the similarity assessment between *stadium* and four entity classes results in values that are greater than or equal to 0.5: *arena* (0.78), *athletic field* (0.62), *tennis court* (0.6), and *construction* (0.5).

The similarity evaluation between *athletic field* and all other entity classes illustrates the asymmetric evaluation of the similarity model. For a symmetric evaluation, the similarity between *athletic field* and *stadium* should be the same as the similarity between *stadium* and *athletic field* (0.62). The similarity between *athletic field* and *stadium*, however, is 0.58.

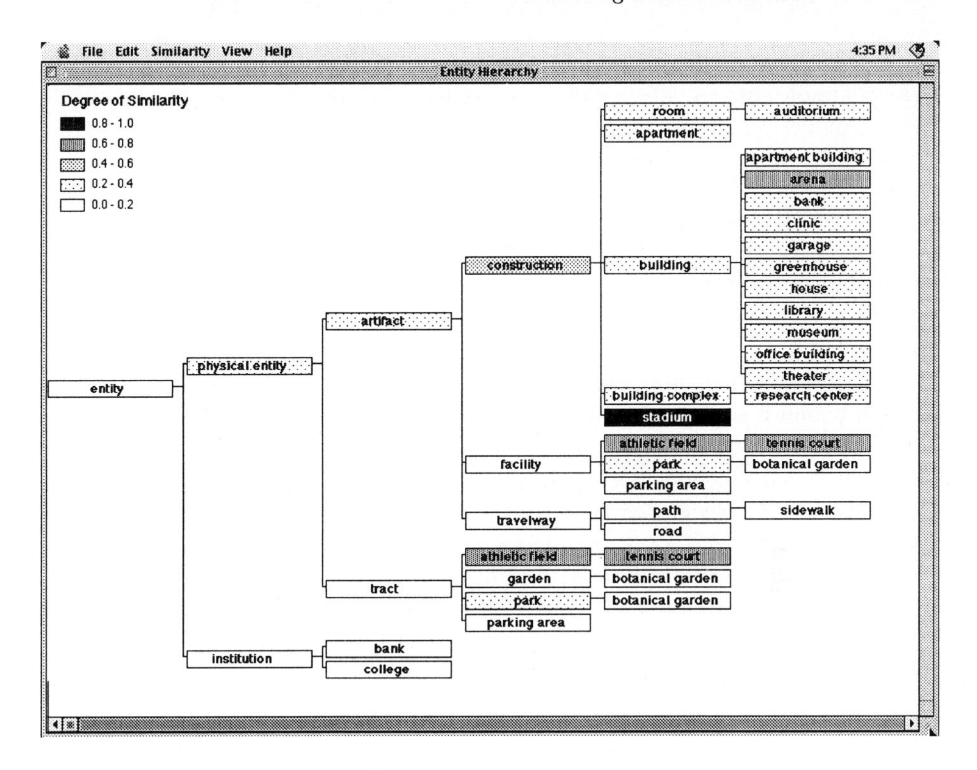

Fig. 3. Similarity assessment: stadium against all entity classes.

6 Conclusions and Future Work

Our model of semantic similarity has a strong basis in linguistics. It introduces synonyms and different senses in the use of terms. It also provides a first approach to handle part-whole relations in the evaluation of semantic similarity. Furthermore, it defines a semantic-similarity function that is asymmetric for classes that belong to different levels of generalization in the semantic network. Although the model is affected by the definition of parts, functions, and attributes, it reduces the effect of the underlying semantic network when compared with many of the semantic distance models.

As defined by our model, the asymmetric weights for the non-common features of each entity class (α, and $1-\alpha$) add up to 1. This means that in total, common and different features have the same weight (i.e., 1). A further refinement can be done to define the weights α and $1-\alpha$ if we consider that, in the assessment of similarity, people may tend to give more importance to the common features (Tverski 1977, Krumhansl 1978).

Global semantic similarity assessment for spatial scenes could also be improved. Our approach can be used to evaluate entity-to-entity similarity to obtain a global optimization of the similarity between two scenes. Problems arise when scenes have different numbers of spatial entities. A study of how much non- common entities affect global similarity assessment will help to obtain a better estimation of the semantic similarity between spatial scenes.

Context has already been suggested to be a relevant issue for semantic similarity (Tversky 1977, Krumhansl 1978) and for interoperability (Kashyap and Sheth 1996, Bishr 1997). We expect to incorporate context, initially through matching a user's intended operations with operations associated with the compared classes, in order to recognize different senses (semantics) of entity classes as well as to be able to define weights that reflect characteristics of a specific application.

Human-subject testing will contribute to testing how closely our model resembles people's similarity judgments. It might also provide new insights about how important common and non-common features are for people.

Finally, a big challenge for our model is to evaluate similarity across multiple knowledge databases or ontologies. When we assumed ontologies for specific domains as customized by users, we found significant differences in the definition of concepts within a single domain. In order to move forward towards the solution of interoperating systems we shall need to account for these differences. We shall have to relax our assumption of a unique ontology, replacing it by a common ontology that integrates multiple and independent domain specific ontologies.

References

1. G. Birkhoff (1967). *Lattice Theory.* American Mathematical Society, Providence, RI.
2. Y. Bishr (1997). *Semantic Aspects of Interoperable GIS.* Thesis Wageningen Agricultural University and International Institute for Aerospace Survey and Earth Science (ITC), Enshede, the Netherlands.
3. L. Cardelli (1984). A Semantics of Multiple Inheritance. in: G. Kahn, D. McQueen, and G. Plotkin (Eds.), *Semantics of Data Types.* Lecture Notes in Computer Science 173:51–67, Springer-Verlag, New York, NY.
4. R. Chaffin and D. Herrmann (1988). The Nature of Semantic Relations: a Comparison of two Approaches. in: M. Evens (Ed.), *Relational Models of the Lexicon: Representing Knowledge in Semantic Network.* pp. 289–334, Cambridge University Press, Cambridge, MA.
5. A. Collins and M. Quillian (1969). Retrieval Time from Semantic Memory. *Journal of Verbal Learning and Verbal Behavior* 8: 240–247.
6. D. Cruse (1979). On the Transitivity of the Part-Whole Relation. *Linguistics* 15:29–38.
7. K. Dahlgren (1988). *Naive Semantics for Natural Language Understanding.* Kluwer Academic Publishers, Norwell, MA.
8. K. Dittrich (1986). Object-Oriented Data Base Systems: The Notation and The Issues. in: K. Dittrrich, U. Dayal, and A. Buchmann (Eds.), *International Worskshop in Object-Oriented Database Systems*, Pacific Grove, CA, Washington, D.C. 2–4, IEEE Computer Society Press.
9. J. Gibson (1979). *The Ecological Approach to Visual Perception.* Houghton Mifflin, Boston, MA.

10. T. Gruber (1995). Toward Principles for the Design of Ontologies Used for Knowledge Sharing. *International Journal of Human and Computer Studies* 43(5/6): 907–928.
11. N. Guarino (1995). Formal Ontology, Conceptual Analysis, and Knowledge Representation. *International Journal of Human and Computer Studies* 43(5/6):625–640.
12. N. Guarino and P. Giaretta (1995). Ontologies and Knowledge Bases: Towards a Terminological Clarification. in: N. Mars (Ed.) *Toward Very Large Knowledge Bases: Knowledge Building and Knowledge Sharing*, Amsterdam, the Netherlands. pp. 25–32, IOS Press, Amsterdam.
13. N. Guarino (1997). Semantic Matching: Formal Ontological Distinctions for Information Organization, Extraction, and Integration. in: M. Pazienza (Ed.) *Information Extraction: A Multidisciplinary Approach to Engineering Information Theory*, Frascati, Italy. pp. 139–170, Springer Verlag.
14. M. Iris, B. Litowitz and M. Evens (1988). Problem of the Part-Whole Relation. in: M. Evens (Ed.) *Relational Models of the Lexicon: Representing Knowledge in Semantic Network.* pp. 261–288, Cambridge University Press, Cambridge, MA.
15. J. Jiang and D. Conrath (1997). Semantic Similarity Based on Corpus Statistics and Lexical Taxonomy. *International Conference on Computational Linguistics (ROCLING X).* Taiwan, pp. 19–35.
16. J. Kahng and D. McLeod (1998). Dynamic Classificational Ontologies: Mediation of Information Sharing in Cooperative federated Database Systems. in: M. Papazoglou and G. Schlageter (Eds.) *Cooperative Information Systems: Trends and Directions.* pp. 179–203 , Academic Press, London, UK.
17. V. Kashyap and A. Sheth (1996). Semantic and Schematic Similarities Between Database Objects: A Context-Based Approach. *The VLDB Journal* 5(4):276– 304.
18. V. Kashyap and A. Sheth (1998). Semantic Heterogeneity in Global Information Systems: The Role of Metada, Context, and Ontologies. in: M. Papazoglou and G. Schlageter (Eds.) *Cooperative Information Systems: Trends and Directions.* pp. 139–178, Academic Press, London, UK.
19. S. Khoshafian and R. Abnous (1990). *Object Orientation: Concepts, Languages, Databases, User Interfaces.* John Wiley & Sons, New York.
20. C. Krumhansl (1978). Concerning the Applicability of Geometric Models to Similarity Data: The Interrelationship Between Similarity and Spatial Density. *Psychological Review* 85(5): 445–463.
21. J. Lee, M. Kim, and Y. Lee (1993). Information Retrieval Based on Conceptual Distance in IS-A Hierarchies. *Journal of Documentation* 49(2): 188–207.
22. D. Lenat and R. Guha (1990). *Building Large Knowledge Based Systems: Representation and Inference in the Cyc Project.* Addison-Wesley Publishing Company, Reading, MA.
23. R. Meersman (1997). An Essay on The Role and Evolution of Data(base) Semantics. in: R. Meersman and L. Mark (Eds.) *DataBase Application Semantics.* Chapman Hall, London, UK.
24. E. Mena, V. Kashyap, A. Illarramendi, and A. Sheth (1998). Domain Specific Ontologies for Semantic Information Brokering on the Global Information Infrastructure. *International Conference on Formal Ontology Information Systems.* Available at `http://ra.cs.uga.edu/publications/pub_ALL.html`.
25. G. Miller (1990). Nouns in WordNet: A Lexical Inheritance System. *International Journal of Lexicography* 3(4): 245–264.
26. G. Miller (1995). A Lexical Database for English. *Communications of the ACM* 38(11):39–41.

27. G. Miller, R. Beckwith, C. Fellbaum, D. Gross, and K. Miller (1990). Introduction to WordNet: An On-Line Lexical Database. *International Journal of Lexicography* 3(4): 235–244.
28. G. Miller and W. Charles (1991). Contextual Correlates of Semantic Similarity. *Language and Cognitive Processes* 6(1): 1–28.
29. R. Rada, H. Mili, E. Bicknell, and M. Blettner (1989). Development and Application of a Metric on Semantic Nets. *IEEE Transactions on System, Man, and Cybernetics* 19(1): 17–30.
30. R. Richardson and A. Smeaton (1996). *An Information Retrieval Approach to Locating Information in Large Scale Federated Database Systems.* Dublin City University, School of Computer Applications, Working Paper CA-0296, available at `http://simpr1.compapp.dcu.ie/CA_Working_Papers/wp96.html#0296`.
31. R. Richardson, A. Smeaton, and J. Murphy (1994). *Using WordNett as a Knowledge Base for Measuring Semantic Similarity Between Words.* Dublin City University, School of Computer Applications, Working Paper CA-1294, available at `http://simpr1.compapp.dcu.ie/CA_Working_Papers/wp94.html#1294.html`.
32. L. Rips, J. Shoben, and E. Smith (1973). Semantic Distance and the Verification of Semantic Relations. *Journal of Verbal Learning and Verbal Behavior* 12: 1–20.
33. E. Rosch and C. Mervis (1975). Family Resemblances: Studies in the Internal Structure of Categories. *Cognitive Psychology* 7: 573–603.
34. A. Sheth (1998). Changing Focus on in Information Systems: From System, Syntax, Structure to Semantics. in: M. Goodchild, M. Egengoher, R. Fegeas, and C. Kottman, *Interoperating Geographic Information Systems.* Kluwer Academic Press (in press).
35. A. Tversky (1977). Features of Similarity. *Psychological Review* 84(4): 327–352.
36. USGS (1998). *View of the Spatial Data Transfer Standard (SDTS) Document.* Available at `http://mcmcweb.er.usgs.gov/sdts/standard.html`. Last modification: Friday, 12-Jun-98.
37. M. Winston, R. Chaffin, and D. Herramann (1987). A Taxonomy of Part-Whole Relations. *Cognitive Science* 11: 417–444.

Proceeding on the Road of Semantic Interoperability – Design of a Semantic Mapper Based on a Case Study from Transportation

Yaser A. Bishr, Hardy Pundt, and Christoph Rüther

University of Münster
Institute for Geoinformatics
Robert Koch Str. 26-28
D-48149 Münster, Germany
{bishr, pundt, ruether}@ifgi.uni-muenster.de

Abstract. Semantic interoperability in GIS is the ability to share geospatial information at the application level. This paper argues that we currently have a relatively clear understanding of what semantic heterogeneity means, what are real world cases that exemplify this heterogeneity, and who to characterize these differences. We believe that the stage is now set for implementing a prototype to resolve semantic heterogeneity. We present here an on going case study and prototype development. This prototype is not aimed to be a definitive solution. Rather, it provides a feedback process to our struggle to seek scientifically sound and technically and commercially viable solutions to semantic interoperability.

1 Introduction

The more the geospatial information communities recognize that the real world is not separated into specific parts that exist independently from each other, the more they require to share information. Advanced technologies for data capture, such as satellites, scanners, automatic digitizing, pen-computer based field data recording, etc. have revolutionized data capture techniques and led to an increase in the availability of digital data. Along with such revolution, new problems arose. For example, information communities find it difficult to locate and retrieve data from other sources, in a reliable and acceptable form for their specific tasks. It is a known fact that the reuse of geodata for new applications is very often a lengthy process. This is due to poor documentation, obscure semantics, diversity of data sets, and the heterogeneity of existing systems in terms of data modeling concepts, data encoding techniques, storage structures, access functionality, etc. [8].

2 Worldwide Awareness

Due to the fact that different geospatial information communities have an increasing need to share spatial information and possibly GIS services, changes to the existing GIS-infrastructure are necessary and currently taking place. The networking between official authorities, industry and academia as well as the provision of geodata to the public are currently changing processes. This affects configuration management, systems engineering and integration and the methodologies and strategies that are applied by those disciplines. An effective management of cross-community technology development and Integration requires new approaches. The OpenGIS Consortium (OGC) introduced some of them, mostly in close cooperation with the GIS industry and universities. Within the framework of the goal setting of interoperable GIS tools the simple feature specification is a start-up to a common agreement between GIS industry and users to overcome non-interoperability in the geometry sector. The idea of semantic translators is an approach aimed at the level of the problem-free exchange of thematic information. Such approaches are currently part of the discussion of a new orientation. They are currently part of the discussion of a new orientation and goal setting of the whole GIS infrastructure [16]. Semantic interoperability is an issue that will play an increasingly important role within this framework.

The OGC has identified the need for open geodata sharing and the exchange of open GIS services. "Openness" requires not only open interfaces and techniques to exchange data between different GIS [14]. The concept of openness has definitively to include semantic interoperability, which is more than pure data transfer. The problem is that the goal of open information sharing is not that easy due to the different meanings and interpretations of data by different information communities. Information sharing between different geospatial information communities is mostly impeded by any of three factors that are summarized in table 1 (see [17]). [5] present some case studies on semantic non-interoperability. Such examples document that pure data transfer makes no sense if the members interpret the data differently. Users often realize, after sharing the data, that they can't use them due to the specific view of the person who recorded them, or, in other words, data transfer was successful, but information was not shared [13].

Table 1. Impediments of information sharing

Impediments
• Ignorance of the existence of information outside one's geospatial information community
• Modeling of phenomena not of mutual interest
• Modeling of phenomena in two representations so foreign to each other that each is not recognized by the other

Geodata have geometric and the thematic aspects. Apart from geometric aspects, for most applications the thematic aspects of terrain description and analysis are of prime importance [2]. To share data and to overcome the mentioned impediments, functionalities are required that, not only deal with the geometric and the thematic aspects, but also with the semantic characteristics of spatial data. A special working group at the Institute for Geoinfromatic, IfGI, is just in the phase of the design of a semantic mapper prototype based on a case study from the area of transportation.

Section 3 outlines some issues that should help to explain the urgent need for technical approaches to overcome semantic heterogeneity. Section 4 describes a case study of existing databases from the area of transportation. This is meant to be the real world application that will be used to implement the semantic mapper. Section 5 explains the details of a conceptualized system architecture. We will also show how a mapping between the different databases describes in the case study could be realized. The work described here is the necessary basis for the implementation of a semantic mapper, or, in other terms, for the "technical" approach to overcome semantic heterogeneity.

3 Approaching Technical Solutions

The cause for semantic non-interoperability is semantic heterogeneity. Classifications of the types of heterogeneity differ. Table 2summarizes some of the existing classifications. The classification by [4] highlights the difference between objects at the conceptual level; i.e., semantics, and their computer representation, i.e., syntax and schema. It makes it possible to focus on these three issues as "independent parts". This means that resolving syntactic and schematic heterogeneity is not a difficult problem if the semantic heterogeneity is resolved. We therefore adopt this classification in this research.

Semantics plays a crucial role when sharing information at the application level. The OpenGIS Consortium emphasizes the "Model of Geographic Information Communities" that are currently not or only partly able to share information [15]. The exchange and transfer of spatially referenced data is possible "once communities have agreed on translations between their different feature definitions" [14]. In many cases, this process has not yet been started. Organizations will still need to negotiate common understandings about the semantics of shared geographic information, to get the most information out of the shared data. Such negotiations need the support of the geospatial communities including GIS users in specific geospatial information communities, GICs, academia and even the GIS industry. The latter should specifically pay attention to the special requirements of users concerning the semantic issues. In addition to the internal level of data structures and record formats, the main challenge to resolve the semantic heterogeneity is to characterize the properties of geographic data needed to ensure that translation and interoperability can be achieved at the semantic level [10].

Sheth argued already in 1991 that a technique is required to support semantic reconciliation, *"...a process or technique to resolve semantic heterogeneity and identify semantic discrepancy, and semantic relativism that supports multiple views or interpretations of the same stored data"* [19]. Contributions to the Interop 1997 in Santa Barbara, California gave a comprehensive overview of the

Table 2. Types of heterogeneity

Type of Heterogeneity	**Description**	**Literature**
Generic	Occurs when different nodes are using different generic models of the spatial information	[22]
Conceptual	Occurs when the semantics of schemas depend upon the local conditions at particular nodes	
Discrepancies in data definition	Equal features are described differently	[11]
Differences in Data Structures	The same information is represented with different data structures in two databases	
Semantic	Occurs mainly due to differences in the context world view of different data users	[3,4]
Syntactic	Occurs as a result of different representations of real world features as fields or as objects in the databases of different users. It is directly related to semantic reference.	
Schematic	Occurs due to different schemata, e. g. the classes, attributes and their relationships, which vary within and across contexts	
Naming	Semantically alike entities in the cognitive content world refer to the same real world fact but have different names	[3]
Cognitive	No common base of definitions of real world facts between two disciplines	

increasing efforts to identify and solve the problems and difficulties that occur in the area of semantic non-interoperability. Different paths to overcome semantic non-interoperability have been proposed. But the conference showed as well, that implementable (and therefore applicable) techniques are still lacking. Concerning such techniques the technique of a "semantic mapper" is meant to be such a tool with a high potential to solve special semantic non-interoperability problems. The main thrust of developing the prototype is the fact that there is no concrete implementations of a semantic mapper known to us. In the sequel we introduce our approach to resolve semantic non-interoperability. We will first introduce a case study where we show how to characterize the semantic differences between two information communities. We then introduce an on going effort in our group to develop semantic mapper between GDF and ATKIS.

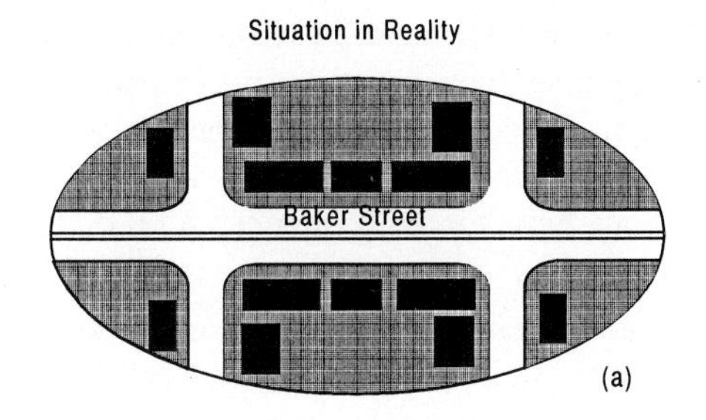

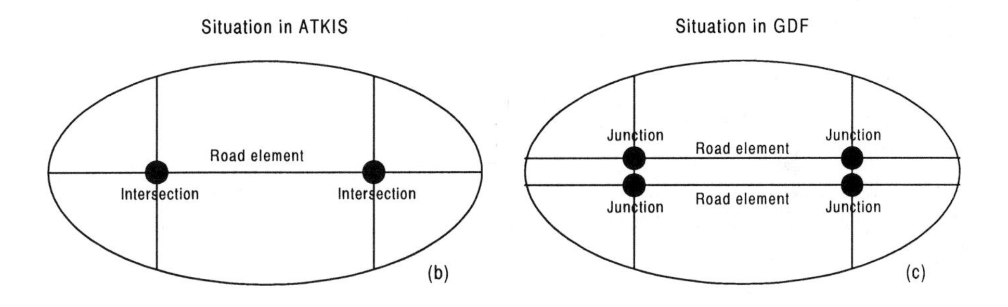

Fig. 1. A two-directions road in the real world (a) is presented as one road element (b) in ATKIS and as two road elements in GDF (c)

4 ATKIS and GDF: A Case Study

Europe has a vast and extensive ground and water transportation network. Several public and private agencies deal with transportation information, e.g., suppliers of data for car navigation systems, logistics transportation, and traffic control, management, and analysis. These agencies usually require transportation information that stretches beyond national borders. For example, traffic management and control agencies often require transportation information collected by mapping agencies.

There are several efforts to standardize transportation definitions and classification, e.g., ATKIS and GDF. Developed between 1989 and 1995 the Authoritative Topographic-Cartographic Information System (Amtliches Topographisch-Kartographisches Information-system, ATKIS) of the federal republic of Germany, is a topographic and cartographic model of reality. It models landscapes within digital landscape maps and the map content within digital map [1].

The Geographic Data Files, GDF, is a European standard released in October 1988 and has gone through several stages of update until 1995 [7] . It aims at providing a reference data model to describe road networks. GDF is created to improve efficiency in the capture and the handling of data for geographic information industry, by providing a model upon which applications can be built, e.g., car navigation, vehicle routing, traffic analysis, etc.

The objectives ATKIS and GDF are different due to differences in the social backgrounds. These differences have created groups called geographic information communities (GICs'). We consider here a German topographic GIC and a pan European traffic management GIC who use ATKIS and GDF standards, respectively. We call them here for convenience, ATKIS GIC and GDF GIC, respectively.

The ATKIS GIC conceptualizes transportation networks as artifacts that are part of landscapes, and ought to be presented in their topographic maps. The GDF GIC conceptualizes transportation networks as a section of the earth, which is designed for, or the result of *any* vehicular movement.

From the GDF GIC point of view, the main purpose of a connection between their information system and the ATKIS information system are to provide the most recent and up to date information about new roads and status, e.g., to provide an online service for car navigation systems. From the ATKIS point of view, the main purposes of a connection between their information system and the GDF information system, is to take advantage of the GDF's traffic flow information and routing information, and provide it for the local applications that adopts ATKIS as their base model and require more information about the traffic flow, direction, rules, etc.

The problem starts by asking the question "does *transportation network* mean the same thing in the two GICs'?".

In GDF, the term road encompasses road, railway, waterways, junctions, rail junctions and water junctions, while in ATKIS waterways are not considered part road.

Roads in the ATKIS GIC refer to ground transportation networks. A road element is the smallest part of road that has a consistent width, i.e., does not change within a certain threshold. In GDF a road network also encompasses ferry connections which are not implied in ATKIS. A road element does not only depend on its width but also on traffic rules. For example a new road element is created in GDF if the direction of flow changes.

Even the term "ferry network" in ATKIS refer to ferryboats, while in GDF a ferry is a vehicle transport facility between two fixed locations on the "road network" and which uses a prescribed mode of transport, for example, ship or train. This definition shows that the term road networks includes waterways.

Considering the ground transportation road network, we find that ATKIS include pedestrian zones, bike roads as part of a road feature, while in GDF, a pedestrian is not part of roads and a bike road is a type of a road network.

Figure 1 further illustrates the difference between the two information community. The *Baker Street* is a two-direction street, in ATKIS it is viewed as one road element that has two intersection points. In GDF the same road is presented as two road elements, one for each direction of traffic flow. If you ask ATKIS GIC about the *Baker Street* you will get one road as shown in Figure 1b. If you ask the same question to the GDF GIC you will get two roads of Baker Street , one for each traffic flow direction, as shown in Figure 1c.

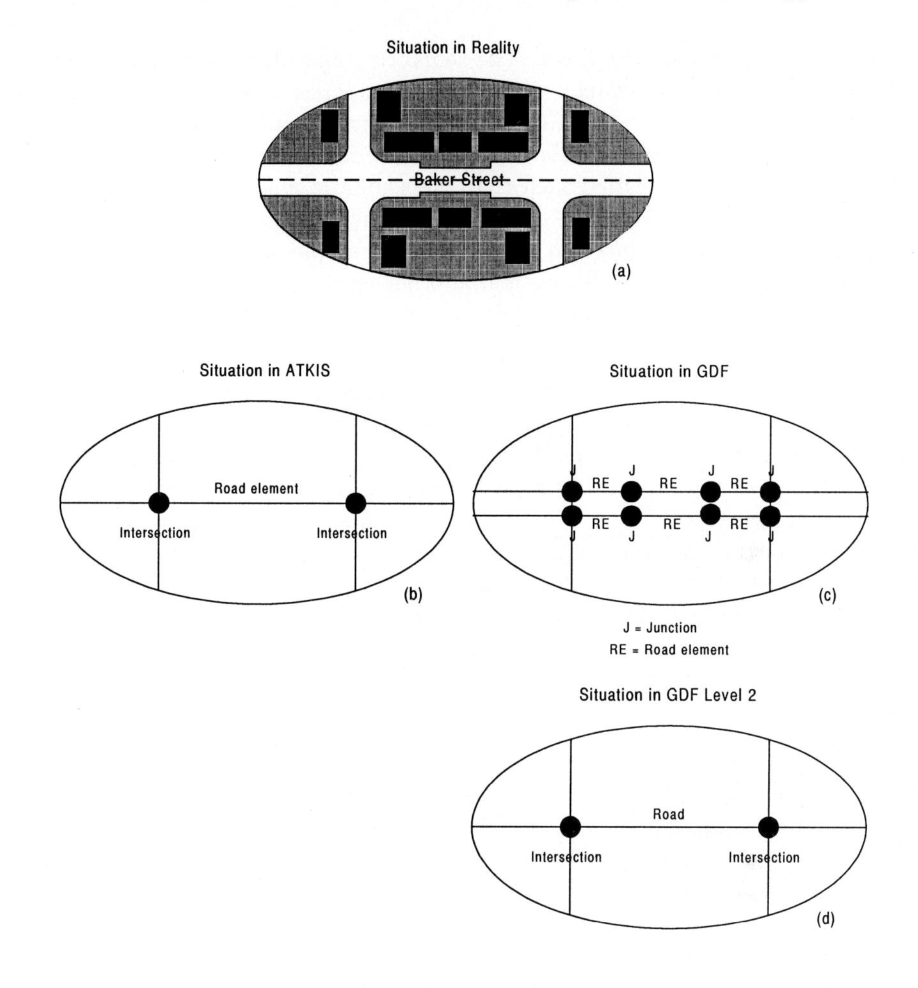

Fig. 2. Intersections and junctions are two different concepts in GDF but not ATKIS

Another example of differences between ATKIS and GDF views are the term "junction" and "intersection". While in ATKIS there is no apparent difference between the two terms, in GDF a junction refers to the end points of a road element and an intersection is the end points of a road as shown in Figure 2. *"Even roads do not always mean roads."*

5 An Overview of the System Architecture

In this section we introduce a system architecture to provide semantic interoperability between heterogeneous systems. The system is currently under development. Figure 1 depicts the two main components of the system and shows where the semantic mapper is situated. The system consists of two main components: semantic mapper and a wrapper. The data source and the client install both the semantic mapper and a wrapper. The semantic mappers at the source and the client are essentially the same. A semantic mapper is developed for a particular application domain, e.g., for transportation, topography, telecommunication. The wrappers, however, are different depending on the underlying database. Finally, the system is designed such that the semantic mapper can communicate with more than one wrapper if the source has more that one database for the application domain, as shown in Figure 3. For example, an information provider can have a database for highways and another for streets of a certain city.

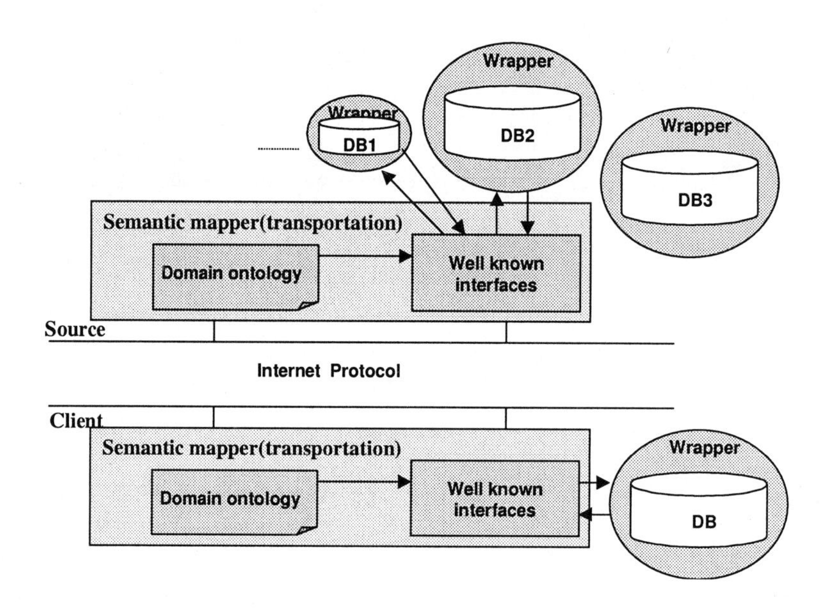

Fig. 3. Architecture of the prototype

To publish data for exchange, a source needs to define what data elements are available and model them as objects using the library of abstract interfaces provided by the semantic mapper. The semantic mapper and the wrapper are called collectively middleware. Middleware provides a unified view and common interfaces, for new applications, of heterogeneous legacy systems while maintaining their autonomy [18, 21]. Middleware typically relies on wrappers that act as an interface between the underlying legacy database and the client.

5.1 The Semantic Mapper

Applications that adopt the semantic mapper see heterogeneous legacy data stored in a variety of data sources as object instances that have well known interfaces. These object instances are provided by the wrapper, as will be shown in the next section. The mapper has two main components: a library of well known interfaces and a domain ontology. The semantic mapper provides five main services.

1. A communication manager between the information provider and the client.
2. Provides a library of abstract interfaces. These interfaces are implemented by the wrapper and allow sending and retrieving objects from and to its underlying database.
3. Wrappers model the underlying database as semantic mapper objects. When these objects are received at the client, the mapper identifies and resolves the semantic and the schematic differences between the client database and the newly received objects.
4. A mechanism that guarantees that each outgoing object from the underlying data source has a universal unique object ID.
5. It cooperates with the wrapper in query planing and execution, where it identifies parts of the query that belong to each wrapper if the query ranges over several wrappers (within the space of one data source).

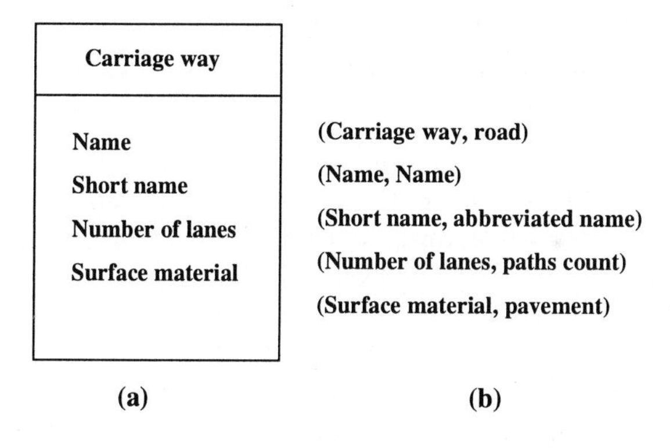

Fig. 4. Ontology is assigned to each outgoing object from the wrapper

5.2 Domain Ontology

In the knowledge representation literature, ontology refers to formal conceptualization of specific domains. The emerging field of ontological engineering has led to several attempts to create libraries of sharable ontology. For example, the KOSMOS project has produced a domain ontology for machine translation from Spanish to English in the filed of economy and cooperate mergers. Information exchange between sources and clients requires a sharing of ontology [4]. An ontology of geographic kinds, of the categories or entity types in the domain of geographic objects, is designed to yield a better understanding of the structure of the geographic world [20]. The ontology publisher develops and maintains domain ontology libraries. The development of ontology and its related issues are outside the scope of this paper and can be found elsewhere [6, 9, 12].

It is important to note here that in our system architecture it is assumed that the ontology is shared and committed by the clients and the sources who are willing to share and exchange their information. The domain ontology library, as an embedded component in the semantic mapper, helps resolve the semantic conflicts between the data source and the client. Each object received by the semantic mapper from the wrapper has an attribute that takes its value from the shared ontology.

Figure 4, shows a class carriage way as represented in the database of the information provider. The class has four attributes. We collectively call classes and the attributes: schema elements. These schema elements are associated to an ontological term brought from the shared ontology, as an ordered pair, as shown in Figure 4 b.

5.3 Constructing the Wrapper

A wrapper is an interface between the underlying database and the semantic mapper. Its main functionality is to provide a protocol for the communication between them. Wrappers provide three main services:

1. Model the instances of the underlying database as objects understood by the semantic mapper. This is achieved by a reference model, provided by the semantic mapper, that defines abstract well-known interfaces and implemented by the wrapper.
2. Participate in the query planning and execution.
3. Implement, together with the semantic mapper, a comprehensive schema for object ID to uniquely identify objects in a heterogeneous distributed environment.

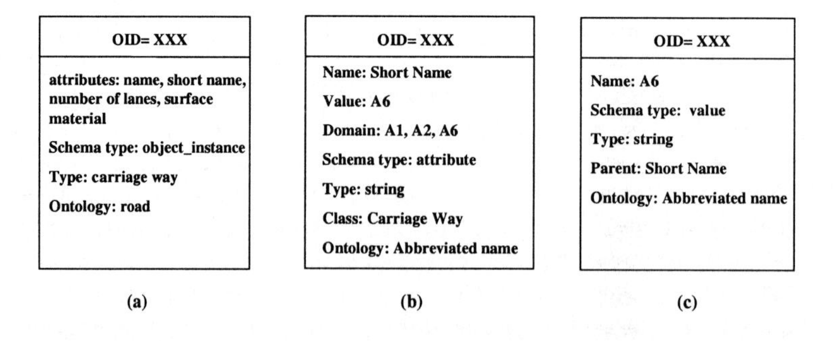

Fig. 5. Objects sent by the wrapper to the semanctic mapper

Figure 5 shows a simple example of three object instances sent by the wrapper to the semantic mapper. The objects are identified by a unique ID. As depicted in Figure 5a, in addition to the attribute that indicates the corresponding ontology term, the object Carriage_way has three attributes.

Attribute_list has the list of the attribute names that belong to that object.

DB_type indicates if the object is originally an attribute or an object instance. Hence, its value can be *attribute* or *object_instance*. In our example the value is object_instance.

Type indicates the type of the object. The object in consideration is originally an instance of a class then its type is that of the class.

In our system, attributes and their domains are also posted to the semantic mapper as object instances as shown in Figure 5 b and c. The attribute short_name is posted as an object instance with a unique ID.

6 Conclusions

Interoperability in general and semantic interoperability will undoubtedly lead to drastic organizational changes in the GI community. In this paper we have shown that semantic heterogeneity has to be resolved first before the syntactic and schematic ones. Semantic differences between ATKIS and GDF were presented. The semantic mapper and the wrapper architecture were presented as our approach to the problem. Both component use the interface technology to communicate with each other. The basic idea of the architecture is that wrappers post objects of their underlying databases, to the mapper as object instances with well known interfaces. The object instances hold detailed information about their semantics and their ontological reference.

Our future task will focus on defining the well known interfaces as well as the relevant information that characterizes object semantics between ATKIS and GDF. Our methodology is to keep our list of the prototype functionality short in order to speed up the feedback cycle. This research paper presents the technical efforts to achieve semantic interoperability. We are also probing the theoretical ground in search for theories to characterize semantic heterogeneity and similarities.

Acknowledgement

This research effort is supported by the seed grant from the National Science Foundation, NSF, the NCGIA/Varenius project.

References

1. ATKIS, dV[1995]: Amtlich Topographisches-Kartographisches Informationssystem (ATKIS). Arbeitsgemeinschaft der Länder der Vermessungsverwaltungen der Bundesrepublik Deutschland (AdV), Bonn. `http://www.tnt.uni-hannover.de/soft/tnt/gis/objkatalog/overview.html`.

2. Bishr Y. and M. Radwan (1996): Design of Federated Databases for Multi-Level Decision Making in Watershed Management. In Proceedings of an International Workshop on New Developments in Geographic Information Systems, March 1996,Milan, Italy, pp. 60 - 68.
3. Bishr, Y. (1998): Overcoming the semantic and other barriers to GIS interoperability. In Vckovski, A. (ed.) Special Issue on Interoperability in GIS. International Geographical Information Science, Vol 12 No 4, Jun 1998, pp299-314.
4. Bishr, Y., (1997). Semantic Aspects of Interoperable GIS. Publisher, The International Institute for Aerospace Survey and Earth Sciences, ITC. ISBN 90 6164 141 1.
5. Bishr, Y., H. Pundt, W. Kuhn and M. Radwan (1998): Probing the Concept of Information Communities - A First Step Towards Semantic Interoperability. Interop '97, Sta. Barbara, CA, (pre-conference version, pp 12 - 14).
6. Chaudhri V. K, A. Farquhar, R. Fikes, P. D. Karp, & J. P. Rice (1998). Open Knowledge Base Connectivity 2.0. Knowledge Systems Laboratory, KSL-98-06.
7. Comité Européen de Normalisation (CEN) (Publisher) [1995]: Geographic Data Files 3.0, TC 278 `http://www.ertico.com/gdf/index.htm`
8. Devogele, Th., Chr. Parent, S. Spaccapietra (1998): On spatial database integration. In Internat. Geographical Information Science, 1998, Vol. 12, No 4, pp 335 - 352.
9. Fikes R., A. Farquhar, & J. Rice (1997). Tools for Assembling Modular Ontologies in Ontolingua. Knowledge Systems Laboratory, KSL-97-03.
10. Gahegan, M. (1997): Accounting for the semantic differences between various Geographic Information Systems. Interop '97, Sta. Barbara, CA, (pre-conference version, pp 18-27).
11. Gangopadhyay, D. and Th. Barslou (1991): On the Semantic Equivalence of Heterogeneous Representations in Multimodel Multidatabase Systems. In Special Issue: Semantic Issues in Multidatabase Systems. SIGMOD Record, Vol. 20, No. 4, December 1991, ACM Press.
12. Guarino, N. and Poli, R (eds). Formal Ontology in the Information Technology. Special Issue of the International Journal of Human-Computer Studies, vol 43, no. 5/6, 1995.
13. Kuhn, W., 1995, Semantics of Geographic Information. GeoInfo Vol. 7. Department of Geoinformation, Technical University Vienna
14. McKee, L. (1997): Computing Concensus. In GIS Europe June 1997, pp 22 - 24.
15. 15. OGC (1998a): The OpenGIS Guide, 3rd edition, by the OpenGIS Consortium Technical Committee, edited by Kurt Buehler and Lance McKee. `http://www.opengis.org/techno/guide/980615/Cover.html`
16. OGC (1998b): The Benefits of OGC Membership. OpenGIS Consortium Inc.
17. OGC (1998c): The OpenGIS Specification Model, Topic 5: The OpenGIS Feature (Version 3). OpenGIS Document Number 98-105.
18. Roth M., Schwarz P. (1997). Don't Scrap It, Wrap It! A Wrapper Architecture for Legacy Data Sources. Jeusfeld M. et. al, (Eds.): VLDB'97, Proceedings of 23rd International Conference on Very Large Data Bases, August 25-29, Athens, Greece. Morgan Kaufmann 1997, ISBN 1-55860-470-7
19. Sheth, A. (1991): Semantic Issues in Multidatabase Systems. In Special Issue: Semantic Issues in Multidatabase Systems. SIGMOD Record, Vol. 20, No. 4, December 1991, ACM Press.
20. Smith, B. and Mark, D (1998). Ontology and Geographic Kinds. 8th Internaitonal Symposium on Spatial Data Handling. Poiker, T. and Chrisman, N. (eds). Vancouver, Canada.

21. Wiederhold, G. (1992): Mediators in the Architecture of Future Information Systems. Computer, March 1992, pp 38 - 49.
22. Worboys, M.F. and Deen, S.M. (1991): Semantic Heterogeneity in Distributed Geographic Databases. In Special Issue: Semantic Issues in Multidatabase Systems. SIGMOD Record, Vol. 20, No. 4, December 1991, ACM Press.

What Are Sports Grounds?
Or: Why Semantics Requires Interoperability

Catharina Riedemann and Werner Kuhn

Institute for Geoinformatics, University of Münster, Robert-Koch-Str. 26-28,
D-48149 Münster, Germany
{riedema, kuhn}@ifgi.uni-muenster.de

Abstract. Beyond technical aspects, interoperability of Geographic Information Systems is widely recognized to include semantic aspects. It is not yet clearly understood, however, which problems in data sharing stem from semantic issues and how they affect people's work. In order to gain a better understanding, this paper takes noise abatement related to sports grounds as a practical example of data sharing and examines in detail the semantic issues arising with the usage of two key terms. The first term we look at is "relevant sports ground". The selection of sports grounds that are considered by noise abatement planning is based on it. The second term is "sports ground" as such. The understanding of it determines how sports grounds are modeled in various data sets that might serve as an information source for noise assessment. We show that the meaning of sports grounds lies in their usage, at least in the case of noise abatement, and that available data sets do not contain the necessary usage information. Furthermore, we show that the available geometric data only become useful when several sources are combined: interoperability is required to map between data and user semantics.

1 Introduction

A great deal of current research attempts to find means for describing the semantics of data in order to support data sharing [8, 5, 10]. This research is motivated by the belief that successful interoperability is not only a technical matter of accessing distributed components, but requires a documented understanding of what the information held by these components means. The work reported here sprang from that same position, but evolved to demonstrate that the reverse claim is true as well: coping with the semantics of different applications requires interoperability. Only the kind of interoperability based on service interfaces can cope with the necessary semantic mappings between user requests and information sources. Data transfers cannot generally satisfy the information needs of applications cutting across information communities, even if the data were complete and with documented unambiguous meanings.

The paper presents a detailed case study of an application taken from ordinary planning practice in Germany. Imagine a local environmental authority that has to assess the noise in an urban area coming from sports grounds. Before actually assessing the impact, the responsible person has to find out which sports grounds represent relevant sources and then build an emission model. Both tasks require a lot of data which the environmental authority usually does not maintain, but which are spread over the sports authority, sports clubs, the surveying and cadastral authority and possibly other authorities having to do with sports grounds, depending on the local administrative organization. This indicates how many different people who have nothing to do with noise reduction issues must be asked for information, how often questions will not lead to the expected answers, how many data sources have to be examined, and how long it can take to merge all the data to the information product the responsible person needs to fulfill the tasks - if it is possible to create such a product at all.

From a semantic perspective, the planners are struggling with the meaning of the term "sports ground" in the context of noise abatement. This meaning rests on the perspective on sports grounds of a person living in their neighborhood, not that of a surveyor, cartographer or land use planner. But also not that of a public transports planner or social worker. Since most available data on sports grounds describe a perspective from surveying, mapping and possibly land use statistics, neither the noise expert nor the transportation planner nor the social worker are likely to find their understanding of sports grounds reflected in them. Consequently, the best means of describing data semantics may not be sufficient to answer their questions.

In an ideal world, these application specialists would be able to query distributed information resources for answers to their questions rather than individual databases for their contents. It would be an agent's task to determine whether and where there are data answering the questions or allowing for inferences to support one or the other hypothesis. In the case of sports grounds, such data may come from use statistics, traffic counts, regulations, surveys, remote sensing images, etc. The crucial challenge that motivates our research is how to bridge the gap between the data and the questions. We take it up by first studying in great detail some specific questions and data sources. The paper first describes the task setting, then discusses the semantics of relevant sports grounds and sports grounds in general, and concludes with observations on the multi-layered nature of such practical semantic problems. The somewhat surprising result is that these kinds of semantic issues are not just aggravated by interoperability, but require interoperability to be dealt with at all.

2 The Task: Noise Abatement

Sports activities make noise. These noise emissions are addressed in the German "Lärmminderungsplanung" (noise abatement planning), an administrative procedure dealing with noise impact (German: Immission) coming from different sources, e. g. traffic, industry, and sports. When assessing the noise emitted from sports grounds, the understanding of terms plays an important role, because it determines which sports grounds have to be considered and how they are modeled.

Noise abatement planning is designed as a small-scale instrument for urban areas. Only those sports grounds that contribute significantly to the noise situation are examined in detail. Consequently, as a first step the following question must be answered: "Which sports grounds are relevant to noise abatement planning?" The term to be examined in this context is *relevant sports ground* (section 3).

When the selected relevant sports grounds are looked at in detail, it is sensible to use existing (digital) information for building the emissions model on which the calculation of noise impact is based, if the data meet acoustic requirements. This leads to the question: "What data sets can be used in modeling sports grounds?" In other words: "Do data sets exist which have the same understanding of a sports ground as an acoustic expert?" The term to be examined in this context is *sports* ground (section 4).

In the following two sections the meaning of the two terms to different groups of people are described. The difficulties arising from different understandings are pointed out as well.

3 Relevant Sports Grounds

The local administration is responsible for noise abatement planning; thus instructions contained in the law must be taken into consideration first. Furthermore, it is useful to have a look at how such a task is executed, because it involves people who did not create the laws and because it usually takes refinement and interpretation to put legal instructions into practice. Table 1 shows how laws and other instructions in the federal state of North Rhine-Westphalia deal with the relevance of sports grounds in noise abatement planning.

Table 1 reveals that there is only one explicit definition of sports grounds available (given by [11], see row 3). The community of noise assessment seems to have a common basic understanding of the term sports ground. In the criteria for the relevance of sports grounds differences show up. The closer it comes to practice more criteria are found and they become more precise, too. In addition, the kind of criteria changes: the law refers to size (certainly taking it as a proxy for aspects of usage), and practice refers mainly to usage. In fact practice does not even mention size, but several more precise aspects of usage. A large number of criteria with high precision might be useful for understanding, but can also prevent information access: people might be overwhelmed by the amount and detail of questions and put them aside until they have enough time to answer them. Or information is gathered on a less detailed level and the questions cannot be answered at all. For a sports ground without scheduled use, e. g., nobody keeps a record of the hours of usage and the number of people present. In this case it becomes necessary to define fewer criteria and dispense with accuracy. Still, basically the same attributes are addressed. One considers, e. g., probability of usage instead of exact training and competition hours (see table 1, rows 5 and 6).

Obviously different answers to the question "Which sports grounds are relevant to noise abatement planning?" will produce different results in impact

Table 1. Relevance of sports grounds in North Rhine-Westphalia according to various sources

Regulation and information source	**Aim; contents; definition of sports grounds**	**Criteria for relevance of sports grounds**
1. Federal law for protection against noise impact [4, §47a])	introduces noise abatement planning; does not mention special noise sources like sports grounds; no definition	no criteria mentioned
2. Instructions for administration concerning the federal law; model [9] and version for North Rhine-Westphalia [12]	make instructions by law more concrete and practicable; mention sports grounds as noise sources; no definition	• larger sports grounds
3. Instruction for protection against the noise of sports grounds [11]	defines a uniform method for the assessment of noise impact caused by sports grounds; assessment procedure; stationary facilities intended for doing sports, including facilities that have close proximity, spatially and operationally	no criteria mentioned
4. North Rhine-Westphalian guide for making noise impact plans [6]	refines legal instructions to put them into practice; no definition	• competitions • preparation of competitions • run by municipalities, clubs, enterprises • considerable noise emission examples: football fields with more than 200 spectators, tennis complex with more than 3 courts

Regulation and information source	Aim; contents; definition of sports grounds	Criteria for relevance of sports grounds
5. Practice, version for sports grounds with scheduled use	carries out noise impact assessment	• outdoors • ball games like soccer or tennis • regular usage • used by clubs • competitions on Sundays between 1 p.m. and 3 p.m. • many spectators • training after 8 p.m. • usage of loud-speakers • residential buildings in the neighborhood (within a radius of 200 m in case of a usage before 8 p.m. or after 10 p.m., otherwise within a radius of 100 m)
6. Practice, version for sports grounds without scheduled use	carries out noise impact assessment	• outdoors • noisy forms of sports • acceptance by population • high probability of usage after 8 p.m. • much usage between 8 a.m. and 8 p.m. (counted in 25, 50 or 75 %)

assessment. This creates a situation which administration must avoid as much as possible. Furthermore, there is the issue of communication. If you have to ask other authorities for data, which happens regularly with noise abatement planning, you must be able to give a precise description of what you want. This requires to be clear in one's own mind about the requirements and able to express them. If you ask for a list of all larger sports grounds and thus take law literally, you might be presented with many grounds that are not used regularly and consequently of no interest. On the other hand, important smaller grounds being used regularly will not appear. Aside from this, appropriate questions must be asked that take into account the availability of information. Otherwise, you may not receive any information at all or only ill-fitting data.

4 Sports Grounds

The "Sportzentrum Roxel", a sports complex in Münster, was chosen as a case study to compare different views on the same real world object. It is a larger sports ground comprising several playing-fields of various sizes dedicated to different forms of sport, a stand for spectators and parking lots which, according to a legal definition [11], belong to the sports ground. The following descriptions only take into account geometry, because the other attributes are too specific to appear in existing data sets whereas geometry can be directly compared.

4.1 Aerial Photograph - The Real World Situation

The aerial photograph in fig. 1 is meant to give an "objective" view of the sports ground, as far as this is possible. There is a large pitch in the center of the complex, the main field, where soccer is played mostly. It is surrounded by track and field areas, and, south of it, a stand for spectators adjoins. In the northeast another large field is located, which is also mainly used for soccer. Between these fields you can see two smaller multipurpose fields (for basketball, handball, and volleyball), which - in contrast to the large fields - are free to be used by everyone and are not subject to any schedule. North of them, there is another track and field area. Ten tennis-courts are situated in the southwest corner and east of them, parking lots can be seen. A beach volleyball field lies next to the parking lots, and east of it there is an indoor swimming pool. The building at the eastern edge is a gym.

4.2 Acoustic Model

Fig. 2 shows how an expert in acoustics models the real world situation. Only those parts are to be found that are important from an acoustic point of view, the sources of noise. Noise can be emitted by players, spectators and cars. But some of the areas, where these sources originate, do not appear in the model due to a low frequency of usage, the form of sports, or because they represent indoor facilities. They are irrelevant for noise abatement planning. The two large fields, the tennis-courts, the stand, and the parking lots are relevant sources that must be looked at.

4.3 Cadastral Data (see [7])

In fig. 3, cadastral data ("Automatisierte Liegenschaftskarte", ALK) of the city of Münster for the same area are shown. Cadastral data are meant to provide information about location, shape and size of parcels as a basis for property documentation and taxation. From this perspective, sports grounds are not an object of primary interest. They belong to the supplementary topography, which is not registered systematically. This is why only few parts of the sports complex appear. There are three objects, each of which represents a generalization of two

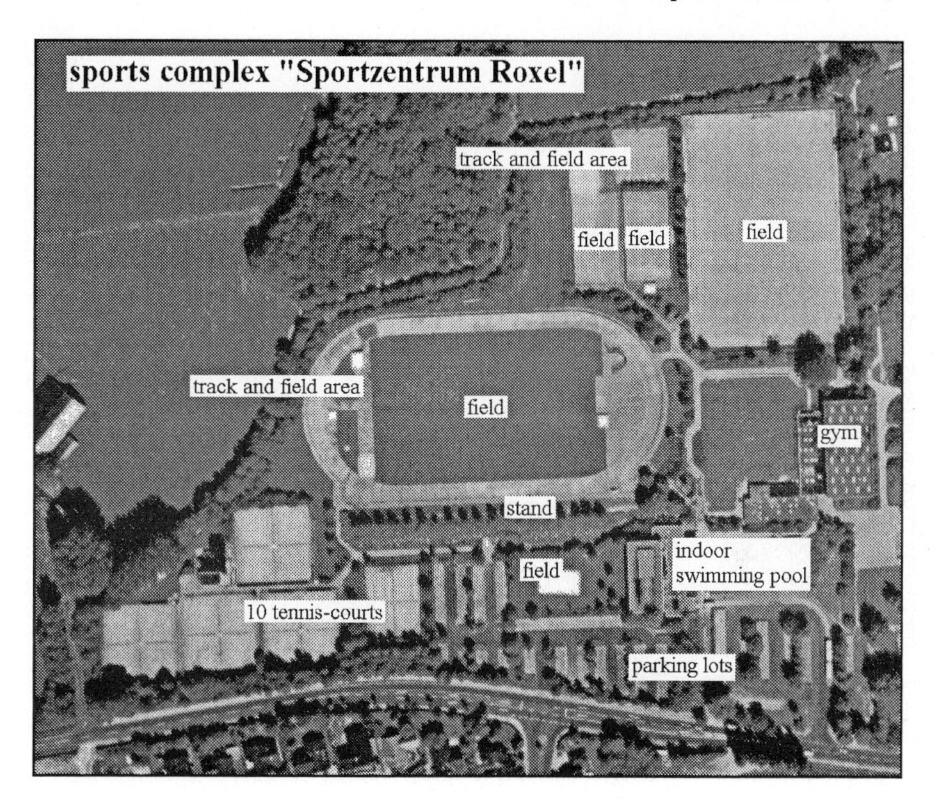

Fig. 1. Aerial photograph of the "Sportzentrum Roxel"

tennis-courts. In contrast to that, the parking lots are modeled in detail. The catalogue of ALK-objects in addition contains small and large fields, which is why we can expect corresponding extensions of the current map at some time, but the catalogue does not provide an object for the stand.

4.4 Topographic Data

The intention of topographic data is to show the surface of the earth and the objects on it. Depending on scale and purpose, the modeling of real world objects in topographic maps varies. Fig. 4 contains two examples for this kind of data. On the one hand, a part of the German base map 1:5000 ("Deutsche Grundkarte 1:5000", DGK 5, here not shown to scale) appears in black lines. This map is available digitally only in raster format. On the other hand, digital vector data of the Authoritative Topographic-Cartographic Information System ("ATKIS") are shown in gray. The contents of ATKIS are comparable to a topographic map with a scale of 1:25000.

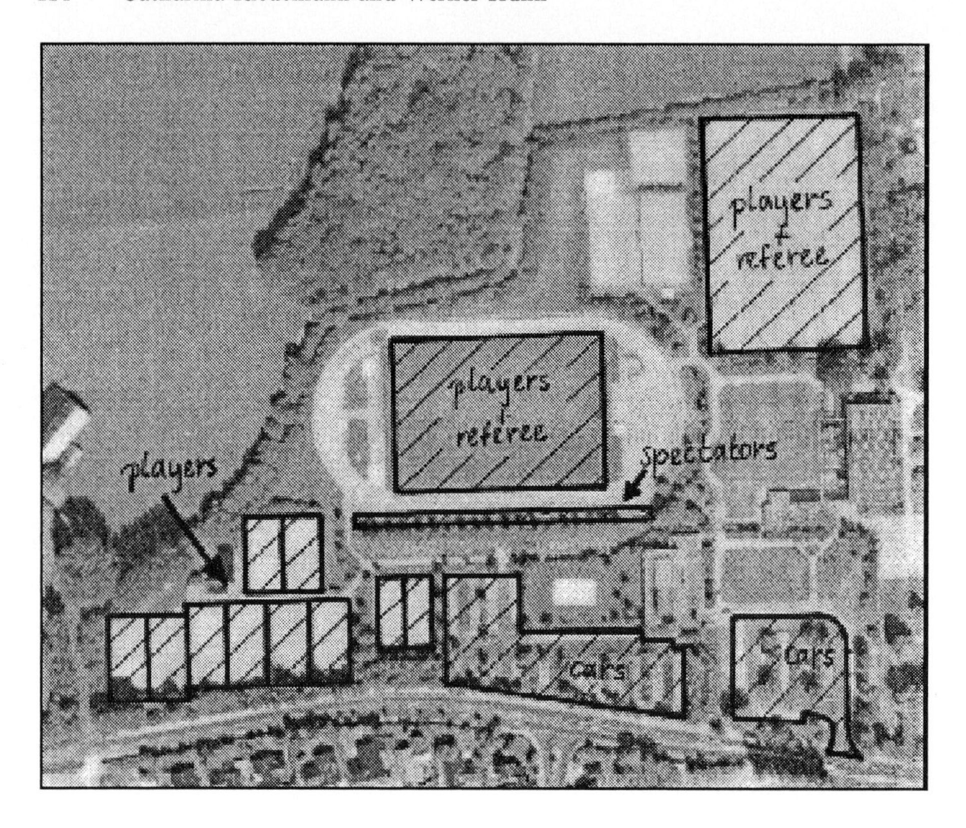

Fig. 2. An acoustic model of the "Sportzentrum Roxel"

DGK 5 (see [1]). In DGK 5, all fields are represented, but there are generalizations: the playing field in the oval area is not a separate object, and two adjacent tennis-courts are represented by one object. The parking lots are generalized as well by depicting just their outlines. Since the outlines are not closed, they merge with the street to one object. The DGK 5 does not have an object "stand".

ATKIS (see [2]). In ATKIS, the whole sports complex is depicted by one large object. ATKIS also comprises objects like playing field, stand and parking lot, which could show more details, but they have not been included here yet. In the future, updates will create a more detailed model of the sports complex in ATKIS.

4.5 Comparison of the Models

The acoustic model requires the existence of certain objects with a certain level of generalization. The large fields, e. g., must be represented by rectangular

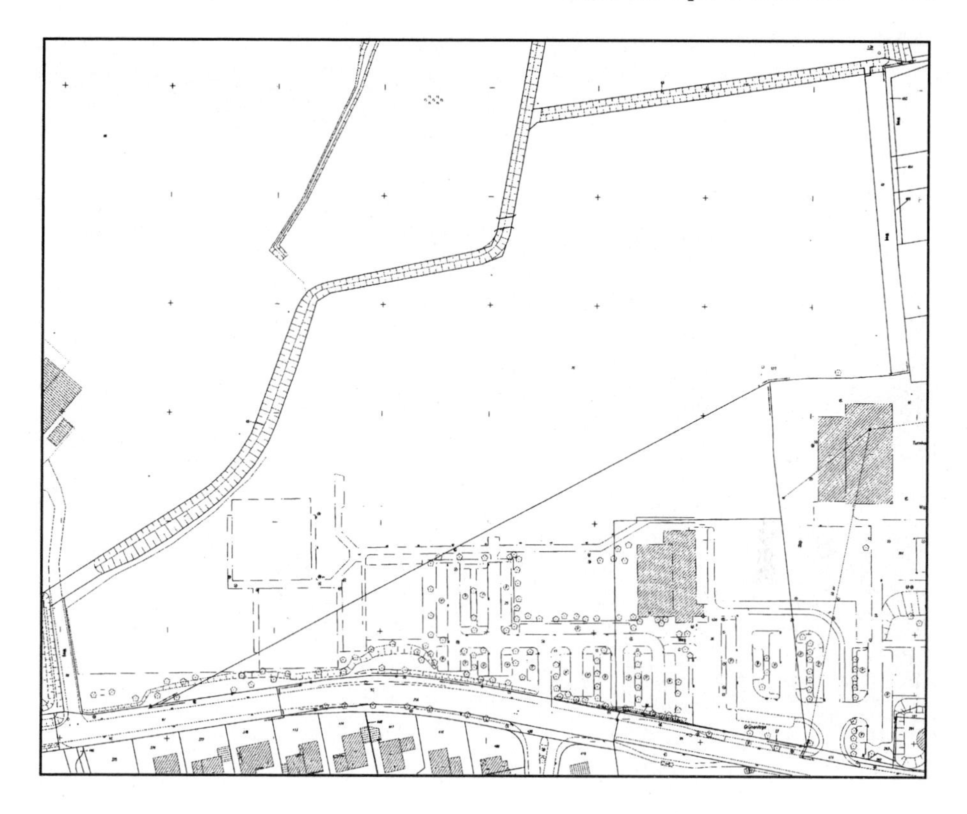

Fig. 3. The "Sportzentrum Roxel" in the ALK

objects. Thus, the generalized oval object in the DGK 5 is not acceptable. The parking lots just require outlines; a subdivision in places to park and areas with bushes and trees between them is too detailed for the purpose of noise abatement planning. In this respect the generalization of the DGK 5 is better than the detailed cadastral data. For the consideration of spectators, which is only necessary for the large fields and not for tennis-courts, there are several possibilities. The easiest way is to assume spectators on the field together with the players and the referee. But this method is only applied for small numbers of spectators. When the number of spectators increases, more exactness is needed and consequently geometry must change. Either the stand must be introduced as an additional object, or the rectangle of the field must be enlarged by several meters at the long sides. This will somewhat improve results near the field.

This means that all data sets show weaknesses and cannot be used alone. In the cadastral data, many objects are missing at this time, whereas parking lots are depicted in too much detail. The DGK 5 is not detailed enough regarding the oval field. ATKIS is lacking any useful object at this time, but it is the only data set which could provide an object "stand" in future.

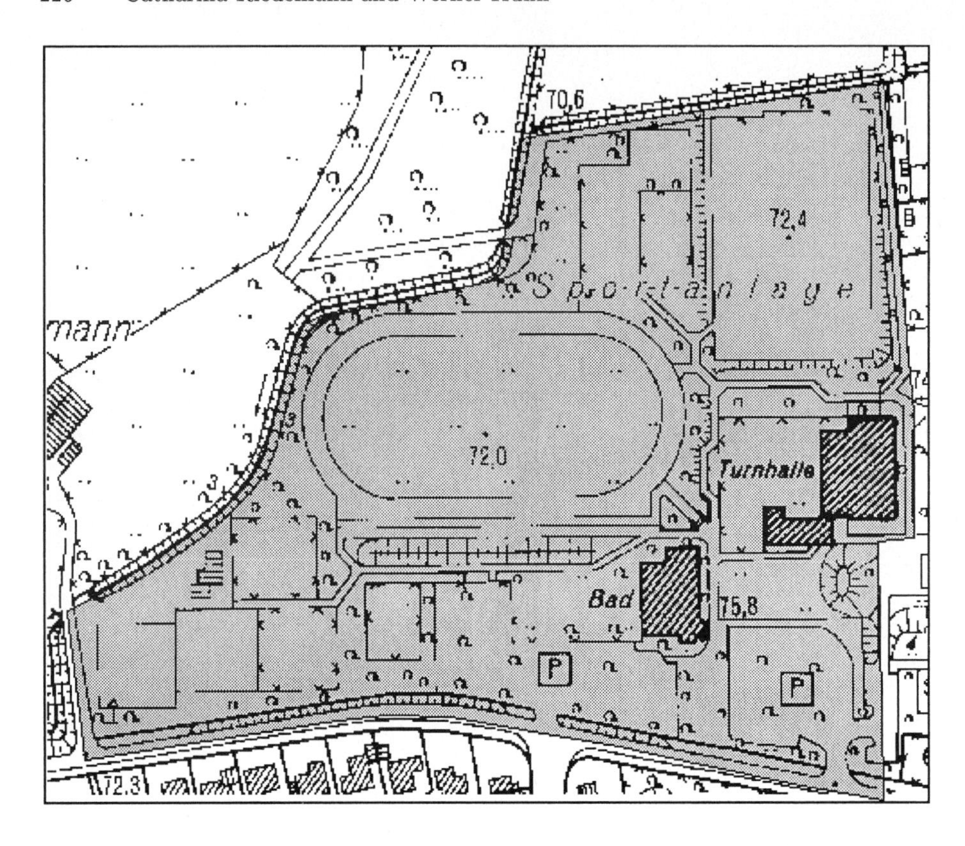

Fig. 4. The "Sportzentrum Roxel" in DGK 5 and ATKIS

All data collections basically concur that there is an area dedicated to sports, but each set has its own perspective differing from the acoustic perspective. Consequently each set provides at best part of what is needed for noise impact assessment. To get an appropriate digital acoustic model, parts of (future) cadastral and future ATKIS data together can be taken as a basis and completed by digitizing supplementary information out of the DGK 5. Currently, existing data are often digitized again, or data not exactly meeting requirements are used, although more appropriate data exist. This occurs, because users are not presented with the data they need and because it is often too complicated to build the necessary data out of (several) existing sets. Support for adapting data to the requirements of users (e. g. dividing the square tennis areas of the cadastral data into two tennis-courts) and merging digital data collections could improve this situation. This would mean, on the one hand, improving the use of available data and, on the other hand, giving access to the most appropriate data and thus making possible the best results for a given task.

5 Conclusions

Our case study on sports grounds has established a series of shortcomings in the current practice of using spatial data. It was originally intended to document specific semantic problems arising in an application picked more or less at random. In this direction, we have at least been able to establish a case of practical needs for semantic information and a detailed account of practical semantic issues arising in a particular case. The paper thereby makes a contribution to the first phase of an ongoing project that collects and analyzes specific examples of semantic interoperability issues (`http://ifgi.uni-muenster.de/english/3_projects/sip/index.html`). The challenges we encountered, however, are far more complex than the kind of semantic differences one might try to resolve by semantic translation [3]. They reveal several layers of technical, organizational and cognitive issues that cannot be resolved by putting more semantics into spatial data models or component interfaces. In conclusion, we attempt to draw up an incomplete list of these issues. Some of them are already well known and documented. Others suggest that information sharing is a much more complex problem than the GI community might have expected. Furthermore, they separate this complexity into several smaller parts. They demonstrate the need for interoperability and the inadequate role of data transfers in applications cutting across traditional boundaries of information communities. The case study established that:

1. Phenomena of interest to "spatially aware professionals", such as sports grounds in the context of noise abatement, are today most often captured in some kind of spatial data.
2. These data are generally held outside the organization of the interested professionals and consequently difficult to access and assess for suitability.
3. Most existing data reflect the particular semantics of the Surveying and Mapping Information Community. It is firmly grounded in planar geometry and has little to say about people, processes of use, or temporal and physical attributes.
4. The questions that can be answered by certain data collections are often several steps of interpretation away from those raised by an application.
5. All data necessarily exhibit certain levels of quality, along the usual axes of accuracy, resolution, completeness, consistency and currency. These different degrees of quality become very obvious when attempting to integrate information from multiple sources.
6. Communication (with people as well as with computer systems) plays an important role in data sharing, because data requests must be unambiguous in order to be successful. Planning tasks mostly involve different information communities, which makes the communication issue even more important.
7. It often takes too much time to find, assess and use data. Especially when the amount of data required is not too large, as in our example, data are rather digitized again for economical reasons.

While it appears too difficult yet to order these issues with respect to their amenability to practical solutions, the list shows that, in order to make spatial information viable, one needs to transcend the limitations of thinking in terms of data collections and data transfers. The needs for spatial information have to be addressed from the user's perspectives and supported by technical as well as institutional means to translate these needs into requests for services that search and exploit data collections. On the road to making this user-driven scenario reality, the establishing of spatial data warehouses by producing some kind of canonical representations of the landscape, documenting them through metadata, and making them accessible by exchange formats is only a beginning. A second step, implemented by OpenGIS®architectures, is to create service interfaces that offer answers to questions and not just excerpts from databases in some cryptic exchange format. A third step, investigated in the research on semantic interoperability is to enrich these interfaces by means to capture more semantics. The real breakthrough to an improved usability of spatial data, however, will only come with a better understanding of the mechanisms needed to map user questions to data through service interfaces. Our research agenda targets the identification and formalization of such semantic mappings [5]. The case study presented here is a first attempt at understanding their nature and requirements.

Acknowledgements

This work is partially supported by a grant from NCGIA's project Varenius. We also would like to acknowledge the support of the following people in preparing this paper and in providing the data: Markus Bellinghoff, Francis Harvey, Ann Hitchcock, Jolanta Krasutzki, Jürgen Müller, Franz-Josef Wenning.

References

1. Arbeitsgemeinschaft der Vermessungsverwaltungen der Länder der Bundesrepublik Deutschland (AdV): Musterblatt für die Deutsche Grundkarte 1:5000. 8th edn. Niedersächsisches Landesverwaltungsamt, Landesvermessung, Hannover (1983)
2. Arbeitsgemeinschaft der Vermessungsverwaltungen der Länder der Bundesrepublik Deutschland (AdV): Amtliches Topographisch-Kartographisches Informationssystem ATKIS. ATKIS-Gesamtdokumentation. Teil D ATKIS-Objektartenkatalog (ATKIS-OK). Landesvermessungsamt Nordrhein-Westfalen, Bonn (1995)
3. Bishr, Y.A.: Semantic Aspects of Interoperable GIS. Ph.D. Thesis. ITC Publication Series, Vol. 56. Enschede (1997)
4. Bundes-Immissionsschutzgesetz (BImSchG). Gesetz zum Schutz vor schädlichen Umwelteinwirkungen durch Luftverunreinigungen, Geräusche, Erschütterungen und ähnliche Vorgänge. BGBl. I (1990) 880
5. Harvey, F. et al.: Sharing Geographic Information: A Central Issue for Semantic Interoperability (submitted to publication)
6. Hillen, R.: Schallimmissionspläne - Basis von Lärmminderungsplänen. Landesanstalt für Immissionsschutz Nordrhein-Westfalen (ed.). LIS-Berichte, Vol. 108. Essen (1993)

7. Innenministerium des Landes Nordrhein-Westfalen (ed.): Vorschriften für die Bildung und Abbildung von Objekten der Automatisierten Liegenschaftskarte in Nordrhein-Westfalen, Objektabbildungskatalog Liegenschaftskataster NRW. Landesvermessungsamt Nordrhein-Westfalen, Bonn (1998)
8. Kuhn, W.: Approaching the Issue of Information Loss in Geographic Data Transfers. Geographical Systems 3 (1997) 261-276
9. Länderausschuß für Immissionsschutz (LAI): Muster-Verwaltungsvorschrift zur Durchführung des § 47a BImSchG. Aufstellung von Lärmminderungsplänen. Ministerium für Umwelt, Raumordnung und Landwirtschaft des Landes Nordrhein-Westfalen (ed.). Düsseldorf (1992)
10. Sheth, A.: Data Semantics: What, Where and How? Paper presented at the 6th IFIP Working Conference on Data Semantics (DS-6), Atlanta, GA (1996)
11. Sportanlagenlärmschutzverordnung (18. BImSchV). Achtzehnte Verordnung zur Durchführung des Bundes-Immissionsschutzgesetzes. BGBl. I (1991) 1588
12. Verwaltungsvorschriften zum Bundes-Immissionsschutzgesetz. MBl. NW (1993) 1472

Identifying Semantically Similar Elements in Heterogeneous Spatial Databases Using Predicate Logic Expressions

Kristin Stock[1] and David Pullar[2]

[1] School of Planning, Landscape Architecture and Surveying
Queensland University of Technology
2 George Street
Brisbane
Qld 4001
kristin_stock@dpa.act.gov.au

[2] Department of Geographical Sciences and Planning
University of Queensland
Brisbane
Qld 4072
D.Pullar@mailbox.uq.eud.au

Abstract. For data to be successfully integrated, semantically similar database elements must be identified as candidates for merging. However, there may be significant differences between the concepts that participants in the integration exercise hold for the same real world entity. A possible method for identifying semantically similar elements prior to integration is based on cognitive science theory of concept attainment. The theory identifies inclusion rules as being the basis for the highest level of concept attainment, once concepts have been attained at lower, perceptive levels. Predicates can be used to combine inclusion rules as a basis for semantic representation of elements. The predicates for different database elements can then be compared to determine the similarities and differences between the elements. This information can be used to develop a set of semantically similar elements, and then to resolve representational conflicts between the elements prior to integration.

1 Introduction

Since the popularization of spatial information systems, a number of researchers have discussed the benefits of spatial data sharing [31]. However, in recent years, data sharing has become increasingly important. This increase has been motivated by growing environmental concerns, pressures on government and the private sector to perform more efficiently, recognition of the synergistic advantages of spatial data and the increasing availability of a wide range of data [33], [21].

Despite agreement that spatial data sharing is an important goal, attempts to achieve such a goal have often been frustrated by heterogeneity in the data.

Data heterogeneity can be classified into schematic heterogeneity, syntactic heterogeneity and semantic heterogeneity [3]. If any of these types of heterogeneity exist, data from different sources cannot be readily integrated for use in problem solving.

Schematic heterogeneity refers to differences in the type of database elements that are used to represent a particular real world entity (for example, attribute, relation or class) [3].

Syntactic heterogeneity refers to differences in the structures of the elements that are used to represent real world entities. For spatial data, structure relates to the type of geometric element (for example, point, line or polygon), and the characteristics of that element. For non-spatial data, structure relates to the representational details of the database element (for example, data type, constraints or domain) [3].

Semantic heterogeneity refers to differences in the definition of concepts and the rules that are used to determine whether a real world entity is an example of a concept. Semantic heterogeneity is the source of most data sharing problems [3], and is the focus of the research described in this paper. Semantically similar database elements must be identified and any heterogeneity resolved before schematic and syntactic heterogeneity can be addressed. Methods for resolution of the latter two types of heterogeneity are provided by [9], [39], [30] and [10].

Semantic heterogeneity between individuals can be significant. The semantics that individuals have for a particular real world entity vary depending on the concepts or categories they use to classify the entities they encounter in everyday life. These concepts differ between individuals depending on education, experiences and theoretical assumptions [29]. For some applications (for example, hospital management systems), a certain level of similarity in world views can be assumed [46]. However, this assumption is not valid for spatial data, as users with a wide range of backgrounds are likely to be interested in spatial data due to its fundamental nature [33].

The OGIS information communities model groups people that have the same semantics for a particular set of concepts together into information communities[1]. Data sharing within information communities is relatively easy because the members of the community use similar concepts. However, if data is to be shared between communities, semantic similarity cannot be assumed. In this case, sharing requires that semantic similarities and differences between elements can be identified and resolved [33].

Methods for determining the semantic similarity of elements proposed to date can be divided into two main groups. The methods in the first group are concerned with the identification of semantic similarity based on representational (syntactic) details. This includes element characteristics [20] ,[7], or at the more complex end of the spectrum, behavior [12], [19]. Elements are considered se-

[1] "An information community is a collection of people...who, at least part of the time, share a common digital geographic information language and share common spatial feature definitions. This implies a common world view as well as common abstractions, feature representations, and metadata." [33]

mantically similar if they have similar representation or behavior respectively. These methods are limited in application in semantically heterogeneous environments because they assume that semantic similarity is correlated with syntactic similarity.

The methods in the second group are more concerned with interrogation of the semantics of elements, and examine definitions [3] or terminological relationships [6], [13] in an attempt to determine semantic similarity. These methods are limited in their ability to identify similarities and differences because they usually assume that a common language is used (that is, a given term has the same meaning across all databases). This assumption is often invalid in the context of spatial data, because the same term may be used by different information communities to mean different things. Similarly, different terms may be used by different information communities to mean the same thing.

This paper describes a new method for identifying semantic similarity of database elements, referred to as the inclusion rules method. The method is an extension of some of the methods that fall into the second group in the previous paragraph, and applies psychological theories of concept attainment to the problem of the representation of element semantics. The method attempts to increase the applicability of the previously proposed methods by reducing reliance on the assumption that participants use a common set of terms and definitions. This goal is particularly important for spatial database integration, because users of spatial data often have widely varying backgrounds and thus different semantics for real world entities. Consequently, this paper focuses on the use of the method for spatial data. However, many other database applications involve a similarly diverse range of individuals or groups, and the method is equally suitable for these applications.

The proposed method allows users to represent the semantics of their own database elements with a predicate that combines several inclusion rules. Inclusion rules are rules that indicate the characteristics that an instance must have to be considered an example of that element. The predicates are then used to determine the similarities and differences between any two database elements. The information about the relationship of the elements provides a tool for resolution of variations and ultimately, element integration, although this is not described in this paper due to space limitations (refer to [47] for details).

The next section in this paper briefly reviews the methods that have been suggested by other researchers, both from conventional and spatial database fields of research, for identifying semantic similarity. Section 3 provides an introduction to the inclusion rules method, including a review of the psychological theory that provides its foundation, and defines a formal language for the inclusion rules method. Section 4 contains an example of the method and its use in representing element semantics and determining the semantic equivalence of database elements.

This paper uses the term *real world entity* to refer to some 'thing' experienced in the real world. Individuals group these real world entities into *concepts* in order to deal with the world. Concepts are defined according to a set of rules that dictate whether an entity is an example of the concept or not. If the experienced real world entity fulfills those rules, it is considered an example of the concept.

In an object oriented database, concepts are implemented as either *classes* or *attributes*, and real world entities as either *objects* of the classes or *values* of the attributes respectively [11], [38]. Similarly, in a relational database, concepts are implemented as either *relations* or *attributes* and real world entities as *tuples* of the relations or *values* of the attributes respectively [11]. In the remainder of this paper, the term *element* is used to refer to any database representation of a concept (that is, class, attribute or tuple), and the term *instance* is used to refer to any database representation of a real world entity (that is, object or value).

2 Review of Integration Methods

As described in the previous section, database integration methods must be capable of identifying both the differences and the similarities between elements. Where elements differ, it must be possible to identify *how* they differ.

The methods proposed by database researchers for identifying semantic similarities and differences can be divided into two groups. The methods in the first group take advantage of the representation of databases in various ways, comparing representations to identify semantically equivalent elements. The methods in the second group attempt to identify semantically equivalent elements by interrogating definitions or relationships between terms. This section provides a very brief review of these methods. The interested reader should refer to the relevant references for more information.

2.1 Representational Methods

The representational database integration methods use the characteristics of each database element to determine semantic equivalence. A common approach is to determine attribute equivalence by comparing structural characteristics including uniqueness, domain, constraints, allowable operations and units [20], and then relation equivalence on the basis of the equivalence of either all the component attributes [7], or only the key attributes [20].

A more general form of this method uses a combination of context and domain to determine semantic equivalence. The context is some representation of the underlying assumptions of the element, and may take the form of the database, a set of named domains or a rule based formal expression [16], [41].

Another method combines usage and access patterns with other representational information to determine semantic equivalence [46].

The use of the representational details to determine semantic equivalence has been applied to the non-spatial components of spatial information systems [31], with the addition of role information. This latter information refers to the use or meaning of the element, and is stored in the data dictionary [32].

Moving to the more complex end of the representational spectrum, some researchers have suggested that element behavior be compared to determine semantic equivalence. A common approach is for behavioral equivalence to be determined by running all operations that are defined on an element on itself, and then comparing the results with the same operations run on the element to which it is being compared [12]. A more sophisticated variation on this approach uses universal algebra to define the set of operations that describe an element in terms of its use or purpose. The method adopts a mathematical device called homomorphism to compare the algebras of different elements and determine whether they are similar [19].

A limitation of the representational methods is their reliance on the ways individuals model their data, or the methods they attach to their elements, both of which will depend on their individual semantics. This approach does not take element semantics into account, meaning that elements that are semantically very different may be considered equivalent.

2.2 Semantic Methods

Two main types of semantic methods have been suggested: those based on terminological relationships and those based on definitions. The former use networks or hierarchies that indicate the relationships (including synonyms and generalizations) between either the names given to database elements [14] or the terms used in element definitions [50]. There is usually a common taxonomy of terms that is used by all those involved in data sharing. The terms used in the local databases are mapped to the common taxonomy, and the links between terms from different databases can be used to determine their semantic similarity [6], [13], [3]. A variation requires the user to define semantic clusters across all databases, which are then used to automatically generate an associative network [42].

The element definitions method involves a simple comparison of the definitions of the elements being compared. This method has been applied to spatial information, using the classification criteria for an element as its definition. However, these definitions are used to resolve schematic heterogeneity rather than determine semantic equivalence [3].

Another method combines terminological relationships and element definitions in a knowledge base. Other types of information like organizational rules may also be included [50], [8], [6], [15].

These semantic methods usually assume that a common language is used (that is, a given term has the same meaning across all databases). Although mappings between the individual's terminology and that of the common language is possible, this only allows a direct one to one mapping between terms, and does not consider the more subtle differences in the classification criteria used for terms by different individuals.

These requirements are limiting for information that has a heterogeneous nature, including spatial information. These limitations are supported by Mark's findings that linguistic labels are limited in their ability to convey semantics [24], and Kuhn's comments about the inadequacy of terms for the representation of meaning [19].

3 The Inclusion Rules Method

The inclusion rules method aims to take advantage of the strengths of the semantic methods, but reduce the need for a common language that limits the application of the semantic methods to information that is used by a number of different information communities (like spatial information). The method does this by applying a cognitive science theory of concept attainment to the problem. This section provides a theoretical background to the method, and gives a formal definition of relevant elements.

3.1 Theoretical Background

Psychological theory asserts that individuals organize the world in terms of categories of things that have similar characteristics or attributes, and that these categories make up the semantics of real world entities from the individual's perspective. The characteristics are selected according to underlying theories that individuals hold about which characteristics are important, and hence may differ from one individual to another [28].

Psychological research has dedicated much effort to the question of how categories (more commonly referred to as concepts), are learnt, and then how they are used to classify newly experienced real world entities. These theories are important for research into database integration, because they provide an indication of how concepts differ between individuals, and following on from this, how it might be possible to translate between concepts held by different individuals so that valid integration is possible.

A number of psychological theories have been suggested in terms of how concepts are attained, many having a number of similarities. The database integration method developed in this research is based on the theory of concept attainment suggested by Klausmeier, Ghatala and Frayer [18]. This theory, or model, is based on several years of experimentation, as well as work by other eminent psychological researchers (including Piaget and Inhelder [36], who have carried out a series of authoritative studies in this area) [18].

Klausmeier et al's model involves a series of levels of concept attainment: the concrete level, the identity level, the classificatory level and the formal level. The concrete level is similar to perception, and is reached when the individual is aware of an entity. The identity level is reached when the individual sees the same entity in different situations, positions or orientations, and realizes that it is the same entity. The classificatory level is reached when the individual groups entities together based on some similarity, but is not able to formalize the criteria for this grouping. Finally, at the formal level, the individual is able to define the group of characteristics that define a concept, and to identify inclusion and exclusion rules in terms of those characteristics [18].

Klausmeier et al's theory suggests that concepts change over time, and may be affected by a number of different external factors. A possible criticism of the application of the the theory to database semantic representation is that the theory applies to the concepts of individuals; but that database elements are more likely to represent the concepts of a group as defined during the database analysis and design process. It is suggested that there is unlikely to be an incompatability between an individual database user's concepts and the concepts of the database itself for the following reasons:

- the database concepts are developed from the concepts of the database users, and since those users have similar experiences with the data and operate in a similar environment, they will probably have similar semantics for the database elements they share [29] and
- it would be difficult for a user to regularly use a database over a period of time that contained concepts that were incompatible with her own without gradually changing her own semantics, or consiously performing a mental translation whenever she used the database.

Since the database concepts are likely to be similar to the concepts of its users, it is appropriate to adopt the individual theory for application to the wider database.

3.2 A Formal Language for the Inclusion Rules Method

In accordance with Klausmeier et al's theory of concept attainment, the inclusion rules method defines the semantics of database elements (which are the implementation of concepts) in terms of formal level rules that the representation of a real world entity must satisfy in order to be considered an instance of the element. These rules are combined into predicates that express the meaning of elements such that each database element is represented by a single predicate. When elements are defined in this manner, the inclusion rules can be used to determine the similarities and differences between the elements in different databases.

For the purposes of defining and illustrating the inclusion rules method, a formal language has been adopted. The formal language has two separate parts. The first part describes how inclusion rules are specified in terms of dimensions and properties. These rules form the foundation for the second part, which describes how rules can be combined into predicates to define element semantics.

For the purposes of this definition, the standard syntactic metalanguage is adopted [49]. Using this metalanguage, the "=" symbol defines a name with its appropriate grammatical structure or well formed formula. A ";" symbol terminates the definition. A "—" symbol indicates alternative choices for the make up of a well formed formula, and any symbol or variable from the language being specified is enclosed in quotation marks.

Rule Definition Klausmeier et al's model specifies that a concept is attained at the formal level when a person can specify the rules that she uses to include or exclude a newly experienced entity from the concept. These rules consist of what Klausmeier et al [18], as well as other researchers (for example, [4]) refer to as dimensions and properties. For example, color is a dimension, and red is a property of that dimension in Klausmeier et al's language [18].

In the formal language of the inclusion rules method, the following variables represent the principles discussed above:

- D_i represents a dimension;
- P_j represents a property (which may be a single value, a range of values or an enumerated set of values) and
- R_k represents a rule.

A symbol [ ] can be interpreted as 'is a property of'.

The syntactic rule for well formed formulas for specifying rules in the formal language is as follows:

$$rule = \text{“}D_i\text{”}\ \text{“[”}\,\text{“}P_j\text{”}\ |\ element\ definition\ \text{“]”}; \quad (1)$$

(element definition will be defined in the next section).

As discussed in the previous section, one of the aims of the method is to remove the requirement for users to have a common language. The inclusion rules method does this in part by defining database elements in terms of inclusion rules that have a standard form (a dimension and property). However, some reliance on language is unavoidable because the dimensions and properties are expressed in language. The method attempts to minimize this reliance in four ways.

Firstly, element semantics are represented using a standard form that removes some of the ambiguities of natural language. Users are confined to representation of rules as dimensions and properties, and to combining these rules using the $\wedge$ and $\vee$ operators. In this way, some of the possible ambiguity is removed.

Secondly, inclusion rules may include not only direct values, but also references to other elements. The property of an element can itself be a predicate. The implementation of this facility is described in more detail in 4.2.

Thirdly, a standard set of dimensions is being developed, using research into semantic theory. Although individuals differ in the dimensions and properties that they user to define concepts, it is thought that a fundamental set of dimensions and properties are used as a source of these definitions. Relationships between the dimensions can be defined to handle redundancy, asymmetry and context [22].

Fourthly, attempts should be made to avoid assumptions about the meanings of terms. Dimensions and properties should be defined in the tradition of scientific research [44]. In this sense, the ability of the method to handle different semantics will depend on the specificity of the inclusion rules.

Element Definition Element definitions are built up using combinations of rules, so no new variables are necessary. Two additional operators are used:

- '$\wedge$ which can be interpreted as 'and" (conjunction), meaning that the element is defined by both the operands and

- '$\vee$ which can be interpreted as 'or" (disjunction), meaning that the element is defined by either or both of the operands.

The inclusion rules method does not use the negation operator. The reason for this omission is that the use of negation in databases causes logical problems. If an element is defined in terms of negative rules (for example the element is not green in color), this implies that all the properties that the element does not have must be specified for the definition to be complete. This would include an infinite list of negative properties, and is clearly impossible. For this reason, it is customary to adopt the closed world assumption. This assumes that if a piece of information is not in the database (in this case, a rule is not in the definition), it is considered to be false. In other words, if it is said that an element is red or yellow, the fact that it is not green, blue or any other color are implied, and the need to explicate these facts is avoided [37], [1].

Well formed formulas for element definition must conform to the following pattern:

$$element = rule \mid element \wedge \ element \mid element \vee \ element; \tag{2}$$

Element definitions are referred to as predicates for the remainder of this paper. Each of these definitions has an implied $\forall x | x \in$ element(element definition), to indicate that each instance of the database element evaluates to true for each rule in the expression.

4 An Example

The inclusion rules method provides a means for undertaking each step in the database integration process. Firstly, the method can be used by participants in data sharing exercises to represent the semantics of the elements in their databases. Following this, the method can be used to determine the semantic equivalence of elements, and thus whether they should be integrated or translated between. Thirdly, the information provided by the method can be used to resolve any schematic or syntactic heterogeneity between the elements that are to be integrated, and to merge the databases into one. This paper is concerned with the first two steps in the process, as described in this section with a running example using two relational databases. A summary of the third stage in the process is provided in [47].

4.1 The Participating Databases

Figures 1 and 2 contain Entity Relationship diagrams for two example relational schemas, together with the relations produced from the diagrams.

The elements in Schema 1 are defined as follows: Relations:

- Block: a physical area of the surface of the earth, defined both in position and extent, with a bundle of property rights attached that are owned by a particular person. A block must exist either wholly or partly within a larger administrative area.

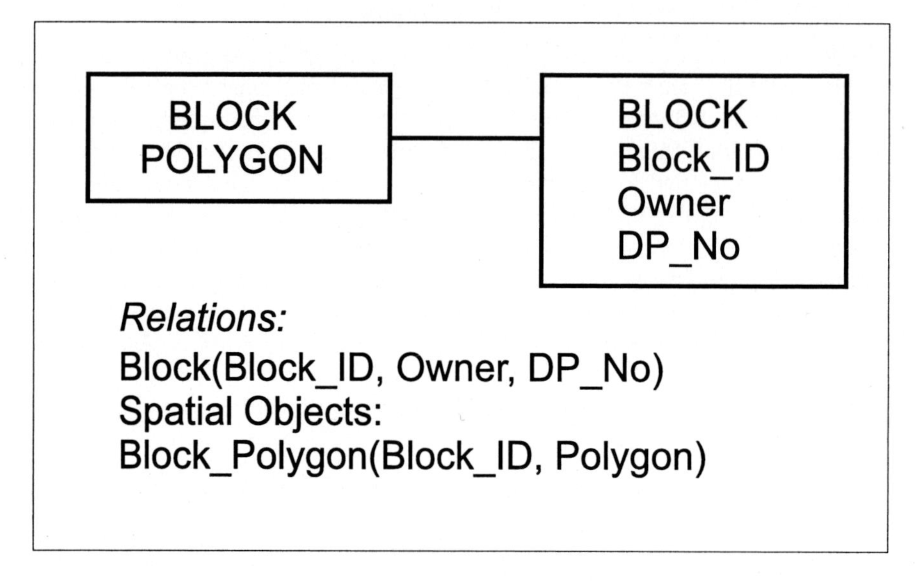

Fig. 1. ER Diagram and Relational Model for Schema 1

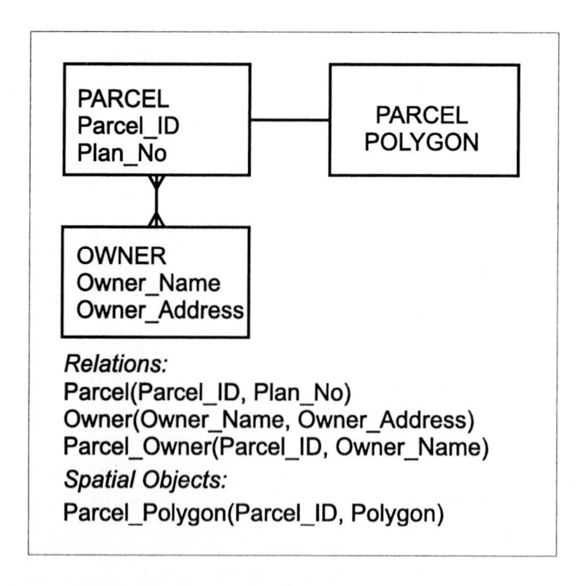

Fig. 2. ER Diagram and Relational Model for Schema 2

Attributes:

- Block_ID: a unique, arbitrarily defined identifier for a block.
- Owner: the person who has the right to use a particular block in accordance with the bundle of rights assigned to the relationship between the block and the person. The owner may only be a single individual.
- DP_No: the number of the plan that created the block and has been deposited with and registered by the jurisdiction's land administration authority.

Spatial Elements:

- Block_Polygon: the physical location and extents of a block, captured using survey plan data entry, using an Australian Map Grid coordinate system defined in meters (Transverse Mercator projection) with an accuracy of +/- 0.2m, oriented to grid north.

The elements in Schema 2 are defined as follows: Relations:

- Parcel: a physical area of the surface of the earth, defined both in position and extent, with a bundle of property rights attached that are owned by one or several people. A parcel must exist wholly within a larger administrative area.
- Owner: the person who has the right to use a particular parcel in accordance with the bundle of rights assigned to the relationship between the parcel and the person.
- Parcel_Owner: the assignment of a particular parcel to the ownership of a particular person (each parcel may have any number of owners, and each owner may own any number of parcels).

Attributes:

- Parcel_ID: a unique, arbitrarily defined identifier for a parcel.
- Plan_No: the number of the plan that created the parcel and has been registered by the jurisdiction's land administration authority.
- Owner_Name: the name of the person who has the right to use a particular parcel in accordance with the bundle of rights assigned to the relationship between the parcel and the person.
- Owner_Address: the street name, number, locality and city of the owner.

Spatial Elements:

- Parcel_Polygon: the physical location and extents of a parcel, captured by digitizing a 1:1000 topographical map, using a local coordinate system defined in feet with a Lambert Conformal Conic projection) with an accuracy of +/- 10m, oriented to true north.

4.2 Representing Database Elements

Before attempts can be made to determine the semantic similarity of database elements, a predicate logic expression must be defined for each element in the participating databases. These expressions consist of rules, combined with the $\wedge$ and $\vee$ operators. Complex expressions can be built up in the usual predicate logic manner.

The rules themselves are contained in a rule repository, which consists of three tables of a relational nature.

1. The Rule Table contains the number of the rule, which is simply the subscript to the "R" variable that represents each rule, and the numbers of the dimension and property that make up that rule.
2. The Dimension Table contains the number of the dimension, which is simply the subscript to the "D" variable that represents each dimension, and a phrase that defines the dimension.
3. The Property Table contains the number of the property, which is simply the subscript to the "P" variable that represents each property, and a data value, range of data values or enumerated set of data values that define the property (a phrase, number, date etc.), or a predicate logic expression that defines an element that represents the property.

For each element in each of the participating databases, the owner of the database must carry out the following steps to define the semantics of the elements in the database:

1. Define a set of rules for the element that dictate whether a newly experienced entity is considered to be an example of that element (that is, define the inclusion rules for the database element), and create an expression to combine the different rules.
2. For each rule:
   - Check whether the dimension is already in the repository. If it isn't, add the dimension to the repository.
   - Check whether the appropriate property value or range is already in the repository.
     - If both the dimension and the property are already in the repository, check to see if there is a rule that matches the two. If there is, place the number of the rule in the predicate. If there isn't, create a new rule that references the appropriate dimension and property, and place its number in the predicate.
     - If the either the dimension of the property (or both) are not in the repository, add them and create a new rule that references the appropriate dimension and property. Place the number of the new rule in the predicate.

An important restriction on the addition of properties to the repository is that circular references must be avoided. Since properties can contain references to other rules, a check must be undertaken to avoid creating logic inconsistencies in the rule definitions. In addition, a constraint on predicates is that they must not contain more than one rule that references any given dimension, as this would indicate a logical contradiction. If an element may have one of several values for a property, these can be included as a property range or enumerated set of values.

Since the repository is extensible by data sharers, it will have a number of versions, and version numbers will need to be maintained. For each database element, the data sharer creates a predicate that references the rules in the repository. In order for database integration to be successful, it is important that participants in a data sharing exercise either use the same version of the repository, or only reference elements that are contained in the versions that are being used by other participants. As the repository becomes large, the need to extend the repository will decrease, and changes will be less likely, thus at this later stage it is more likely that database integration can occur successfully even if the versions used by the participants are slightly different.

The Appendix A contains the three rule repository tables for the example introduced in Section 4.1. The predicates for each element in the schemas are as follows: Schema 1:

- Block = $R_1 \wedge R_2 \wedge R_3 \wedge R_4$
- Owner = $R_5 \wedge R_6$
- DP_No = R_7
- Block_Polygon = $R_8 \wedge R_9 \wedge R_{10} \wedge R_{11} \wedge R_{12}$

Schema 2:

- Parcel = $R_1 \wedge R_2 \wedge R_3 \wedge R_{13}$
- Owner = $R_5 \wedge R_6$
- Plan_No = R_7
- Owner_Name = $R_5 \wedge R_6$
- Owner_Address = $R_{14} \wedge R_{15} \wedge R_{16}$
- Parcel_Polygon = $R_{17} \wedge R_9 \wedge R_{18} \wedge R_{19} \wedge R_{20} \wedge R_{21} \wedge R_{22} \wedge R_{23}$

Although most of the properties listed in this example contain single values, there are some properties that include predicates as references to other elements. For example, rules 5 and 17 both include a dimension that provides a reference to the semantic aspects of the spatial element being defined. Rule 16 also includes a reference to the owner of a block as a means of indicating the importance of an address. In a real life situation, many more of these references would be used to avoid undefined terminology (for example, street, residential area and administrative area would be defined with predicates).

The use of the $\vee$ operator is relatively unusual. This is because definitions of elements do not often rely on two alternatives that refer to an entirely different dimension. It is more common for alternatives to refer to a single dimension, in which case they can be included as an enumerated set (for example, see Rule 4).

4.3 Identifying Semantic Similarity

Up to this point, the aim of identifying the semantic equivalence of elements has been defined as identifying database elements that represent the same real world entity. In the case where two database elements represent exactly the same real world entity (in terms of the users' semantics of that entity), they should be represented by the same predicate. However, it is often the case that two elements may have slightly different semantics, but the user would still like to integrate them. Thus for the purpose of schema integration, semantic similarity is defined by the integrator and the purpose of the integration. If slight differences are of no consequence for the purpose to which the data will be put, semantic similarity will have a more generous definition than for other purposes.

Using the inclusion rules method, it is possible to integrate any two database elements. This integration creates a database element with the semantics of the rules that the two integrated predicates share. If they do not share any rules, the created element will be infinitely general (that is, it will contain any real world entity). The more rules the two elements share the more specific the integrated element will be. The level of generality of the integrated elements will depend on the user's requirements, so the inclusion rules method offers user input to allow control over the process.

Each element in one database must be compared to each element in the other database in order to determine whether the two elements are semantically equivalent within the tolerances of the human integrator. This comparison relies on the predicates that represent the elements as a representation of element semantics. For the remainder of this discussion, the two predicates are referred to as P_i and P_j.

The first step in comparison of the predicates is to convert them to conjunctive normal form. Conjunctive normal form (CNF) is a form where an expression is the conjunction of a number of conjuncts. Each of these conjuncts is a single variable or a disjunction of disjuncts, each of which is a single variable. Every expression has an equivalent in CNF. [23] discusses methods for converting predicate expressions into CNF. Once P_i and P_j are in CNF, they can be directly compared. Each rule in P_i should be classified as one of the following:

- a shared conjunct, being a conjunct, whether a single rule or a disjunction, that also appears in P_j;
- a shared disjunct, being a disjunct of a disjunction that is contained in either a conjunct or a disjunction in P_j;
- an unshared conjunct, being a conjunct, whether a single rule or a disjunction, that does not appear in P_j or
- an unshared disjunct, being a disjunct of a disjunction that is not contained in either a conjunct or a disjunction in P_j.

This classification is used to determine how similar two elements are semantically. Shared conjuncts are considered a stronger indication of similarity than shared disjuncts, because they are compulsory rather than optional. For example, an element that is (red and square) is more similar to an element that is (red and round) than it is to an element that is ((red or old) and round). Similarly, unshared conjuncts are considered a stronger indication of difference than are unshared disjuncts.

The classification method described above requires that a judgment is made regarding similarity or difference of two rules. If two rules have the same number, they are obviously the same, but if they have different numbers, they may still be similar. Thus any rules that do not have the same number must be examined in more detail to determine their possible similarity. A first step in this more detailed examination involves comparison of the dimensions. If these dimensions are different, the rules are also different, so they can be classified accordingly. If the dimensions are the same (but the rule number different), the property values must be compared.

The compared property values will not be identical (because this would mean that the rule repository contains identical rules), but the values in one rule may share some common values with the values in the other rule. In this case, the decision regarding whether the rules should be classified as similar (and thus shared) or different (and thus unshared) depends on the property type and the rule comparison strategy. The rule comparison strategy can be biased towards similarity or difference, depending on the requirements of the integration or translation exercise. The similarity bias assumes similarity if the relationship is unknown and requires only an overlap between property values in order for two rules to be considered similar. In contrast, the difference bias assumes difference if the relationship is unknown, and requires one value to be a subset of the other for two rules to be considered similar.

The classification process is carried out in both directions. That is, P_i is compared to P_j, and then P_j is compared to P_i.

The classification discussed above is used in order to determine the degree of similarity between two elements by calculating a ratio of similarity to difference between two predicates. This is based on the principle that more similar elements will probably have more rules in common. A calculated value termed the comparison ratio is used to encapsulate this similarity. The comparison ratio (CR) is a simple ratio relating the number of rules that the two element predicates share to the number of rules by which they differ. The shared and unshared disjuncts are given a lesser value than the shared and unshared conjuncts for reasons discussed above. A value of about half may be appropriate, but this may also be set by the user:

CR = number of shared conjuncts + (number of shared disjuncts / 2) : number of different disjuncts + (number of different disjuncts / 2) for either the P_i to P_j comparison or the P_i and P_j comparison, whichever is larger.

This definition of the CR assumes that rules are of equal weight. If the rules that the two predicates differ by are much more important than those that it shares, the elements represented by the predicates may seem much more similar than they really are. For this reason, the inclusion rules method allows the definer of an element semantics (in the form of a predicate) to assign weights to the different rules in the predicate. If such weights are assigned, the definition of CR becomes:

CR = $\sum$ weights of shared conjuncts + ($\sum$ weights of shared disjuncts / 2) : $\sum$ weights of different disjuncts + ($\sum$ weights of different disjuncts / 2) for either the P_i to P_j comparison or the P_i and P_j comparison, whichever is larger.

CR values can be interpreted as follows:

- CR = 1:1 - the similarities and differences are approximately equal;
- the first value in the ratio is much larger than the second value - the elements represented by the predicates are similar and
- the first value in the ratio is much smaller than the second value - the elements represented by the predicates are not very similar.

The CR values for each pair of elements can be used to automatically determine the semantic equivalence of elements. The user can specify a similarity threshold, and if the ratio is greater than the threshold values, the elements concerned are considered semantically similar, and are integrated or translated between. For example, if a user specifies a similarity threshold of 1:1, any pairs of elements that have more similar rules than different rules will be considered semantically equivalent (if weights are attached, this may not necessarily mean that the number of similar rules will be greater than the number of different rules).

Table 1. CRs for Element Pairs

Schema 1(top), Schema 2(side)	**Block**	**Owner**	**DP_No**	**Block_Polygon**
Parcel	3:1	0:4	0:4	0:5
Owner	0:4	2:0	0:2	0:5
Plan_No	0:4	0:2	1:0	0:5
Owner_Name	0:4	2:0	0:2	0:5
Owner_Address	0:4	0:3	0:3	0:5
Parcel_Polygon	0:8	0:8	0:8	2:6

The CRs for the pairs of elements in Schema 1 and 2 are shown in Table 1. Using a similarity threshold of 1:1, the set of semantically equivalent elements is as follows: {1:Block, 2:Parcel} {1:Owner, 2:Owner} {1:Owner, 2:Owner_Name} {1:DP_No, 2:Plan_No}

The case of the polygons is one that might benefit from the application of weights. Although the polygons differ in a number of representational details, the polygons are linked to elements in their respective schemas that are considered semantically equivalent (that is, Block and Parcel). Thus it may be appropriate to give the rule that results from the combination of R_8 and R_{17} a greater weight. If this rule is given a weight of 5, a CR of 6:6 for the Block_Polygon - Parcel_Polygon pair is the result. This CR is within the similarity threshold, so that pair will be added to the set of semantically equivalent elements.

At this stage, the set of semantically equivalent elements that has been determined by the process described in this section can be presented to the user for alteration or confirmation as desired. This represents a significant reduction in effort relative to the entirely manual method that has sometimes been used. The attention of the user can be directed to particular data sets for confirmation.

From this point, the conflicts between the semantically equivalent elements must be resolved, and then elements can be merged. Details of this process are provided in [47].

5 Conclusion

The inclusion rules method as described above provides a means for the representation of the semantics of database elements. Although this method is not completely isolated from the language and semantics of individuals, the method reduces reliance on assumptions of similarity by incorporating aspects of Klausmeier et al's [18] model of concept attainment. This includes specification of dimensions and properties that are used to define a concept, as well as the types of relationships that exist between concepts.

Despite the persistent, if reduced, level of reliance on language, the method is considered useful in providing a formalized approach for expressing not only the differences between elements, but also the specific ways in which they differ in terms of dimensions and properties. This information can then be used to identify semantically similar database elements with a reduced requirement for human interaction.

Ongoing research is testing the use of the method in real world situations to determine its practicality, as well as its ability to handle the wide range of semantic heterogeneity that exists in the spatial user community.

The example provided in Section 4 indicates that the inclusion rules method is capable of identifying semantically similar elements that contain variations in database representation. This suggests that the method may be a useful tool for database integration and semantic translation of spatial data from different information communities.

References

1. Apt, K.R. and Bol, R.N.: Logic Programming and Negation: A Survey. Journal of Logic Programming. **19-20** (1994) 9–71
2. Batini, C., Lenzerini, M. and Navathe, S.B.: A Comparative Analysis of Methodologies for Database Schema Integration. ACM Computing Surveys. **18 (4)** (1986) 323–364
3. Bishr, Y.: Semantic Aspects of Interoperable GIS. Unpublished PhD Thesis, ITC, The Netherlands (1997)

4. Bourne, L.E.: Knowing and Using Concepts. Psychological Review. **77 (6)** (1970) 546–556
5. Breitbart, Y., Olson, P.L. and Thompson, G.R.: Database Integration in a Distributed Heterogeneous Database System. In: A.R. Hurson, M.W. Bright and S.H. Pakzad (eds): Multidatabase Systems: An Advanced Solution for Global Information Sharing IEEE Computer Society Press, Los Alamitos, California. (1994) 231–240
6. Bright, M.W., Hurson, A.R. and Pakzad, S.: Automated Resolution of Semantic Heterogeneity in Multidatabases. ACM Transactions on Database Systems. **19 (2)** (1994) 212–253
7. Chatterjee, A. and Segev, A.: Data Manipulation in Heterogeneous Databases. SIGMOD Record. **20 (4)** (1991) 64–68
8. Collet, C., Huhns, M.N. and Shen, W.: Resource Integration Using a Large Knowledge Base in Carnot. Computer. **24 (12)** (1991) 55–62
9. Dayal, U. and Hwang, H.: View Definition and Generalization for Database Integration in a Multidatabase System. IEEE Transactions on Software Engineering. **10 (6)** (1984) 628–645
10. Deen, S.M., Amin, R.R. and Taylor, M.C.: Data Integration in Distributed Databases. In: A.R. Hurson, M.W. Bright and S.H. Pakzad (eds): Multidatabase Systems: An Advanced Solution for Global Information Sharing. IEEE Computer Society Press, Los Alamitos, California. (1994) 255–259
11. Elmasri, R. and Navathe, S.: Fundamentals of Database Systems. The Benjamin/Cummings Publishing Company Inc, Redwood City, California. (1994)
12. Fang, D., Hammer, H. and McLeod, D.: The Identification and Resolution of Semantic Heterogeneity in Multidatabase Systems. In: A.R. Hurson, M.W. Bright and S.H. Pakzad (eds): Multidatabase Systems: An Advanced Solution for Global Information Sharing. IEEE Computer Society Press, Los Alamitos, California. (1994) 52–59
13. Fankhauser, P. and Neuhold, E.J.: Knowledge based integration of heterogeneous databases. In David Hsiao, Erich Heuhold and Ron Sacks-Davis (eds): Interoperable Database Systems (DS-5). Proceedings of the IFIP WG2.6 Database Semantics Conference on Interoperable Database Systems. Lorne, Victoria. North-Holland, Amsterdam. (1993) 155–175
14. Fankhauser, P., Kracker, M. and Neuhold, E.J.: Semantic vs. Structural Resemblance of Classes. SIGMOD Record. **20 (4)** (1991) 59–63
15. Hammer, J. and McLeod, D.: An Approach to Resolving Semantic Heterogeneity in a Federation of Autonomous, Heterogeneous Database Systems. International Journal of Intelligent and Cooperative Information Systems. **2 (1)** (1993) 51–83
16. Kashyap, V. and Sheth, A.: Semantics-based Information Brokering. Proceedings of the 3rd International ACM Conference on Information and Knowledge Management Gaithersburg, Maryland, USA. (1994) 363–370
17. Kim, W. and Seo, J.: Classifying Schematic and Data Heterogeneity in Multidatabase Systems. Computer. **24 (12)** (1991) 12–18
18. Klausmeier, H.J., Ghatala, E.S. and Frayer, D.A.: Conceptual Learning and Development. Academic Press, New York. (1974)
19. Kuhn, W.: Defining Semantics for Spatial Data Transfers. In Thomas C. Waugh and Richard G. Healey (eds): Advances in GIS Research: Proceedings Volume 1, Sixth International Symposium on Spatial Data Handling. Edinburgh, Scotland. (1994) 973–987

20. Larson, J.A., Navathe, S.B. and Elmasri, R.: A Theory of Attribute Equivalence in Databases with Application to Schema Integration. IEEE Transactions on Software Engineering. **15 (4)** (1989) 449–463
21. Laurini, R.: Distributed Databases: An Overview. In The AGI Source Book for Geographic Information Systems The Association of Geographic Information, London. (1995) 45–55
22. Leech, G.: Semantics: the Study of Meaning. Penguin Books, Middlesex. (1981)
23. McKeown, G.P. and Rayward-Smith, V.J.: Mathematics for Computing. Macmillan Press, London. (1982)
24. Mark, D.M.: Toward a Theoretical Framework for Geographical Entity Types. In A.U. Frank and I. Campari (eds): Spatial Information Theory: Theoretical Basis for GIS. Springer Verlag, Berlin. (1993) 270–283
25. Mark, D.M., Egenhofer, M.J., Rashid, A. and Shariff, M.: Toward a Standard for Spatial Relations in SDTS and Geographic Information Systems. Proceedings Volume 2, GIS/LIS '95 Annual Conference and Exposition. Nashville, Tennessee, USA. 14-16 November 1995. (1995) 686–695
26. Mark, D. and Frank, A.: Concepts of Space and Spatial Language. Proceedings of Autocarto 9, Ninth International Symposium on Computer-Assisted Cartography. Baltimore, Maryland. 2-7 April 1989. (1989) 538–556
27. Mark, D.M. and Frank, A.U.: Experiential and Formal Models of Geographic Space. Environment and Planning B: Planning and Design. **23(1)** (1996) 3–24
28. Medin, D.L. and Wattenmaker, W.D.: Category cohesiveness, theories and cognitive archaeology. In Ulric Neisser (ed): Concepts and Conceptual Development: Ecological and Environmental Factors in Categorization Cambridge University Press, Cambridge. (1987) 25–62
29. Moore, G.T.: Theory and Research on the Development of Environmental Knowing. In Gary T. Moore and Reginald G. Golledge (eds): Environmental Knowing: Theories, Research and Methods Dowden, Hutchinson and Ross, Stroudsburg, Pennsylvania. (1976) 138–163
30. Motro, A.: Superviews: Virtual Integration of Multiple Databases. IEEE Transactions on Software Engineering. **13 (7)** (1987) 785–798
31. Nyerges, T.: Schema integration analysis for the development of GIS databases. International Journal of Geographical Information Systems. **3 (2)** (1989) 153–183.
32. Nyerges, T.: Information Integration for Multipurpose Land Information Systems. URISA Journal. (1989)
33. OGIS Project Technical Committee The OpenGIS Guide. Edited by Kurt Beuhler and Lance McKee. Open GIS Consortium, Wayland. (1996)
34. OpenGIS Consortium.: The Open GIS Specification Model. Topic 5: The OpenGIS Feature. OpenGIS Project Document Number 98-105. (1998)
35. Ozsu, M.T. and Valduriez, P.: Principles of Distributed Database Systems. Prentice Hall, Englewood Cliffs, New Jersey. (1991)
36. Piaget, J. and Inhelder, B.: The Child's Conception of Space. Routledge and Kegen Paul, London. (1956)
37. Reiter, R.: On Closed World Data Bases. In H. Gallaire and J. Minker (eds): Logic and Databases. Plenum Press, New York. (1978)
38. Rumbaugh, J.: OMT: The object model. Journal of Object Oriented Programming. **8 (1)** (1995) 21–27
39. Saltor, F., Castellanos, M.G. and Garcia-Solaco, M.: Overcoming Schematic Discrepancies in Interoperable Databases. In David Hsiao, Erich Heuhold and Ron Sacks-Davis (eds): Interoperable Database Systems (DS-5). Proceedings of the IFIP

WG2.6 Database Semantics Conference on Interoperable Database Systems. Lorne, Victoria. North-Holland, Amsterdam. (1993) 191–205
40. Seligman, L. and Rosenthal, A.: A Metadata Resource to Promote Data Integration. http://www.nml.org/resources/misc/metadata/proceedings/seligman/seligman.html. 25 September 1996. (1996)
41. Sheth, A.P., Gala, D.K., and Navathe, S.B.: On Automatic Reasoning for Schema Integration. International Journal of Intelligent and Cooperative Information Systems. **2 (1)** (1993)23–50
42. Sheth, A. and Kashyap, V.: So Far (Schematically) yet So Near (Semantically). In David Hsiao, Erich Heuhold and Ron Sacks-Davis (eds): Interoperable Database Systems (DS-5). Proceedings of the IFIP WG2.6 Database Semantics Conference on Interoperable Database Systems. Lorne, Victoria. North-Holland, Amsterdam. (1993) 283–312
43. Sheth, A.P. and Larson, J.A.: Federated Database Systems for Managing Distributed, Heterogeneous and Autonomous Databases. ACM Computing Surveys. **22 (3)** (1990) 183–236
44. Simon, J.L. and Burstein, P.: Basic Research Methods in Social Science. McGraw-Hill, New York. (1985)
45. Spaccapietra, S. and Parent, C.: Conflicts and Correspondence Assertions in Interoperable Databases. SIGMOD Record. **20 (4)** (1991) 49–54
46. Srinavasan, U.: A Framework for Conceptual Integration of Heterogeneous Databases. Unpublished PhD Thesis, University of New South Wales. (1997)
47. Stock, K.: The Representation of Geographic Object Semantics Using Inclusion Rules. Paper presented at GIS/LIS '98, Fort Worth, Texas, 10-12 November 1998. (1998)
48. Urban, S.D. and Wu, J.: Resolving Semantic Heterogeneity through the Explicit Representation of Data Model Semantics. SIGMOD Record. **20 (4)** (1991) 55–58
49. Woodcock, J. and Loomes, M.: Software Engineering Mathematics. Addison-Wesley, Reading, Massachusetts. (1988)
50. Yu, C., Jia, B., Sun, W. and Dao, S.: Determining Relationships among Names in Heterogeneous Databases. SIGMOD Record. **20 (4)** (1991) 79–80

Appendix: Example Rule Repository

Table 2. Rule Table

Rule Number	Dimension Number	Property Number
1	1	1
2	2	2
3	3	3
4	4	4
5	2	5
6	3	6
7	2	7
8	5	8
9	6	9
10	7	10
11	8	11
12	9	12
13	4	13
14	2	14
15	10	15
16	11	16
17	5	17
18	7	18
19	8	19
20	12	20
21	13	21
22	14	22
23	9	23

Table 3. Dimension Table

Dimension Number	Dimension
1	Main object ingredient
2	Object purpose
3	Source of authority
4	Spatial relationship
5	Object defined by spatial object
6	Spatial object type
7	Capture method
8	Coordinate system
9	Accuracy
10	Format
11	Allows communication with
12	Measurement unit
13	Projection
14	Orientation

Table 4. Property Table

Property	Description
1	Earth
2	Definition of the position and extents over assigned land use rights
3	The legally recognized land registration authority for the jurisdiction
4	Wholly or partly within a larger administrative area
5	Reference to the rights of land holder to use land in accordance with specified relationships
6	Legal name as shown on a birth certificate or similar document
7	Recording of the legal survey that creates and defines a unit of earth in diagrammatic form
8	$R_1 \wedge R_2 \wedge R_3 \wedge R_4$
9	Polygon
10	Survey plan data entry
11	Australian Map Grid
12	+/- 0.2m
13	Wholly whithin a larger administrative area
14	Facilitation of postal communication
15	Street number, street name, residential area
16	$R_5 \wedge R_6$
17	$R_1 \wedge R_2 \wedge R_3 \wedge R_{13}$
18	Digitizing from 1:1000 topographic map
19	Local
20	Feet
21	Lambert Conformal Conic
22	True North
23	+/- 10m

Designing a Mediator for Managing Relationships between Distributed Objects

Hanna Kemppainen

Finnish Geodetic Institute,
Department of Cartograhy and Geoinformatics,
Geodeetinrinne 2,
FIN-02430 Masala, Finland
hanna.kemppainen@fgi.fi

Abstract. The paper discusses issues related to the design of an integrating mediator that facilitates the use of distributed and heterogeneous data resources. The architecture for such a mediator is outlined in the paper. The implementation of relationships between distributed objects is presented. The design of a mediator is based on the CORBA software model that considers relationships between distributed objects as first-class objects with state and operations rather than as stored links.

1 Introduction

High-speed networks facilitate the exchange of data and services between information systems, and developments in distributed software architectures make it possible for heterogeneous software systems to interoperate. Complementary advances should be obtained in the field of data integration (not only system integration) and especially maintenance of integrated data sets. Mediators have been suggested to serve as modules between user applications and data resources, capturing tasks required to overcome difficulties introduced by huge volumes of data, heterogeneities among data resources and mismatch between data values [15].

The integration problem domain involves bringing together data items describing some characteristic of a real-world entity emanating from different data sources. The system integration problem domain is mostly addressed under the title 'interoperability' [1], which requires agreements on how software systems connect themselves to a computer network and allow interaction between components through well-defined interfaces. Interoperability through a common communication protocol rather than through a common data format [14] constitutes a starting point for an integrating mediator design in this paper.

One aspect of importance when integrating data from autonomous databases is how to guarantee maintenance of the integrated data set with respect to changes that occur in the component data sets. Since the contents of information sources may change frequently, an application dependent on such resources needs to be aware of such changes.

The contribution of the work is to identify issues related to the design of an integrating mediator, deriving from experience gained in a GIS integration project. This work also demonstrates the use of existing techniques developed in the information technology community by implementing relationship objects in a distributed environment. The implementation is based on both OpenGIS Simple Feature specification [11] and OMG's CORBA relationship service specification [9].

The paper is organized as follows. Section 2 describes the organizational setting for this work. Section 3 explains what an integrating mediator is supposed to do. Section 4 describes the implementation of relationships between distributed objects and section 5 discusses the meaning and implications of such relationships. Section 6 summarizes the topics discussed in this paper.

2 The Case: Institutional Setting for Data Integration

A case study was recently undertaken to gain information on factors contributing to the mismatch between independently established and maintained building data [4]. The institutional setting for the study consists of 1) local authorities recording building permit data, 2) the Population Register Centre database registering different types of data concerning buildings and residences and 3) the building data stored in the topographic database. The building data of interest here consist of data describing the location of buildings, expressed as a reference either to a coordinate system or to road networks (addresses). A problem often encountered in data integration has been mismatch in data values: applications combining map databases with data from the population register often fail to perform the task automatically because coordinate values recorded for buildings do not fit together (Fig. 1).

Integration of building data is performed primarily by matching coordinate values. Another alternative for performing the match is the building identifier, an official identifier assigned to each building requiring a building permit. The building identifier is one independent of any software system implementation. Its value is a string of characters consisting of various pieces of information that

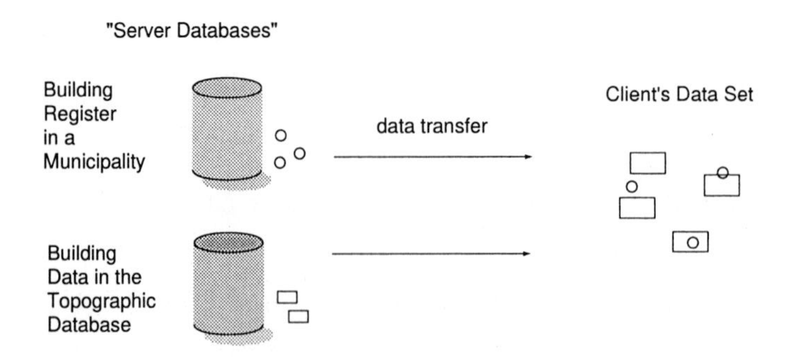

Fig. 1. Database integration

describe the location of the building with respect to administrative and real estate areal divisions. Since the topographic database does not contain building identifiers, the corresponding information can be derived from the cadastral map.

Building data are originally gathered in municipalities during the various stages of the follow-up process related to construction activity. A new building is assigned with an identifier, expected location data and address when the building permit is applied for. The municipality informs the Population Register Centre of the new building permit application and provides the new building data.

The occasional low quality of the coordinate data obtained from the municipalities is mostly due to the fact that not all municipalities have adequate resources and means for gathering the location. It is envisaged here that such municipalities could access existing map databases through the Internet, using web techniques [5] to obtain reference data for determining the location data for the new building. Such reference data sets are already available in Finland through the Net [7].

3 Integration Problem Domain

It is envisaged here that compatibility of data values emanating from different databases could be improved by the use of software that aids in accessing data from various databases, establishing interdependencies between data items through relationships, and monitoring the validity of such relationships.

Mediator is a software module that exploits encoded knowledge about some sets or subsets of data to create information for a higher layer of applications [15]. What exactly an integrating mediator is supposed to do is further explained in this chapter. The design of the mediator is based on distributed object technology concepts which are also shortly explained.

3.1 Integration and Mediators

The tasks required for performing data integration comprise:

- Accessing and retrieving data from multiple heterogeneous resources, using data export and import services [12] by the target and source information communities [10]
- Abstracting and transforming retrieved data into a common representation and semantics, thereby homogenizing any differences between the underlying data representations [3] and information content [2]
- Combining the data items according to matching values.

Given a set S of local database objects O_{Li}, the mediator should be able to create an integrated object R_{i} that holds information from some elements of S. The mediator accesses O_{Li} and imports them to the mediator's address space, creates appropriate relationship objects R_{i}, and monitors the state of each R_{i} with respect to changes occurring in O_{Li}.

O_{Li} are integrated to form R_{i} if they fulfill a condition that has been associated with the definition of R_{i}. The relationship object R_{i} is created by the mediator if the required condition is fulfilled. The tasks required from our integrating mediator are summarized as:

1. Creation of the relationship object
   - determine the condition to be fulfilled by the component objects (e.g. building geometries emanating from different databases must coincide)
   - retrieve local objects
   - determine whether the condition between retrieved objects is fulfilled
   - create relationship object
2. Monitoring of the relationship object
   - local objects should notify the mediator when a change has occurred in them
   - determine whether the condition still holds after change has occurred in the component

3.2 Distributed Object Technology Supporting Integration

A mediator of the type discussed above can be based on existing software architectures designed for integrating heterogeneous, autonomous and distributed computing resources. Middleware technologies facilitate integration of components (objects) in a distributed system by providing a run time infrastructure of services for components to interact with each other despite differences in underlying communications protocols, system architectures, operating systems and other application services [13].

An example of middleware technology is CORBA [8]. A CORBA object system is a collection of objects that isolates the requesters of services (clients) from the providers of services by a well-defined encapsulating interface. Clients request services by issuing requests for services that are implemented on the server, and called using an object reference. CORBA relationship service architecture [9] provides a mechanism for associating isolated objects and managing their creation, navigation and destruction. A mechanism for associating independently implemented and maintained objects is used as a basis for an integrating mediator.

The CORBA relationship service defines relationships as first-class objects, i.e. objects that have a state, and that may have operations. A relationship object groups together CORBA objects that are related to each other and defines the semantics of the relationship in terms of the degree of the relationship, the types of object that are expected to participate in the relationship and cardinality constraints. A *role* represents a CORBA object in a relationship and provides mechanisms for navigating the relationship [9].

The relevant portion of the CosRelationships module [9] for purposes of the present work is given below. The client acquires the roles involved in a relationship with the aid of the *named_roles* attribute defined in the relationship interface. The method *get_other_related_object* defined in the role interface provides access to the related object. Another role object involved in a particular relationship is accessed by the *get_other_role* method.

```
interface Relationship : CosObjectIdentity::IdentifiableObject {
  readonly attribute NamedRoles named_roles;
};
interface Role {
  readonly attribute RelatedObject related_object;
  RelatedObject get_other_related_object (in RelationshipHandle,
                                          in RoleName target_name);
  Role get_other_role (in RelationshipHandle rel,
                       in RoleName target_name);
};
```

4 Implementing Relationships between Building Objects

4.1 An Integrating Mediator

This section describes a simple implementation on which more functionality aiding in the management of integrated objects can be built. An integrating mediator is envisaged as an application that accesses distributed building object implementations, performs matching between object values to identify the objects that are to be associated with the relationship service and aids in maintaining the relationship (Fig. 2). The relationship object integrates object values from the component objects and thus provides an integrated view of the distributed building objects to the client.

An abstract service domain has been assumed here as the basis for the work. The mediator invokes services on remote objects (registered in the network and thus made available to clients) to acquire their values. Services required to implement such a system include access to the remote object, exporting data from the

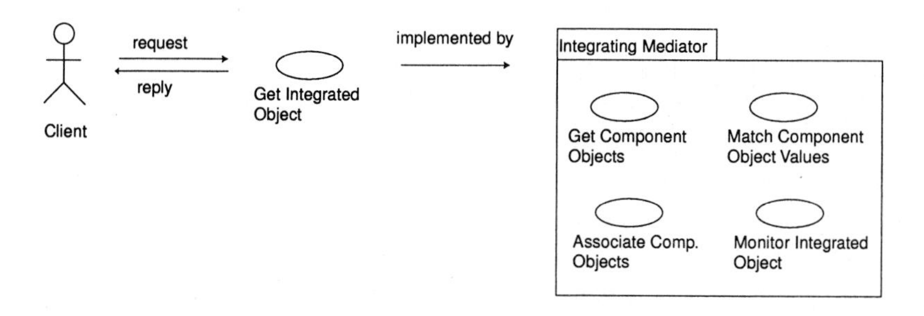

Fig. 2. Integrating mediator and the software package for its implementation

remote object and importing data to the mediator's address space. The integrating process finds a match between object values and establishes the relationship service.

The following sections in this paper describe how the relationship service is established in this setting. The objects to be associated by the relationship service are modelled as OpenGIS simple features [11]. Java classes were implemented to instantiate the following object types:

- Factory objects for creating feature types and features [11]
- Features [11]
- Role and relationship factories [9]
- Roles and relationships [9].

4.2 Related Objects without Relationship Service

The objects to be associated by the relationship service are digital representations of real-world buildings. The representations conform to the following IDL interface specifications:

```
interface TopoBuilding : OGIS::Feature {};
interface MunBuilding : OGIS::Feature {};
```

TopoBuilding and *MunBuilding* object types represent the building data stored in the topographic database and the building permit database, respectively, as was explained in Sect. 2. The OGIS module specification is given in [11].

In the event that the features were modelled as those not inheriting from the OpenGIS Feature, the integrated feature schema capturing the properties central to the current integration task would be:

```
interface TopoBuilding {
  readonly attribute string BuildingId;
  readonly attribute string Use;
  readonly attribute OGIS::Geometry PolygonGeometry;
};
interface MunBuilding {
  readonly attribute string BuildingId;
  readonly attribute string Address;
  readonly attribute OGIS::Geometry PointGeometry;
};
```

A client that collects data to implement an integrated building object in, say, a population register application, might have the following local model of a building:

```
interface PopBuilding {
  readonly attribute string BuildingId;
  readonly attribute string Address;
  readonly attribute OGIS::Geometry Geometry;
};
```

The purpose of the client application is to access available objects of *TopoBuilding* and *MunBuilding* types and find a match between either *BuildingId* value or between *PointGeometry* and *PolygonGeometry* values. Once a match is found, the client copies values to the *PopBuilding* object from *TopoBuilding* and *MunBuilding* objects. If there were no relationship service, i.e. no integrating mediator available, the client would make requests for objects implementing the given interfaces and perform the integration in the local application address space.

4.3 Associating Building Object Implementations Using the CORBA Relationship Service

CORBA Relationship Service Architecture models relationships as first-class objects (with identity), whose components are named roles. Roles represent related objects, in this case, *TopoBuilding* and *MunBuilding* types. Relationship objects are designed and maintained by the integrating mediator for the purposes of providing the client with an integrated view of the isolated building objects. The client accesses the integrated relationship object and is not responsible for performing the access and matching of values of the component object itself. Separate component object implementations hold implementation-dependent identities and have a property value that can be used for matching the objects (*BuildingId* attribute).

The component objects, *TopoBuilding* and *MunBuilding* objects are related by the mediator to form a relationship object of the *IntegratedBuilding* type given below. The object simply associates *TopoBuilding* and *MunBuilding* objects. Roles corresponding to these objects are named *TopoRole* and *MunRole.*

```
interface TopoBuilding : OGIS::Feature {};
interface MunBuilding : OGIS::Feature {};
interface IntegratedBuilding : CosRelationships::Relationship {};
```

Table 1 lists the types of object that are involved in the resulting relationship object. The relationship semantics is given by cardinality constraints saying that no role object may exist without belonging to a relationship object; on the other hand, the role object is associated with exactly one relationship object.

4.4 Registering Object Implementations with the ORB and Naming of the Objects

The central component in OMG's CORBA is the ORB (Object Request Broker), which makes it possible for objects to communicate with each other with

Table 1. Roles Participating in the Integrated Building Relationship

Related object	Role type	Cardinality (min,max)
TopoBuilding	TopoRole	(1,1)
MunBuilding	MunRole	(1,1)

no prior knowledge about the location or implementation details of the other object. The objects that are registered with the ORB can be accessed by resolving the names bound with the objects. Names are structures consisting of an identifier and 'kind' describing the type of object. The client needs to know about the semantics associated with each name [9]. The Naming Service provides the principal mechanism through which most clients of an ORB-based system locate objects that they intend to use. When given an initial naming context, the client can navigate through the naming contexts and retrieve lists of the names bound to that context.

A simple naming scheme was used here: the relationship object name identifier contains the *BuildingID* value for clients to identify the correct building. Designing an appropriate naming scheme for the environment [6] is a key question when designing an integrating mediator.

5 Relationships: Meaning and Implications

The purpose of the integrated relationship object (*IntegratedBuilding*) is to associate existing building objects for prospective clients to access their values in an integrated fashion. A client is assumed to be interested in viewing and copying values of the integrated object. The mediator provides services that provide the client with the necessary information on the object so that the client can use it.

5.1 Client Scenarios

The client knows of relationships (integrated objects) through the names that have been bound with the naming context in the ORB and their semantics through a catalogue service associated with the mediator. The primary purpose of the catalogue service in this context is to support identification of registered relationship objects.

A client may simply view the implemented relationship (and its values) or copy the values to the client's local data store. From the point of view of client accessing an object whose state is in agreement with real-world entities, it is of relevance to think of the validity of the object. When the client accesses the relationship object through the mediator it is always up to date (as to the mediator's knowledge). If the client stores the relationship locally, the object no longer remains under the control of the mediator, and state changes in the relationship are not mirrored in the client database.

The client navigates through the relationship object with the aid of roles which can be accessed by the relationship object's *named_roles* method. Those clients having an object reference to an object may wish to know whether that object is related to another object. This can be accomplished by determining all the registered relationship objects, roles participating in them and comparing the object references of the objects represented by roles.

5.2 Modelling and Management of Relationships

Once a relationship object has been created, it is possible that the related component objects change in a way that makes the association no longer valid. And even though the association does not lose its validity, it may be of importance to inform the client about the change. To design the activities of the mediator service, a model of the relationship life cycle is required on which to base their management. Based on the object life cycle model, event and triggering mechanisms monitoring object state changes can be designed.

Data management consists of the activities of defining, creating, storing, maintaining and providing access to data and associated processes in one or more information systems [12]. We may assume the existence of a data management environment in terms of relationship controller services responsible for

- Adding and modifying relationship definitions
- Adding, modifying and deleting relationship instances.

In the case examined here, possible changes in real-world buildings that must be accounted for include:

- Change in the attribute value that was used in matching the component objects
- Change in existence of the component object.

For example, the implementation described in Sect. 4.3 used the building identifier value to match component objects. The value is subject to change when changes occur in real estate division (e.g. two real estates are merged). The implementation also used the building identifier as a basis for creating names of the objects, the knowledge of which the client is assumed to rely on when accessing them. Another type of change that can occur is one in which the real-world building ceases to exist, thereby making existence of the whole relationship object meaningless.

6 Summary and Conclusion

We have reviewed issues related to integration of data emanating from various sources and outlined functionality of an integrating mediator that integrates data and manages the integrated data set. An experiment has been carried out in which the CORBA relationship service and OpenGIS simple feature specification was used as the basis for designing the mediator.

This work has been motivated by experience gained when carrying out an integration task in which building data were integrated. Without the mediator service, users have to invoke various applications when integrating data sets. This is not considered to be appropriate. It has been demonstrated here that distributed object technology offers techniques that can be used to build applications that provide the end user with a more integrated view of autonomous data resources.

This paper has demonstrated the use of the CORBA relationship service in forming associations between distributed objects. The next step in managing the implemented relationship objects would be to use event services to monitor and invoke further actions when encountering state changes in related objects. One key problem area that was identified is the naming scheme that needs to be established to identify objects between organizations, a problem area that is currently being discussed in the OpenGIS community.

7 Acknowledgements

The author wishes to express her gratitude to the anonymous referees whose constructive criticism provided new insights into the issues represented in the paper. The discussions with Tapani Sarjakoski and Lassi Lehto are also gratefully acknowledged. Several other colleagues from our group at the FGI also helped me during the earlier stages of the work.

References

1. Brodie, M.L., Ceri, S.: On Intelligent and Cooperative Information Systems. International Journal of Intelligent and Cooperative Information Systems **1**, 2 (1992) 1-35
2. Devogele, T., Parent, C., Spaccapietra, S.: On Spatial Database Integration. International Journal of Geographical Information Systems, **12**, 4 (1998) 335-352
3. Gahegan, M. N.: Specifying the Transformations within and between Geographic Data Models. Transactions in GIS, **1**, 2, (1996) 137-152
4. Kemppainen, H.: Integration and Quality of Building Data. Notices of the Finnish Geodetic Institute, **18**, (1998) In Finnish
5. Lehto, L.: Java/CORBA Integration - A New Opportunity for distributed GIS on the Web. 1998 ACSM Conference Proceedings, Baltimore, MD (1998) 474-481
6. Lynch, C.: Identifiers and Their Role In Networked Information Applications. ARL Newsletter **194**, (1997) `http://www.arl.org/newsltr/194/identifier.html`
7. Map Site, National Land Survey of Finland. `http://www.kartta.nls.fi/index\_e.html`
8. Object Management Group (OMG): The Common Object Request Broker: Architecture and Specification, Revision 2.2, February 1998
9. Object Management Group (OMG): Common Object Services Specification, Revised Edition, November 1997 update

10. OpenGIS Consortium: The OpenGIS Guide, Introduction to Interoperable Geoprocessing and the OpenGIS Specification. Third Edition June 3, 1998
11. OpenGIS Consortium: OpenGIS Simple Features Specification For CORBA. Revision 1.0, March 18, 1998
12. RMDM. ISO/IEC.: Reference Model on Data Management.
ISO/IEC 10032:1995.
13. Rymer, J.: The Muddle in the Middle. Byte Magazine **21**, 4 (1996) 67-70.
14. Vckovski, A.: Interoperable and Distributed Processing in GIS. Taylor & Francis, London (1998)
15. Wiederhold, G.: Mediators in the Architecture of Future Information Systems. The IEEE Computer Magazine, March 1992 38-49

Geospatial Mediator Functions and Container-Based Fast Transfer Interface in Si^{3}CO Test-Bed*

Shigeru Shimada[1] and Hiromichi Fukui[2]

[1] Hitachi, Ltd., Central Research Laboratory,
1-280 Higashi-koigakubo Kokubunji City Tokyo 185-8601, Japan,
shimada@crl.hitachi.co.jp,
http://koigakubo.hitachi.co.jp
[2] Keio University,Faculty of Policy Management,
5322 Endo Fujisawa Kanagawa 252-8520, Japan,
hfukui@sfc.keio.ac.jp,
http://www.sfc.keio.ac.jp/\homedirhfukui/index.html

Abstract. In order to improve the spatial information infrastructure in Japan, we have organized SI^{3}CO (Spatial Information Infrastructure Interoperability Consortium), and we are newly developing a Japanese interoperable test-bed based on OGIS. In this system, we propose the new three tier model which is composed of web clients, legacy database wrappers, and GSM (Geo Spatial Mediator). Especially GSM locates between client and wrappers, and can compensate spatial objects. Moreover, we propose container-based fast transfer interface of spatial objects as for the CORBA implementation.

1 Introduction

As well known already, many NSDI (National Spatial Data Infrastructure) activities or projects have started in the world. As the example in the United States, OGC (OpenGIS Consortium) WWW Mapping SIG is promoting OGIS (Open Geodata Interoperability Specification) based interoperable test-bed project now [1]. On the other hand, GIPSIE (GIS Interoperability Projects Stimulating the Industry in Europe), which is newly organized as the industry-university joint project, started to support the European standard interoperable GIS [2].

1.1 GIS Situation of Japan

In contrast to these activities in the world, the importance of GIS was not enough considered by the Japanese central government traditionally. But 18

* We would like to thank the SI^{3}CO members for developing this system. There are 29 contributors from two universities (Tokyo and Keio), eight companies (Asia, Falcon, Hitachi, IBM, Kokusai, NEC, Oki, and Pasco) and NSDIPA. There are too many people to list here, but we would like to thank these contributors.

inter-ministerial liaison committees have recently started taking advantage of the Hanshin-Awaji earthquake disasters. Moreover, the importance of interoperability of legacy geospatial databases is not acknowledged yet.

1.2 Project Motivation

We will therefore construct a new geospatial information infrastructure by using distributed object technology connecting various information communities. First, we create the spatial object model that enables interoperable utilization of geospatial data supported by different information communities. Second, we develop a distributed object environment which enables information supply to web terminals by interoperable utilization of legacy geospatial databases supported from each information communities. Our motivation for these activities is how to construct the middle ware for the GIS software environment. This middle ware should provide the following functions.

- Effectively sharing and circulating the geospatial information based on simple feature specifications.
- Quick and high-quality user interface
- Easy interoperability between legacy databases

2 The Japanese Interoperable GIS Test-Bed

With the aim of providing the above-mentioned functions, we organize SI^{3}CO (Spatial Information Infrastructure Interoperability Consortium) with two universities, eight companies, and NSDIPA (National Spatial Data Infrastructure Promoting Association in Japan). And we are newly developing a Japanese interoperable GIS test-bed based on OGIS. This project is promoted as a national project and is financed by the quasi-governmental organization IPA (Information Technology Promotion Agency).

2.1 Purpose of System Development

The application fields for geospatial information in Japan are split into two categories: consumer use or professional use. A consumer application includes the road-route guidance system known as car navigation. This system usually utilize the middle-range precision road map of the 1/25,000 - 1/10,000 scale. And these road maps are stored on CD-ROM and are accessed from a stand-alone-type navigation system.
On the other hand, professional applications include municipal administration systems such as disaster protection systems and environmental preservation systems, Another professional applications are the facility management systems in utility companies. These applications usually utilize the high-precision residential maps of the 1/500 - 1/2,500 scale.

This test-bed project handles professional use of the geospatial information infrastructure shared among various divisions. And the purpose of test-bed construction is to confirm whether is it possible to construct a practical use system by utilizing high-precision geospatial data and the system based on the OpenGIS simple feature specification.

2.2 Problems in System Development

The FGDC (Federal Geographic Data Committee) demonstration system, which is now being developed by the WWW Mapping SIG, adopts a two-tier model. In this system, legacy databases managed by different GIS vendors such as Bentley, Intergraph, ESRI, and ORACLE are integrated by three kind of DCP (Distributed Computing Platform): CORBA, OLE/DB, and SQL. And this system enables overlay display on web terminals via Open Map IDL [3] [4] [5]. The data handled by this system is composed of relatively rough-scaled geospatial data such as municipal boundaries, road routes, tin, and coverage [6].
On the contrarily, Japanese domestic requires not only high precision geospatial data but also utility data such as electric power lines and water supply and sewage lines. In such cases, the following problems arise.

1. Geographic data and facility data are usually supplied from the each different legacy database. And two kinds of object difference can occur even in simple overlay processing. One is physical difference such as coordinate rotation or aberration; the other is semantic difference such when the same object is different even though the name title is the same.
2. The construction cost of high-precision geospatial data is very high, so it will be impossible to maintain a wide area. A compensation technique is therefore required, in which areas lacking high precision are compensated seamlessly by rough-precision data.
3. The common clearing house function mainly supports searching network addresses of web sites. The properties of these web sites are matched with user selected properties. But when each spatial object contains physical or semantic differences, these clearing house functions are not sufficient. So an autonomous trading function must search more profitable web sites if the obtained and evaluated spatial object contains large differences.

2.3 System Organization Concepts

In order to solve the above mentioned problems effectively, we propose a new three-tier model composed of web clients, legacy database wrappers, and a GSM (Geo Spatial Mediator). The GSM locates between clients and wrappers, and it compensates spatial objects. The block diagram of our proposed three-tier interoperable system is shown Fig.1. The bottom level is the wrapper part which converts the retrieval result from legacy databases into spatial objects based on the OpenGIS simple feature spec., the middle level is GSM which composes tree kinds of spatial object processing functions such as composer, compensator and trader; and the highest level is the web clients which support high-speed display of spatial objects.

Portrayal Model Correspondence If we compare our proposed architecture with the portrayal model adopted by the WWW mapping SIG, the result is as follows. The portrayal model clusters the pipeline process into two groups: one is the process from legacy database retrieval to feature extraction as one component; and the other is the process from feature transportation to graphical display as three-detail components. On the other hand, our proposed three-tier model clusters pipeline processes into two groups: one is the former process as two components and the other is the latter process as one component. The latter display process is regarded as the deep client system. For implementing GSM, it will be possible to support any level of interfaces from the thin client level to the deep client level depending on user demands. This user adaptable function is supported by GSM's active retrieval mechanism. And detail process flow of this mechanism is shown Fig.2. However, it will be possible that, the filter component of the portrayal model can be treated as a more precise process in the case of intersecting GSM.

2.4 GSM (Geo-spatial Mediator)

The characteristics of our proposed system function as the middle level of GSM. Generally speaking, the mediator role under the heterogeneous database retrieval environment has been already formulated by various studies. For example, TSIMMIS project has offered a data model and a common query language that are designed to support the combining of information from many different sources [7] [8]. But this formulation is not enough concrete for the geospatial information processing.
To cope with this situation, our proposed GSM architecture supports spatial object processing functions specified by the OpenGIS Spec. Topic 12: "The OpenGIS Service Architecture" except for image processing to solve the above problems of first and second. For example, GSM partially implements "Geospatial Coordinate Transformation Services" and "Geospatial Feature Manipulation Services". And GSM supports trader functions, which can search web sites, based on meta-data in order to solve problem of third. Moreover, these functions are organized each are split into three categories.
Control of distributed retrieval If GSM receives a retrieval demand from web terminals, it orders the trader to search web sites which hold the most suitable spatial objects. Then GSM controls the compensator to check the difference between the composed spatial objects. And GSM actually processes the obtained spatial object archives such as retrieval, composition, and conversion. GSM has following two functions..
Spatial object trading Search the location of legacy databases for the most suitable spatial object according to the trader graph which the server holds.
Spatial object compensation If the application object is composed of plural spa-

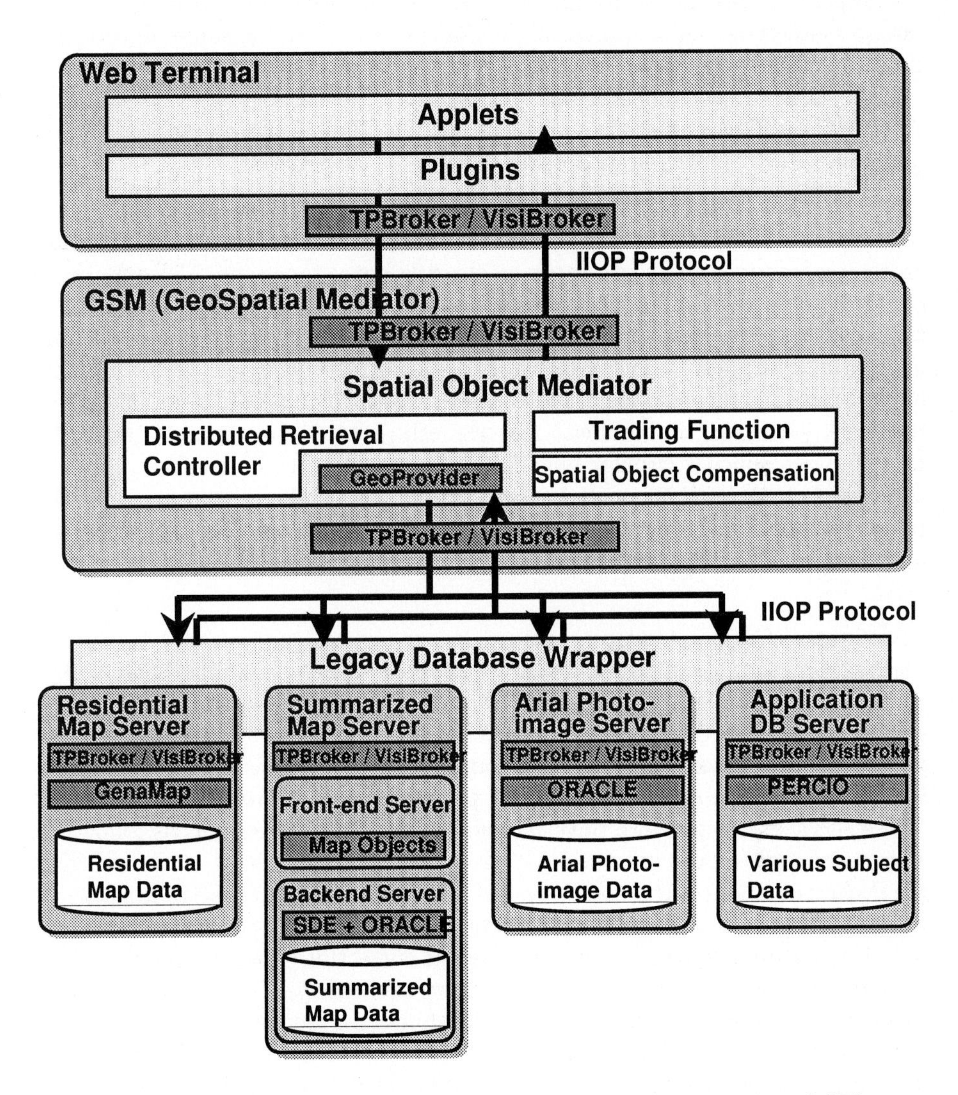

Fig. 1. Proposed three-tier model for Japanese test-bed based on OGIS

Map Objects is a trademarked object oriented GIS environment produced by ESRI.
SDE is acronym of spatial data engine produced by ESRI.
GenaMap is a trademarked GIS produced by GenasysII Corp.
GeoProvider is a interoperable GIS tool produced by Hitachi Software Engineering Co., Ltd.
Object Spinner is a trademarked ODBMS produced by NEC Corporation.
VisiBroker is a trademarked CORBA produced by Imprise Corporation.
TPBroker is a trademarked CORBA, of which object-transaction management is added to VisiBroker, produced by Hitachi, Ltd.

tial objects simply, physical differences caused by the difference between coordinate systems as well as semantic differences caused by the same objects with different names still exist. So GSM is equipped with compensation functions to regulate these object differences.
Among these GSM functions, we focus on the characteristic functions, trader and compensator, as explained below.

Geospatial Object Trading Function We developed meta-data oriented searching functions similar to common clearing house functions. These functions can grasp information about where a spatial object is and what kind it is. In this case, the searching function must consider the special relationships among geospatial information. These special relationships result from the scale hierarchy or the difference of media style such as vector or image. So it is necessary to develop following hierarchical search method based on a trader graph. We therefore developed a circular information infrastructure. This infrastructure is composed of network connected servers, in which GSM is the center and this builds an individual community. The trading function is the core technology for autonomous search between GSMs.

Geospatial Object Compensation In our test-bed system, we try to access the high-precision geospatial data, so we suppose that various differences will be contained in spatial objects. By using interoperable composition methods, these spatial objects are directly supplied from legacy databases. Spatial objects are optimized for ordinary usage in each community. Realtime compensation mechanisms of spatial objects are therefore prepared as one service menu of GSM. These services are split into physical and semantic compensations as follows.
Physical compensation
As already mentioned, there are physical differences between spatial objects supplied from legacy databases. These differences are caused by coordinate rotations or shifts. Compensation is thus composed of correcting processes for two kinds of errors. One is the transformation error between geodetic coordinates and orthogonal coordinates. And the other is the shift error among coordinate systems with different scales. The latter case is mainly caused by offset of the origin coordinates which belong to the Japanese standard 19 coordinate systems.
The positional distortion can be compensated by adding transformation process to standard coordinate systems or standard geometrical models. This compensation process is specified in OGC spec. 12: "Geospatial Coordinate Transformation Services".
Semantic compensation
In order to compose spatial objects while keeping semantic consistence, semantic conversion based on the definition of terminology and schema structure used in each community is needed. In our test-bed system, semantic conversion of structure means that feature names, properties, and structure between spatial databases are compensated as shown in Fig.3. This figure shows the feature correspondence between the residential map, the summarized map, and the dis-

aster prevention. We thus achieved compensation processing and based on the OGC service architecture spec., such as "Feature Manipulation Service", "Feature Generalization Service", and "Feature Analysis Service".
This semantic compensation is expected to become more complicated, but it will become more domain specific.

3 Implementational Situation

Our test-bed system is now implemented by CORBA based on a three-tier model. In this implementation, we utilize VisiBroker which is CORBA 2.0 based on ORB products supported by Inprise Corp. Regarding the ORB products, there are already many products announced, such as IONA's Orbix, and we adopt VisiBroker in our implementation. This is because VisiBroker is adopted as the IDL-Java mapping in OMG, and moreover, as the standard CORBA interface in Netscape Corp.
The greatest improvement in this implementation is the proposed new method for transferring features. The OpenGIS Simple Feature Specification for CORBA specifies the method as for transferring features in each feature unit. But transfer speeds are very slow when transfering heavy geospatial data loads. So we propose "ContainerFeatureCollectionSet Interface" which enables extremely fast transfer of a set unit of features by one IIOP protocol communication.

3.1 ContainerFeatureCollectionSet Interface

First, we specify ContainerFeatureCollectionSet in order to express the set of features. The top level content of internet geospatial-data format (IGF [1]) is the collected unit of a plural common layer. That is, collection of the same kind of features as a layer and collection of layers as a container unit. Second, we explain the two newly supported operations from the ContainerFeatureCollectionSet interface as follows.

- Spatial retrieval operation (`get_ContainerFeatureCollectionSet_by_geometry`): Describe the retrieval area as the retrieval condition and get the set of features contained in the specified area.
- Property retrieval operation (`get_ContainerFeatureCollectionSet_by_property`): Describe the retrieval condition for the attribute information and get the set of features which hold the properties matched with the retrieval condition.

The following IDL program is a concrete description of the above mentioned ContainerFeature-CollectionSet.

[1] IGF is acronym of internet geospatial-data format settled by Hitachi Software Engineering Co., Ltd.

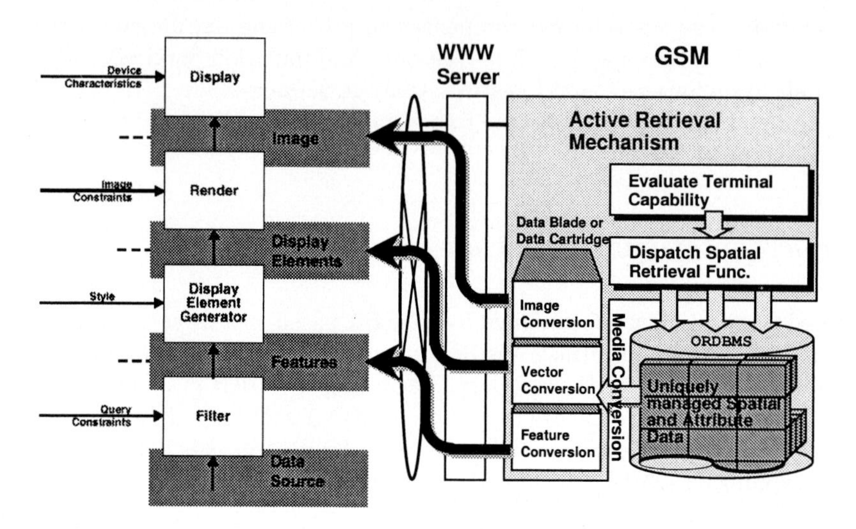

Fig. 2. Process flow of GSM's active retrieval mechanism. If the terminal capability is sent to the GSM, it is evaluated and dispatched spatial retrieval functions Then retrieved data is actively converted into various forms by the media conversion function

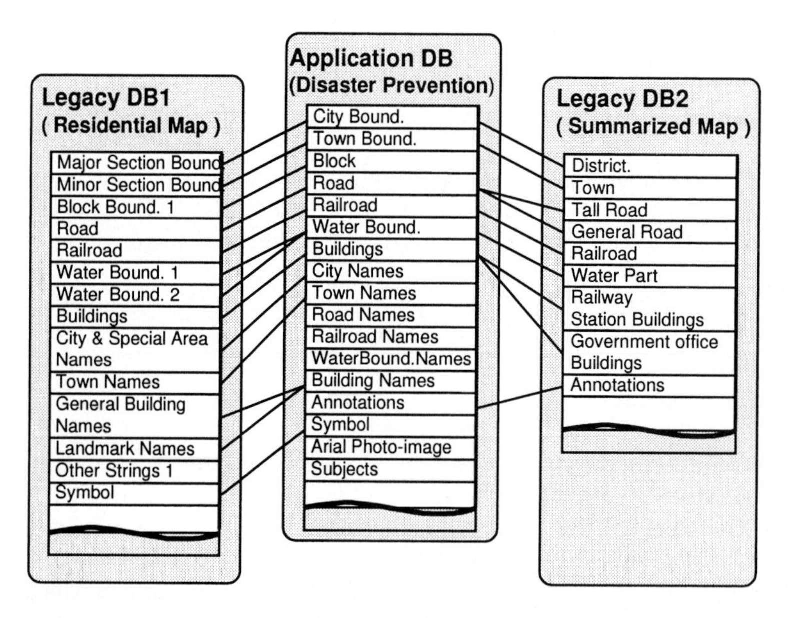

Fig. 3. Diagram of feature correspondence among DB communities. Semantic structure conversion is compensation of feature name, properties and structure among different communities. In this example, features defined on legacy databases such as residential map and summarized map must be converted into newly defined features on application system such as disaster prevention.

```
//ContainerFeatureCollectionSet InterFace
module CFCSIF{

  typedef string Istring;                //Extracted portion of OGIS CORBA spec.
  typedef sequence <Istring> IstringSeq; //
  typedef sequence <octet> OctetSeq;     //Following is SICO supplement

  struct RetrievedFeatureType {
      Istring ftype_name;                //Name of Feature Type
      Istring condition_prop_name;       //Name of Condition
      any   prop_value;       //Value of Property
      Istring condition;      //Condition (eq,ne,ge,le,gt,lt,like,etc.)
  };

  typedef sequence<RetrievedFeatureType> RetrievedFeatureTypeSeq;

  //Interface Definition
interface ContainerFeatureCollectionSet {
  //Spatial Retrieval Operation
    void get_ContainerFeatureCollectionSet_by_geometry(
      in Istring coordinate_system, //Kind of Coordinates  "GEOGCS" or "PROJCS"
      in Istring coordinate_name,   //Name of Coordinates "BESSEL" or "JA19-9"
      in short unit,                //Orthogonal:Unit (1x10n cm),Geodetic:null
      in double x,                  //Center of X Coordinate
      in double y,                  //Center of Y Coordinate
      in double w,                  //Width
      in double h,                  //Height
      in long scale,                //Display Resolution
      in RetrievedFeatureTypeSeq  ftype_name_list,//Name of FeatureType + Condition List
      out OctetSeq  geodata         // IGF Typed Geospatial Data
    );
  //Property Retrieval Operation
    void get_ContainerFeatureCollectionSet_by_property(
      in RetrievedFeatureType ftype,    //Name of FeatureType+Retrieval Condition
      in Istring   request_prop_name,   //Name of demanded properties
      out OctetSeq  geodata             //IGF Typed Geospatial Data
    );
  };
};
```

3.2 Web Client Implementation

To widen the market of business applications without compromising the ability of application programs, we construct the common portion of various applications as the plug-ins and construct their different portions as the application applets.

Application applets. For example, we suppose that these applications are utilized in a disaster prevention system used in municipal government. And we implement a "Refuge planning system", which needs a middle-scale range and wide area maps, and "Dangerous place management system", which needs narrow but precise-scale range maps, such as Java applets.

Plug-ins. We developed the following functions which enables high speed display and scrolling on the web terminals. And these operations are effective even for high precision maps:

Scroll function: Scroll quantity management, pixel thinning during fast scrolling, estimate scroll direction, and interlocking scrolling of two windows.

Zoom in/out: Automatic selection of graphic information fit for display capabilities.

Layer on/off control: Control function of layers according to applet demand, and on/off control of display according to display limit under the selected scale conditions.

Split control of display: Display management that splits display into two windows, displays two kind of maps in the same area, and merges them into one window.

3.3 GSM Implementation

Physical compensation mechanism of GSM is implemented as uniform geometrical processing composed of coordinate transformation based on the property of legacy databases. On the other hand, semantic compensation must be implemented based on the OpenGIS service architecture, but the details of this specification are not settled yet. So, in this case, we implement the function which conceptually adjusts geo-spatial objects on the disaster prevention application area. We implement the "Feature Type Translation Function", which translates the name and the structure of geo-spatial objects, and the "Composition Function" which relates and composes objects semantically the same as each other.

Feature-type translation function. This function absorbs schema differences between legacy databases by interoperabl transforming geo-spatial object names and structures between each database. The OpenGIS service architecture shows the concept of object translation and translates and transforms object schema between each community. But in the real world, object relationships are not a simple one-to-one relation, but they are M-to-N relations composed of "is-a" or "part-of" relationships. In order to transform these M-to-N relationships, we implement the feature type name translate function and the data structure conversion function.

In this translate function, the object name is managed by the geospatial object relationship table which enables the name used in each databases to be obtined, and any object names from this table to be easily accessed. On the other hand, in the data structure conversion function, the name of the object is also accessed from the geo-spatial object relationship table and the structure of the accessed objects is converted into the corresponding structures according to the feature type management table.

Composition function. Geo-spatial objects, which are acquired from plural legacy databases, are not directly supplied to the applications. But the composed results are supplied. And they are obtained from the following processing. First relationships are resolved from synonyms between geometrical data and attribute data which are both held by geo-spatial data. Second, the new geo-spatial objects are composed of relationships and they are supplied to the application.

Judgement of synonyms between objects is based on the following operations, and the target of these operations is the geometrical and attribute data held by geo-spatial objects.

- Synonym judgement based on searching for geometrical similar processes.
- Synonym judgement based on matching text string partial process.

3.4 DB Wrapper Implementation

We implemented four database wrappers, and each wrapper is developed by different venders. Properties of the geo-spatial database and native GIS supported by each venders are summarized in the following table.

Table 1. DB Wrapper and Assigned Vendors

DB-Name	Scale	Contents	GIS&DBMS	Assigned Vendor
Residential Map	1/1,500	Digital Map related with Householder Name	Original DBMS	Oki Electric Corp.
Summarized Map	1/25,000	Outline Map composed of Boundaries, Roads and so on.	MapObjects+SDE ORACLE	PASCO Corp.
Aerial Photo	1/7,500	Orthographic Aerial Photo-Image	ORACLE	Kokusai-Kogyo Corp.
Application Map	any	Various Subjects focused on Disaster Prevention	ObjectSpinner (ODBMS)	NEC Corp.

These retrieved results from legacy databases are wrapped into the simple feature level geo-spatial objects, and they are transferred to the upper level of GSM via the IIOP protocol by the previously mentioned ContainerFeatureCollectionSet interface.

4 Conclusions

We have developed the SI^{3}CO architecture as the Japanese interoperable GIS test-bed system based on the Open GIS simple feature specification. And we proposed the new three-tier model composed of web clients, legacy database wrappers, and GSM. In GSM, we try to accomplish trading and semantic compensation in geospatial infrastructures. Moreover, these architectures are now being implemented by CORBA. And we have successfully implemented a massively fast transfer protocol of spatial objects named "ContainerFeatureCollectionSet Interface".

In future, we try to realize an object transaction mechanism for distributed databases and an agent based asynchronous object concurrency control, by adopting the function of geospatial object repository in GSM.

References

1. Open GIS Consortium, Inc.: http://www.opengis.org/
2. GIPSIE: http://gipsie.uni-muenster.de/index.html
3. Open GIS Consortium, Inc.: OpenGIS Simple Feature Specification for CORBA Rev.0 (1997)
4. Open GIS Consortium, Inc.: OpenGIS Simple Feature Specification for OLE/COM Rev.0 (1997)
5. Open GIS Consortium, Inc.: OpenGIS Simple Feature Specification for SQL Rev.0 (1997)
6. Kenn Gardels: Open GIS and On-Line Environmental Libraries. SIGMOD Record, Vol. 26, No. 1 (1997) 32–38
7. Y. Papakonstantinou, S.Abiteboul, H. Garcia-Molina: Object Fusion in Mediator Systems. VLDB'96 Conference Record, (1996)
8. Hector Garcia-Moria, et al.: The TSIMMIS Approach to Mediation: Data Models and Languages. Journal of Intelligent Information Systems, (1997)

Plug and Play: Interoperability in CONCERT

Lukas Relly and Uwe Röhm

Database Research Group, Institute of Information Systems,
ETH Zentrum, 8092 Zürich, Switzerland
{relly|roehm}@inf.ethz.ch
http://www-dbs.inf.ethz.ch/

Abstract. In order to make database systems interoperate with systems beyond traditional application areas a new paradigm called "exporting database functionality" as a radical departure from traditional thinking has been proposed in research and development. Traditionally, all data are loaded into and owned by the database, whereas according to the new paradigm data may reside outside the database in external repositories or archives. Nevertheless, database functionality, such as query processing and indexing, is provided exploiting interoperability of the DBMS with the external repositories. Obviously, there is an overhead involved having the DBMS interoperate with external repositories instead of a priori loading all data into the DBMS. In this paper we discuss alternatives for interoperability at different levels of abstraction, and we report on evaluations performed using the CONCERT prototype system making these cost factors explicit.

1 Introduction

Todays Database Management Systems (DBMS) make the implicit assumption that their services are provided only to data stored inside the database. All data has to be imported into and being "owned" by the DBMS in a format determined by the DBMS. Traditional database applications such as banking usually meet this assumption. These applications are well supported by the DBMS data model, its query and data manipulation language and its transaction management. Advanced applications such as GIS, CAD, PPC, or document management systems however differ in many respects from traditional database applications. Individual operations in these applications are much more complex and not easily expressible in existing query languages. Powerful specialized systems, tools and algorithms exist for a large variety of tasks in every field of advanced applications requiring these systems to interoperate and make their data available to other systems.

Because of the increasing importance of advanced applications, DBMS developers have implemented better support in their systems for a broader range of applications. Binary Large Objects provide a kind of low-level access to data and allow individual data objects to become almost unlimited in size. Instead of storing large data objects in BLOB's, some newer systems such as ORACLE

(Version 8) and Informix (Dynamic Server with Universal Data Option) provide the BLOB interface also to regular operating system files. Because the large objects in any of these two options are uninterpreted, database functionality for this kind of data is only very limited. In order to better support advanced applications, the standardization effort of SQL3 specifies, among others, new data types and new type constructors. Most recently, SQL3 and object-orientation have fostered the development of generic extensions called datablades [8], cartridges [11], and extenders [7]. They are based on the concept of abstract data types and often come with specialized indexing.

Although they provide better support for advanced applications, however, except for the file system case, they all have the same fundamental deficiencies: First, it is the DBMS together with its added extensions that prescribes the data structure and data format of the data to be managed. The consequence is that all complex specialized application systems and tools must be rewritten using the data structures enforced by the DBMS, or at least complex transformations must take place to map the DBMS representation into the application representation. Second, the DBMS owns the data. All data has to be physically stored inside the DBMS requiring to possibly load gigabytes of data into the database store.

These observations led to a radical departure from traditional thinking as it is expressed in [19]. In the CONCERT project at ETH, we focus on *exporting database functionality* by making it available to advanced applications instead of requiring the applications to be brought to the DBMS. In [14] and [15], we presented the concepts needed to enable the DBMS to interoperate with external data repositories exporting its functionality to data stored outside the DBMS. Query processing and indexing is performed by generic methods of physical database design invoking operations of user-defined abstract (external) data types. In this paper, we identify different levels of abstraction, at which interoperability for query processing can take place, and we present performance measurements identifying the costs required for the additional flexibility. With the exception of [18] we are not aware of other work that deals with external data and related performance measurements.

This paper is organized as follows: Section 2 introduces the two possible levels of abstraction, at which interoperability of the CONCERT query engine with external repositories can take place. Section 3 discusses the lower level of abstraction exploiting the integration of abstract data objects in the Database Kernel. In Section 4 we present CONCERT's object manager "Harmony", which adds higher level query processing capabilities. Section 5 concludes.

2 The CONCERT Architecture

In CONCERT, interoperability can take place at two different levels of abstraction corresponding to the CONCERT system architecture. CONCERT consists of a database kernel system and a generic object manager. Figure 1 gives an overview of the CONCERT architecture. Interoperability at the kernel level is performed on an object by object basis while interoperability at the object manager level is based on accessing collections of objects.

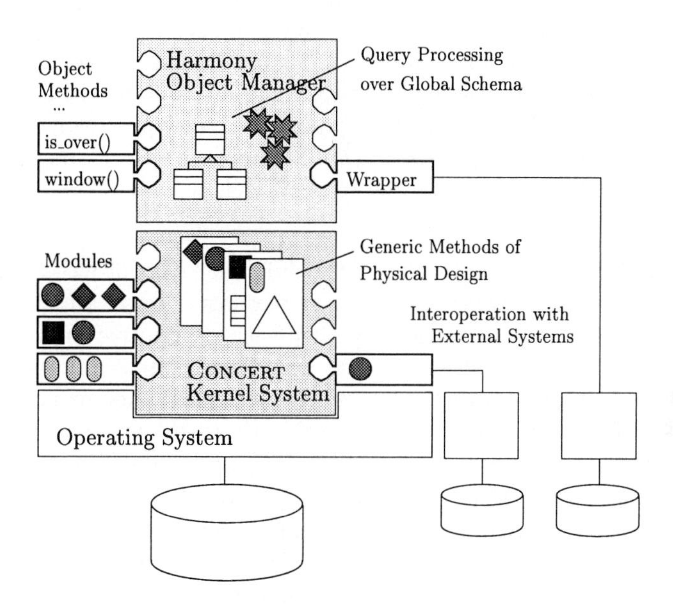

Fig. 1. Overview of the CONCERT system architecture

The CONCERT kernel system provides low-level database functionality such as storage management, low-level transaction management, and predicate and projection evaluation on single collections as basic query processing capability. It's role is comparable to System R's RSS [1], the Starburst Kernel [9] or the DASDBS Kernel [16]. In order to make data management efficient, the kernel is *tightly bound to the underlying operating system* providing multithreading and exploiting efficient secondary storage access using a memory-mapped buffer [2]. With respect to interoperability, the kernel's capability exporting physical database design making it available for external systems is important. The kernel implements *generic methods of physical design* such as a Btree index, an inverted file index, an Rtree-like spatial index. In contrast to traditional systems, these indexes are not connected to a predefined type system, rather they rely on object properties. These object properties are made available through *modules* that can be plugged into the kernel providing access to external objects through method invocation. The CONCERT kernel and its interoperability evaluation is presented in Section 3.

The Harmony object manager sits on top of CONCERT's kernel system. At this level, CONCERT abstracts from single data objects. Harmony provides declarative access to collections of abstract objects according to the *global schema*. The main concept of this layer is a collection, which represents a homogeneous set of data objects. Its elements are accessed through an iterator interface. Furthermore, Harmony defines a query algebra over the collections. Each collection of the object manager is in fact a queryable collection: it offers an `evaluate` method, through which a *declarative query* can be executed on its elements resulting in a new collection representing the result set. The global data model is object-oriented. The user can define *object methods*, whose object code is plugged into the object manager. Further, Harmony can interoperate with external systems through *wrappers*. They allow to export the query processing functionality of CONCERT to external repositories. The Harmony object manager and its interoperability evaluation is presented in Section 4.

Example: To illustrate the interplay of the different layers of CONCERT, we look at the following simple example: Imagine a geospatial image archive. A satellite periodically generates new images together with descriptive information (e.g., current date and time, satellite position, etc.). These satellite images are stored in a huge tape archive. Storing the image triggers the creation of a corresponding new image object in CONCERT, which holds the meta information about the image and its position in the tape archive – but without the image data itself.

First, we ask for the titles of all images. This query will be mainly processed inside the CONCERT kernel. The abstract image objects are of the plugged-in type `SatelliteData` (see Appendix), which has the general form `RECORD ( SCALAR, SPATIAL, SPATIAL )`. Harmony scans the image objects, asking for their first component, which the global schema declares as `title` attribute (see Appendix). The kernel calls the concept-typical method `SUB_OBJECT()` on the corresponding plugged-in type `SatelliteData`, and returns the extracted subobject. This is in fact a `string`. Harmony generates the result collection of string values. No further processing is needed.

Second, we want to display a part of an image, whose title we know. Harmony now scans the image objects, asking for all such objects, where the first component holds the specified string-value (i.e., is the search title). This can again be evaluated inside the CONCERT kernel. On the retrieved abstract `SatelliteData` Harmony now executes the user-defined `window()` method. This method extracts the part of the image we want to display. In order to do so, it falls back on the concept-typical methods `SPLIT()` and `COMPOSE()` of the SPATIAL component of each `SatelliteData`. These methods are provided by the kernel module `SatelliteImage`. They retrieve the needed parts of the satellite image from the external tape archive. This is an example for interoperability at the storage system layer.

Third and last, there might exist a web page, on which weather data about Europe is published. Now we are interested in all satellite images of Europe, which are taken at a the same time, as this published weather information. This means, we need interoperability at the object manager layer. Harmony has to access the external web server via a corresponding wrapper. This wrapper transforms the contents of the page into a collection of weather data entries. Harmony afterwards joins these entries with the image objects stored inside CONCERT according to the time attributes.

3 An Abstract-Object Kernel System

Under the term "Object-Relational", database systems have become popular, that allow extensions to their kernel system. Such extensions are called blades, cartridges, or alike. They allow the DBMS to be extended by *application specific types* and *access methods.* While implementing new types is relatively easy, new access methods is not. The new access method has to cooperate with the various components of the DBMS, such as concurrency control, data allocation, query evaluation, and optimization. This requires substantial knowledge of the DBMS internals. In contrast, the CONCERT kernel offers a limited, built-in set of physical design mechanisms in form of generic, trusted DBMS code provided by the DBMS implementor. Physical design is performed through relating new types to the fundamental concepts of the built-in physical design mechanisms.

3.1 Concepts of Physical Database Design

In [17], Stonebraker introduced the idea of a generic B-Tree that depends only on the existence of an ordering operation to index arbitrary data objects. Our CONCERT approach generalizes this idea by identifying *all relevant concepts of physical database design* and expressing them by the so called *concept typical* operations required to implement them over external data. The data objects are treated as abstract data types (ADT) in CONCERT, and physical database design is performed based on the operations of the ADT only. These ADT's are user-defined and their methods are dynamically linked to the kernel at run time. In order to implement search tree access methods, a generic search tree approach (similar to GiST [6]) can be used as it integrates nicely into the CONCERT framework.

The physical design concept behind Stonebraker's generic B-Tree is that of data objects having a *scalar* property. Therefore, we call it the `SCALAR` concept and its concept typical *ordering operation* `COMPARE`. The comparison operation is sufficient to instantiate a generic B-Tree index without any further knowledge of the data objects. A second concept called `RECORD` concept allows to identify components of objects. A data object might be decomposed into object parts. This is exploited for example in a relational context as vertical partitioning. Its concept typical operation is the decomposition of objects into object parts called `SUB_OBJECT`. A third fundamental concept of physical database design is the one found for example in the information retrieval context, where objects are organized according to sets of object properties such as the index terms of the document object. The concept typical operations of this concept are the ones iterating over the set of properties. We therefore call it the `LIST` concept. The iteration allows the properties to be entered for example into an inverted file index. Finally, the last concept of physical database design, we identified in CONCERT is the one concerned with spatially extended objects and is therefore called the `SPATIAL` concept. It is used for expressing space–subspace relationships as they appear in GIS and CAD systems, but also in temporal applications.

These concepts are the means for interoperability at the CONCERT kernel level as discussed in Section 2. From a physical design point of view, storing the satellite images of our sample application in a tape archive while storing the corresponding metadata in the CONCERT storage component corresponds to a vertical partition of the image objects. The concept typical operation required is the decomposition of the RECORD concept. Therefore, the Application Programmer can make the fact of the image residing in the tape archive known to the kernel via the RECORD concept by implementing the concept typical operation SUB_OBJECT. This operation is responsible for accessing the image on the archive.

In addition to the four concepts SCALAR, RECORD, LIST and SPATIAL and their concept typical operations, three fundamental operations are required for all abstract objects. They are needed to pass abstract objects across system internal interfaces. If for example an abstract object is inserted into a Btree index, the object has to be recursively passed through the nodes of the tree. The kernel has to be able to COPY an object, which is the first and most important operation, that any object in CONCERT has to provide. Depending on the usage of the object, copying can be performed in different ways. If the object is to be passed to a function call within the same process, a shallow copy might be appropriate. If the object has to be stored in a database disk page, a full copy is required. In addition, this copy has to be linearized into a single continuous address space. CONCERT allows the copy operation to be driven by a set of copy flags making such distinctions. While the copy operation is specific for each object type, memory allocation has to be performed by the generic database code. In order to get to know the resource requirements, the COPY_SIZE operation has to be provided by each object enabling the generic kernel to perform the necessary allocations. In order to actually perform a copy operation, additional resources such as temporary memory, network connections, file handles or alike might possibly be required. These resources have to be freed once the copy is no longer needed by the database. Therefore, the third operation required for all generic objects is the DELETE_AUX operation. As a consequence, the usual steps passing abstract objects around is performed as shown in Figure 2.

```
s     := COPY_SIZE (o, copy_flags);
new_o := allocate (s);
COPY (o, new_o, copy_flags);
.. do something with  new_o ..
DELETE_AUX (new_o);
```

Fig. 2. Steps required to move an object around

The Appendix shows the interface definition of the CONCERT kernel concepts and their concept typical operations. It is beyond the scope of this paper to give full details here. More information on CONCERT concepts in particular and the CONCERT kernel system in general can be found in [2, 4, 14] and [15].

3.2 Performance Evaluations for Interoperability at Kernel Level

It is clear, that the interoperability flexibility available in the CONCERT kernel allowing the kernel to access data from remote repositories has its price. This is not specific for CONCERT but rather inherent to any interoperable system. Because interoperability in the kernel is done at a very low level of abstraction, it is very efficient and measuring the local overhead therefore gives a minimal lower bound for the overhead to be expected in any interoperable system.

Passing objects as parameters of procedure calls accross kernel modules is often used in kernel systems. Therefore, the main reason for the low-level interoperability cost in CONCERT is the fact, that the algorithm in Figure 2 involving method calls to abstract object is executed frequently. In systems with hard-coded object types, passing them as parameters of procedure calls can be specialized for the supported types and therefore can be coded more efficiently. Measuring the overhead of the generic algorithm compared with the hard-coded gives a good indication of the low-level interoperability cost.

We identify three typical cases for base type objects:

- The object type is a built-in type (such as `longint` or `float`) of the compiler, that the database system is compiled with. Copying the object can be done using compiler-generated object assignment code. This is the best possible case for a hard-coded system and the most advantage compared with the generic case is expected. However, for these types, data independence can not be guaranteed, as the type representation is compiler and hardware dependent.
- The object type is compiler independent, but of simple structure and of fixed size. Most standard database types in traditional database systems are of this category, such as `INTEGER`, `NUMBER`, `CHAR(n)`. In some systems, aggregations of simple types such as `ROW` types in SQL3 fall into this category as well. Because their size is known a priori, and their representation is a continuous byte sequence, copying these objects corresponds to a simple `memcopy` operation.
- The object type is of variable size, such as `VARCHAR`, `BLOB` or aggregations of object types. Their object size varies from object to object. Therefore, the object size has to be determined at run time and space allocation has to be performed dynamically.

We do not discuss object type with complex structure, because there is virtually no cost difference between hard-coded and abstract objects.

Using the CONCERT kernel system we measured the three typical cases comparing hard-coded with abstract objects using generic algorithms. Figure 3 summarizes the results showing the copy time in nanoseconds on two different SUN Solaris system architectures.

It does not surprise that moving more complex objects around is much more expensive than simple, small ones. It is clear that the more recent system architecture (UltraSparc 1) is substantially faster than the older one. From these

	SparcCenter 2000		UltraSparc 1	
	hard-coded	generic	hard-coded	generic
int (32bit)	446	2108	153	840
NUMBER	1345	2877	719	1389
VARCHAR	11363	13199	3539	4196

Fig. 3. Comparison copying hard-coded objects versus abstract objects using generic algorithms (execution time in ns)

measurements, we see that the interoperability costs, that is the difference between the hard-coded version and the generic version, is especially high for very simple objects (with an interoperability overhead of much more than 100%). The overhead is much smaller for larger objects (the VARCHAR object had an average length of 150 characters resulting in an overhead of approximately 15%). For even larger and more complex objects, the overhead is only a few percent. While the first case due to the of lack of data independence is not very relevant for database interoperability, already in the second case the overhead is in the order of improvement of one hardware generation. Furthermore, the local interoperability cost for small objects is very small compared with the overall system cost. We conclude that *building a system capable of low-level interoperability using generic instead of hard-wired algorithms has only a minimal impact on the overall system performance: the rapid hardware development makes low-level interoperability affordable.* In the next section we concentrate on the higher level aspect of interoperability.

4 An Abstract-Object Query System

The usual understanding of interoperability between systems is quite narrow, addressing only SQL-interoperability. Commercial "SQL middleware" like the Informix Enterprise Gateway Manager, Oracle Transparent Gateways, or Sybase' OmniSQL Server rely on a declarative surface of object managers. This means, they presume a declarative interface like ODBC or JDBC, which is already capable of executing SQL. A precondition, which is not feasible for non-database systems. Therefore, we will not discuss this, in fact, third level of interoperability in this paper.

We rather address interoperability of the layer between the kernel system and the (declarative) user interface. CONCERT's object manager only relies on single-scan collection interfaces. Especially, it does not require any further query capabilities of the storage system. However, if certain repositories provide such capabilities, they can be exploited. The object manager itself adds complete query functionality to the underlying storage kernel and exports a declarative interface to subsequent system layers.

An additional important feature of CONCERT's object manager is its distributed peer-to-peer query execution. While the kernel offers typical storage system functionality for a single node system, the object manager layer is capable of distributing its query execution among several sites with CONCERT instances. As underlying infrastructure the CORBA middleware standard [10] is used. CORBA specifies a communication infrastructure for arbitrary heterogeneous components of distributed systems. Furthermore, it defines a set of basic system services and standard components. One system service is of particular interest for CONCERT: the *Object Query Service (OQS)*. As core of CONCERT's object manager, we have designed and realized an implementation of the OQS, called *Harmony*.

4.1 The Harmony Query Service

Harmony's query algebra operators are implemented by means of physical query operations. E.g., the *join* operator might actually be executed as simple *nested loop join*, or more sophisticated algorithms like a *merge join* or *bind join*. Harmony evaluates a query by a sequence of such physical query operations which are partially ordered: an *execution plan of the query*. An execution plan has a tree structure, each node representing one physical operation. Edges connect the nodes according to their partial ordering. Each edge therefore represents the dataflow between subsequent operations. The leaves stand for the data sources needed for query evaluation. The root node represents the result of the query.

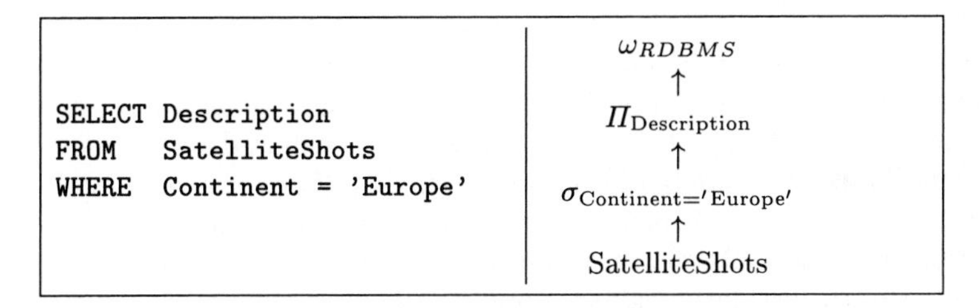

Fig. 4. Example query and corresponding execution plan in Harmony.

An efficient implementation of the OQS is not straightforward. To come up with a realization of a CORBA-based query service whose performance is competitive, we have deployed classical database concepts, notably dataflow evaluation, bulk-transfer and intra-query parallelism: internally, Harmony evaluates a query in a dataflow manner according to an execution plan. As introduced in Section 2, each operation of the plan corresponds to a *QueryableCollection* object instantiated before evaluating the query. These objects actually implement the query algebra operators of Harmony. In addition to the usual operators, the query algebra includes *meta operators*: *wrap*, *send* and *receive*. Meta operators do not change the data stream, but perform some control function.

Send / Receive The *send/receive* operators model the asynchronous set-oriented data transfer between sub-plans. They are used for performance optimizations of distributed query processing and of interoperating queries, as we will show in Subsection 4.3. For a more in-detail description and evaluation of send/receive see [13].

Wrap wrap (ω) abstracts from the underlying storage systems and therefore allows to integrate different storage systems on the logical level. In order to interoperate with a certain external system, an instance of the *wrap* operator for this storage system must be created. These instances correspond to the concept of *wrappers* [5]. Available wrappers are configured via Harmony's global schema (cf. Appendix). Their code is dynamically loaded at runtime into Harmony as soon as the first access to a collection at the corresponding repository is required.

A Harmony wrapper transforms the interface of the underlying storage system into the collection/iterator interface needed by Harmony for query evaluation. This includes type conversion into the CORBA type system. As Harmony provides all further query operations, a simple wrapper does not need to implement own query functionality. The only functionality a wrapper must provide is to access data collections of its source. In the simplest case, this means the ability to scan the source. A wrapper of a more sophisticated system may publish its query capabilities. Harmony can then decide to delegate whole sub-queries to such a wrapper, which will translate it into and execute it exploiting the local query language.

We distinguish three different types of repositories which can be accessed by Harmony:

- The CONCERT kernel storage system. This is the internal interface to CONCERT's own kernel system. The major aspect here is the integration of the kernel's abstract object types into the CORBA-based query service. This affects the type system and the user-defined methods, as we indicated in the satellite image example before. A further discussion follows in Subsection 4.2.
- A data repository. As mentioned in the introduction, the main idea of CONCERT is to export database functionality to external systems. At the level of the object manager this is possible by providing a wrapper for such a simple repository, e.g., an GIS system or a file system. Via the wrapper, CONCERT is capable of evaluating declarative queries on data managed by such repositories. Furthermore, externally stored data can be combined (joined) with objects managed by CONCERT itself (as shown with the third example query of Section 2). A similar approach is taken by Microsoft with OLE DB [3, 12], but only for relational data and without an own storage system.
- A database system. A special case of interoperability with external storage system is the integration of full-fledged database systems. First, such systems have own query capabilities which should be exploited by the corresponding wrapper. Second, the interface level at which Harmony now interoperates with the external system, is typically one above the layer of the object manager. E.g., with a relational database, the wrapper must use the Embedded SQL API, which is already a declarative interface. As state-of-the-art database system do not provide lower APIs, interoperability is simply not possible at a lower abstraction layer.

4.2 Abstract-Object Types in Harmony

As Harmony relies on the CORBA standard, it has to type the data with respect to the CORBA type system. Here, we have to distinguish two different abstraction levels of data access in Harmony. So far, we concentrated on the higher level of collections of objects. All data access and query capabilities of Harmony are defined with respect to abstract (queryable) collection and iterator interfaces. These are CORBA objects, defined in CORBA IDL and fully embedded in the middleware infrastructure.

At the abstraction level below, we are interested in the member objects of Harmony's queryable collections. As they may be of arbitrary type, Harmony employs the dynamic typing capability of CORBA, its *any* type. The actual content type of an *any* value is determined at runtime. For example, the data produced by the ω_{RDBMS} node in Figure 4 corresponds to tuples of the form `(Description VARCHAR(100))`. The wrap operation maps these data items into the CORBA type `sequence<any>`. At runtime, it contains values of type `string`.

In order to benefit from the abstract object types of the kernel, we extended CORBA's own dynamic type to include CONCERT's build-in types. For the CORBA system this is a user-defined *opaque* type, which Harmony provides the code for. This means, Harmony can also exploit the capabilities of concept typical operations. Especially, this allows to combine interoperability at the physical and the logical level. The Harmony object manager can evaluate a query over collections of CONCERT objects, which are actually stored outside the kernel and accessed by the concept typical operations.

4.3 Performance Evaluations for Interoperability at Object Manager Level

In this subsection we are interested in quantifying the costs of interoperability. It is clear, that any access through the wrapper interface of Harmony must be more expensive than direct access to the underlying storage system. Interoperability does not come for free.

Beside the performance difference of a three tier (client – Harmony– repository) against a two tier system approach (client – repository), another major performance bottleneck must be considered: as mentioned before, the Harmony object manager is capable of distributed execution of an query by exploiting a CORBA infrastructure. As any middleware system, CORBA introduces some overhead in order to provide the location and implementation transparency one expects from such a system. E.g., it has to marshal/demarshal each value transmitted via the middleware. This additional data conversions certainly affect the performance of a middleware-based system.

The answer Harmony offers for this issue are its *send/receive* meta operators introduced above. These operators allow for intra-query parallelism and optimized bulk-data transfer between different query operations. While the receive operator consumes the set of intermediate results it got from the corresponding set operator, the subsequent query operators can produce the next partial result in parallel. For further details see [13].

As a starting point to quantify the interoperability overhead of Harmony, we measured the access times for data access to a result set of 30000 string values stored in a relational database. We chose a database system, so that we could easily obtain a reference result. Therefore, we executed the SQL query shown in Figure 4 via an Embedded SQL/C program. The query was executed in two different system configurations: with the client and server on the same local machine and separated via a local area network (LAN). The results are compared to the execution times of Harmony. First, we submitted the same query to the RDBMS via Harmony's RDBMS wrapper and retrieved the result set. Second, we introduced a send/receive pair above the wrap-operation, so that the external database produced its result items concurrently to the further processing in Harmony. For our experiments, the Harmony client retrieved the query results in chunks of 100 values. This is the same number of items as the ESQL/C program retrieved with one of its array fetches. The runtimes are presented in Figure 5.

	RDBMS	**Harmony**	
		wrap only	with send/receive
local	26.22s	391.63s	32.04s
LAN	24.19s	366.45s	29.98s

Fig. 5. Local/remote data access times (seconds) for CONCERT and a RDBMS.

This experiment clearly shows a massive interoperability overhead for the "naive" execution of the query with Harmony. Without deploying any query optimization techniques, data access through the Harmony wrappers is about 15 times slower than direct access to the RDBMS (first two columns in Figure 5). This drastically changes, if we introduce intra-query parallelism in Harmony: data access through Harmony is now only about 20% slower as a direct database access via Embedded SQL/C. This result shows, that *medium-level interoperability is affordable, as long as it is combined with proven query optimization techniques.*

5 Conclusions

In this paper, we showed, how interoperability between heterogeneous storage systems is achieved in our CONCERT system. We discussed two different levels of interoperability: between kernel systems, and on the object manager level. The main motivation is to be able to export database functionality to external systems.

This is achieved in a very flexible way: the kernel allows to plug-in modules with user-defined types, which only have to provide the fundamental concept-typical operations. Above the kernel, the Harmony object manager includes a wrap-operator in its query algebra. Instances of this operator, so-called wrappers, allow to interoperate with external repositories at the query-processing level. Their only required functionality is a simple-scan interface. Extensions at all levels are implemented in a plug and play like fashion exploiting the operating system capability of dynamic linking.

We also presented first results of a cost evaluation of interoperability with CONCERT at the different levels. From our experiments, we can learn two conclusions: First, the interoperability overhead increases with the abstraction layer at which interoperability takes place. Second, the overhead can be minimized by exploiting well-known database optimization techniques, like bulk-transfer and intra-query parallelism in the case of Harmony. Naive interoperability is very costly. An efficient implementation of an interoperability interface needs careful design and conscious deployment of proven database techniques.

The ideas presented here are part of ongoing work. We are primarily interested in further evaluating the interoperability costs at the different abstraction levels, especially in comparison with spatial data access modules of commercial systems. We also plan to investigate, how external data can be accessed fully dynamically via wrappers without prescribing the data format.

References

1. M. M. Astrahan, M. W. Blasgen, et.al. System R: Relational approach to database management. *ACM Transactions on Database Systems*, 1(2):97–137, June 1976.
2. Stephen Blott, Helmut Kaufmann, Lukas Relly, and Hans-Jörg Schek. Buffering long externally-defined objects. In *Proceedings of the Sixth International Workshop on Persistent Object Systems (POS6)*, pages 40–53, Tarascon, France, September 1994. British Computer Society, Springer-Verlag.
3. Jose A. Blakeley. Data access for the masses through OLE DB. In *Proceedings of the 25th ACM SIGMOD Conference on Management of Data, Montreal, Canada*, June 1996.
4. Stephen Blott, Lukas Relly, and Hans-Jörg Schek. An open abstract-object storage system. In *Proceedings of the 1996 ACM SIGMOD Conference on Management of Data*, June 1996.
5. H. García-Molina, J. Hammer, K. Ireland, Y. Papakonstantinou, J. Ullman, and J. Widom. Integrating and accessing heterogeneous information sources in TSIMMIS. In *AAAI Spring Symposium on Information Gathering*, 1995.
6. Joseph M. Hellerstein, Jeffrey F. Naughton, and Avi Pfeffer. Generalized search trees for database systems. In *Proceedings of the 21st International Conference on Very Large Databases*, pages 562–573, September 1995.
7. Ibm corporation. `http://eyp.stllab.ibm.com/t3/`.
8. Informix corporation. `http://www.informix.com/informix/products/options/udo/datablade/dbmodule/index.html`.
9. Bruce Lindsay, John McPherson, and Hamid Pirahesh. A data management extension architecture. In *Proceedings of the 1987 ACM SIGMOD International Conference on Management of Data*, ACM SIGMOD Record, pages 220–226. IBM Almaden Research Center, San Jose, 1987.
10. Object Management Group. *The Common Object Request Broker: Architecture and Specification*, 2.0 edition, July 1995.

11. Oracle corporation. `http://www.oracle.com/st/cartridges/`.
12. Stephen Rauch. Talk to any database the COM way using the OLE DB interface. *Microsoft Systems Journal*, July 1996.
13. Uwe Röhm and Klemens Böhm. Working together in harmony — an implementation of the corba object query service and its evaluation. To appear in: *Proceedings of the 15th International Conference on Data Engineering*, Sydney, Australia, March 1999.
14. Lukas Relly, Hans-J. Schek, Olof Henricsson, and Stephan Nebiker. Physical database design for raster images in CONCERT. In *5th International Symposium on Spatial Databases (SSD'97)*, Berlin, Germany, July 1997.
15. Lukas Relly, Heiko Schuldt, and Hans-J. Schek. Exporting database functionality — the CONCERT way. *Data Engineering Bulletin*, 21(3):43–51, September 1998.
16. Hans-Jörg Schek, H.-B. Paul, M.H. Scholl, and G. Weikum. The DASDBS project: Objectives, experiences, and future prospects. *IEEE Transactions on Knowledge and Data Engineering*, 2(1):25–43, March 1990.
17. Michael Stonebraker. Inclusion of new types in relational database systems. In *Proceedings of the International Conference on Data Engineering*, pages 262–269, Los Angeles, CA, February 1986. IEEE Computer Society Press.
18. Hans-Jörg Schek and Andreas Wolf. From extensible databases to interoperability between multiple databases and GIS applications. In *Advances in Spatial Databases: Proceedings of the 3rd International Symposium on Large Spatial Databases*, volume 692 of *Lecture Notes in Computer Science*. Springer Verlag Berlin Heidelberg New York, June 1993.
19. Avi Silberschatz, Stan Zdonik, et.al. Strategic directions in database systems – breaking out of the box. *ACM Computing Surveys*, 28(4):764–778, December 1996.

Interface Definition of the Kernel Concepts

CONCEPT UNKNOWN	
COPY_SIZE	object, copy_flags
COPY	source, target, copy_flags
DELETE_AUX	object

CONCEPT SCALAR ISA UNKNOWN		
COMPARE	o1, o2	$\rightarrow \{-1, 0, 1\}$

CONCEPT RECORD ISA UNKNOWN		
SUB_OBJECT_SIZE	object, component	$\rightarrow \mathbb{N}$
SUB_OBJECT	object, component	$\rightarrow$ part

CONCEPT LIST ISA UNKNOWN		
OPEN	object	$\rightarrow$ cursor
FETCH	cursor	$\rightarrow$ element
CLOSE	cursor	

CONCEPT SPATIAL ISA UNKNOWN		
OVERLAPS	object1, object2	$\rightarrow$ SCALAR_HASH
SPLIT	object	$\rightarrow$ { object }
COMPOSE	{ object }	$\rightarrow$ object
APPROX	{ object }	$\rightarrow$ object

Example Kernel Modules

```
// "StringType" and "DateType" are predefined as SCALAR and
// SPATIAL concepts
CREATE CONCEPT SatelliteImage AS
```

```
       SPATIAL
  WITH OVERLAPS := SI_Overlaps.so
       SPLIT    := SI_Split.so
       COMPOSE  := SI_Compose.so
       APPROX   := SI_Approx.so

CREATE CONCEPT SatelliteData AS
       RECORD ( StringType, DateType, SatelliteImage )
  WITH SUB_OBJECT_SIZE := SD_GetSubObjSize.so
       SUB_OBJECT      := SD_GetSubObj.so
```

Example Global Schema

```
repository WeatherServer := WebPageWrapper.so
{  URL := "http://www.weathernet.com/..."  };

// object types "Image", "Coordinate" and "Time" are predefined
class SatelliteImage ( extent SatelliteImages )
{
  attribute string title;
  attribute Time   date;
  attribute Image  picture;

  Image   window  ( x, y, w, h : integer ) := SI_window.so;
  boolean is_over ( pos : Coordinate )     := SI_translate.so;
};
class WeatherMeasurement@WeatherServer ( extent Measurements )
{
  attribute Time       when;
  attribute Coordinate where;
  attribute int        rainfall;
  attribute int        airPressure;
  ...
};
```

Functional Extensions of a Raster Representation for Topological Relations

Stephan Winter and Andrew U. Frank

Dept. of Geoinformation
Technical University Vienna
Gusshausstr. 27-29, 1040 Vienna, Austria
{winter,frank}@geoinfo.tuwien.ac.at
http://www.geoinfo.tuwien.ac.at

Abstract. Topological relations are not well defined for raster representations. In particular the widely used classification of topological relations based on the nine-intersection [8, 5] cannot be applied to raster representations [9]. But a raster representation can be completed with edges and corners [14] to become a cell complex with the usual topological relations [16]. Although it is fascinating to abolish some conceptual differences between vector and raster, such a model appeared as of theoretical interest only.
In this paper definitions for topological relations on a raster – using the extended model – are given and systematically transformed to *functions* which can be applied to a regular raster representation. The extended model is used only as a concept; it need not to be stored. It becomes thus possible to determine the topological relation between two regions, given in raster representation, with the same reasoning as in vector representations. This contributes to the merging of raster and vector operations. It demonstrates how the same conceptual operations can be used for both representations, thus hiding in one more instance the difference between them.

1 Introduction

Topological relations are not well defined for raster representations. In particular the widely used classification of topological relations based on the four- and nine-intersection [8, 5] cannot be applied to raster representations [9]. This is due to the topological incompleteness of a raster: it consists, in the field view, of (open) two-dimensional cells only. In contrast, vector representations consist also of one- and zero-dimensional elements, used for the representation of boundaries, which close two-dimensional point sets and demarcate from their exterior. Boundary constructions in raster representations require the use of raster elements [13], although they are two-dimensional by nature. Two-dimensional boundaries contradict to topology, so they cause some well-known paradoxes.

Kovalevsky has suggested that the raster can be completed with edges and nodes to become a full topological model [14]. In this representation, called here a *hybrid* raster, topological relations are defined equivalent to a vector representation [16]. But the hybrid raster appeared as of theoretical interest only, mostly due to its additional and redundant memory requirements (see Section 2.3).

Here detailed definitions for topological relations on a raster – using the hybrid raster representation – are given and then systematically transformed to yield functions which can be used in a convolution operation applied to a regular raster representation. Hereby the hybrid raster is only used as a concept. It need not be stored and is only partially constructed during the execution of a determination of a topological relation. It becomes thus possible to determine the topological relationship of two regions, given in raster representation, by the four- or nine-intersection.

A formal approach is used to understand the structure and the theory of an extended raster representation and its application for topological relations. The specification is written in a functional language. Pure functional languages [1], like Gofer [11], provide a useful separation of specification and implementation [10]. With executable specifications, the result is a provable code – in syntax as well as with test cases – with a clear semantic. Furthermore, such a specification is basis for iterative optimization; e.g. the Gofer code published here[1] was optimized in several cycles of improvements. The value of such formal specifications is recognized more and more. So Dorenbeck and Egenhofer presented a formal specification of raster overlay, with a generalization for polygons [3]. We also specify an overlay, but of an extended raster, deriving the same behavior of raster and vector representations.

This contributes to the merging of raster and vector operations. It demonstrates how the same conceptual operations can be used for both representations, thus hiding in one more instance the difference between them.

The paper is structured as follows. In Section 2 previous work is collected, regarding topological relations between regions, and hybrid raster representation. In Section 3 the raster representation is extended to a hybrid raster, and the combination of two raster images is presented to determine a four-intersection. It is also discussed how to optimize computations. An example in Section 4 shows the advantage of an executable specification. Finally a discussion sums up the results and perspectives (Section 5).

2 Previous Work

2.1 Topological Relations

Egenhofer proposed a representation of topological relations between point sets, based on the intersection sets of such point sets [6, 7, 5]. Point sets in $\mathbb{R}^2$ refer to Euclidean topology, with the Euclidean distance as a metric. The metric is needed to define a boundary of (open) sets. Distinguishing the interior $\mathcal{X}^\circ$, the boundary $\partial\mathcal{X}$ and the exterior $\mathcal{X}^c$ of a point set X, two point sets $\mathcal{A}$ and $\mathcal{B}$ may

[1] The complete code is available at our web-page.

have nine intersection sets, which form a partition of the plane. For describing topological properties the size of the intersection sets is irrelevant, only being empty or not is characterizing.

For regular closed and singular connected sets – *simple* regions – even four intersection sets are sufficient, because the omitted five intersection sets do not vary. The sets can be ordered in a 2 × 2-array, the four-intersection **I4**:

$$\mathbf{I4} = \begin{pmatrix} \mathcal{A}^\circ \cap \mathcal{B}^\circ & \mathcal{A}^\circ \cap \partial\mathcal{B} \\ \partial\mathcal{A} \cap \mathcal{B}^\circ & \partial\mathcal{A} \cap \partial\mathcal{B} \end{pmatrix} \tag{1}$$

The *nine-intersection* contains the other five sets, too. – Eight relationships between two simple regions can be characterized using this schema (Table 1).

Table 1. The eight distinct four-intersections for simple regions, and the names of the characterized topological relations.

$\begin{pmatrix} \emptyset & \emptyset \\ \emptyset & \emptyset \end{pmatrix}$	$\begin{pmatrix} \emptyset & \emptyset \\ \emptyset & \neg\emptyset \end{pmatrix}$	$\begin{pmatrix} \neg\emptyset & \neg\emptyset \\ \neg\emptyset & \neg\emptyset \end{pmatrix}$	$\begin{pmatrix} \neg\emptyset & \emptyset \\ \emptyset & \neg\emptyset \end{pmatrix}$
DISJOINT	MEET	OVERLAP	EQUAL
$\begin{pmatrix} \neg\emptyset & \neg\emptyset \\ \emptyset & \neg\emptyset \end{pmatrix}$	$\begin{pmatrix} \neg\emptyset & \emptyset \\ \neg\emptyset & \neg\emptyset \end{pmatrix}$	$\begin{pmatrix} \neg\emptyset & \neg\emptyset \\ \emptyset & \emptyset \end{pmatrix}$	$\begin{pmatrix} \neg\emptyset & \emptyset \\ \neg\emptyset & \emptyset \end{pmatrix}$
COVER	COVEREDBY	CONTAIN	CONTAINEDBY

The found relationships were investigated and applied to spatial reasoning [4, 12], with the interest to speed up spatial queries in GIS or in AI. They are an important improvement of vector representations, which base on point sets in $\mathbb{R}^2$.

2.2 Topological Relations and Raster Representations

A raster representation is a two-dimensional array of elements with integer coordinates. Interpreting the raster elements as fields – instead of lattice points –, the raster is a regular subdivision of space into squares of equal size, *resels* (short form for 'raster elements'). – For the general principle it doesn't matter how the raster is implemented (see e.g. [15]). But a comparison to vector representations directly shows that only (open) two-dimensional elements exist, and one- and zero-dimensional elements are missed. Boundaries of regions cannot be defined by infinite balls as in the Euclidean space.

This problem was treated so far in two ways:

- omitting boundaries, having only regions as open sets, as it is done in region based reasoning methods (e.g. [2]);

– defining substitutes for one-dimensional boundaries, using raster elements and any arbitrary neighborhood definition [13].

The first solution allows the application of the nine-intersection only for its two-dimensional intersection sets, i.e. the region *interiors* and *exteriors*, which yields a subset of the relationships in Euclidean space [17]. The resulting four-intersection may not be mixed up with the four-intersection defined with boundaries (Eq. 1).

The second solution generates two-dimensional boundaries, resel chains or bands, either as interior boundaries or as exterior boundaries. Two-dimensional boundaries contradict to topology, so they cause some well-known paradoxes [13]. The result of intersecting the sets of interior, boundary and exterior raster elements depends heavily on the definition of the boundary (interior or exterior). Even worse is the possibility of more than the eight four-intersections described in Table 1, simple regions presumed [9]. These intersections have no common-sense meaning; they appear as variations of the eight presented intersections and need special care.

2.3 The Hybrid Raster Representation

Kovalevsky has suggested that the raster can be completed with edges and nodes to become a full topological model, to be precisely: an abstract cell complex [14]. The only specialty of this cell complex is its regular structure (Figure 1). Generally all elements of a cell complex are called (2D-, 1D-, 0D-)cells, but we will speak in the following of two-dimensional *cells* – identical with the resels in raster –, one-dimensional *edges* and zero-dimensional *nodes*. The union of edges and nodes will be called the *skeleton* of the cells.

Fig. 1. A (regular shaped) cell complex, replacing a raster element of usual raster representations in the hybrid raster: each cell is closed by four edges and four nodes.

In this representation, topological relations regard again to Euclidean space, and the four- or nine-intersection can be applied in full accordance to vector representations [16]. In vector representations these tests are expensive, requiring polygon intersection. In a hybrid raster representation the tests are simple to evaluate: two hybrid rasters (of the same resolution, same size and common origin), labeled by three values for interior, boundary and exterior, are overlaid by $\wedge$ (equivalent to $\cap$ in set denotation). Then the nine possible combinations can be accumulated in a histogram. Binarizing the histogram ($= 0$, > 0) yields the nine-intersection. In a more sophisticated algorithm one would consider the dimension of the intersection sets, and reduce the overlay to the cells of the relevant dimension.

Winter presented also a data structure to store and access the cells and their skeleton. If the raster is of size $n \times m$, additional elements in a hybrid raster are $(n+1) * m$ horizontal edges, $n * (m+1)$ vertical edges, and $(n+1) * (m+1)$ nodes: the required memory space is of order 4 higher than for the raster. Another critical point of such data structures is the considerable amount of index transformations for each access.

However, if the hybrid raster is used only to represent regions – as raster does –, and no lines or points, then the additional elements of the hybrid raster become totally redundant to the cells. The skeleton can be renounced from explicit storage, applying dependency rules instead, which work locally. This paper will investigate these ideas, using a functional approach to specify semantically the rules and their application.

3 Topological Relations in a Functional Extended Raster Representation

The determination of the nine-intersection is simple in a raster representation, if the topologically completed raster is used (Section 2.3). But this does not seem practical, as the model includes not only the cells, but also the edges and the nodes; this would quadruple the storage requirement and also make computation four times longer. We will develop now a functional extension of the raster that fulfills all the conditions of a hybrid raster virtually, without explicit representation. The functions are specified in Gofer [11].

3.1 Specification of a Hybrid Raster in Natural Language

The hybrid raster representation can be computed from the regular raster representation, i.e. the necessary information is already contained in the raster, and all additional elements are redundant.

Assume an arbitrary region – without loss of generality let us confine ourselves to simple regions – given as the set of resels with value 'Region', and the background resels have the value 'Empty'. These two values are mapped to the Boolean values true and false, to allow the regular logical operations.

Cells: Cells are identical to resels. △

Vertical edges: A vertical edge belongs to the interior of the region, iff the adjacent left and right cells are labeled as 'Region'. It belongs to the exterior of the region, iff the adjacent left and right cells are labeled as 'Empty'. It belongs to the right boundary, iff the adjacent left cell is 'Region' and the right cell is 'Empty', otherwise it belongs to the left boundary (Figure 2). △

Fig. 2. Classification of edges by the two adjacent cells (bright cells are outside of the region, dark cells are elements of the region). An edge belongs (a) to the exterior or (b) to the interior, if both raster elements are homogenous, (c) and (d) to the boundary, if the values of the raster elements are different.

Horizontal edges: A horizontal edge belongs to the interior of the region, iff the adjacent upper and lower cells are labeled as 'Region'. It belongs to the exterior of the region, iff the adjacent upper and lower cells are labeled as 'Empty'. It belongs to the lower boundary, iff the adjacent upper cell is 'Region' and the lower cell is 'Empty', otherwise it belongs to the upper boundary. △

To distinguish the orientation of the boundary is not necessary in the context of this paper. But it could get importance in other tasks like line following.

Nodes: A node belongs to the interior of the region, iff all four adjacent cells are labeled as 'Region'. It belongs to the exterior of the region, iff all four adjacent cells are labeled as 'Empty'. Otherwise the node belongs to the boundary (Figure 3). △

Fig. 3. Classification of nodes by the four adjacent raster elements (bright resels are outside of the region, dark resels are elements of the region). A node belongs to the exterior, if all resels are outside (a), to the interior, if all resels are element of the region (b), and to the boundary if the four resels are not homogenous; the given examples (c, d) are not complete.

To transform the rules into a formal language, an identification of each single elementsis required. We define the following index schema:

Cell index: Cells are indexed in the regular way of resels. △

Vertical edge index: Vertical edges are indexed with the same index as the cell to their left. △

Horizontal edge index: Horizontal edges are indexed with the same index as the cell above. △

Node index: Nodes are indexed with the same index as the cell left above. △

Figure 4 shows that this indexing schema is indeed complete for the Euclidean plane and gives for each element of the representation a unique index. However, any subset of the plane will miss the edges and nodes at the left and upper border by this indexing schema. For that reason it is presumed that the subsets are chosen with a border of at least one resel width ('Empty') around the represented region.

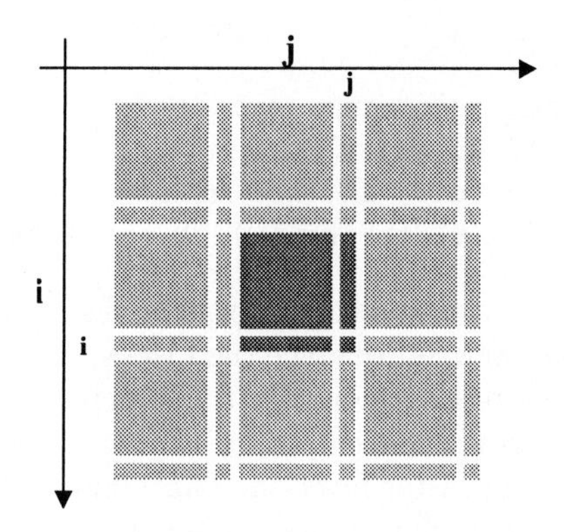

Fig. 4. Indexing schema for egdes and nodes.

3.2 Specification of a Hybrid Raster in a Functional Language

In Gofer an array can be realized as a class of `{bounds, [index := value]}`, where `bounds` are the lower and upper limit of indices, and the remainder is a list of associations between an index and a value. In the context of this paper arrays are two-dimensional and rectangular, indices are integer tuples, and the type of cells is Boolean:

```
instance Arrays (Int,Int) Bool
```

Let us extract an arbitrary 2-by-2 sub-array from a binary raster image, by applying the class method `getSubMat`:

```
get22Mat image i j = getSubMat image ((i,j),(i+1,j+1))
```

The sub-array contains the four resels (i, j), $(i + 1, j)$, $(i, j + 1)$, and $(i + 1, j + 1)$. In the following, they are referred to as patterns `cIJ`, `cEast`, `cSouth`, and `cSouthEast`, cf. Figure 5.

All the following functions map this sub-array onto a Boolean. They represent the rules of Section 3.1:

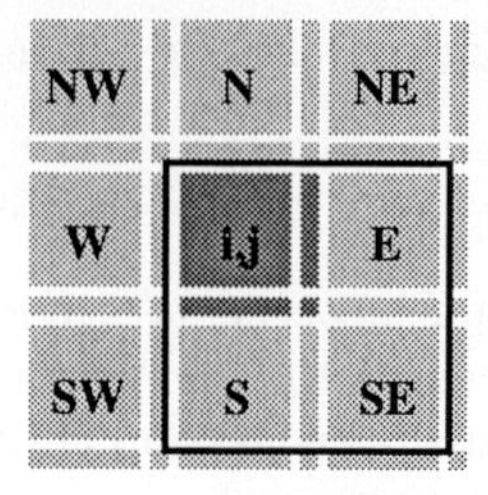

Fig. 5. The names of resels/cells in a window at cell (i, j).

```
cInterior mat22 = cIJ
cExterior mat22 = not (cInterior ar)
vInterior mat22 = cIJ && cEast
vExterior mat22 = (not cIJ) && (not cEast)
vBoundary mat22 = not (vInterior ar) && not (vExterior ar)
```

and so on for horizontal edges and for nodes. The result confirms or falsifies the rule name; e.g. if `vBoundary` returns true then the vertical edge (i, j) belongs to the boundary. While the resels are binary, the skeleton elements are ternary.

With the functions above the elements of a hybrid raster can be derived from a raster on demand at any raster position. That ability allows the construction of the hybrid raster on the fly during the overlay of two raster images. No storage of the results is specified for the functions.

3.3 Determination of Topological Relations

Testing for the intersection between boundary and interior of two simple regions determines their topological relation. In a hybrid raster, the tests must be repeated for cells, for horizontal and vertical edges, and for nodes. With regard to the limited dimension of some intersection sets, some of these tests can be neglected.

For two hybrid raster images (of the same resolution, the same orientation, and the same origin), only cells intersect with cells, only edges intersect with edges, and only nodes intersect with nodes. That is a consequence of the regular decomposition of the plane, and exceeds the usual properties of vector representations. Taking advantage from these properties, the four intersection sets of Equation 1 can be reformulated as:

```
ii_intersect a b = (&& (cInterior a) (cInterior b))
ib_intersect a b = (&& (vInterior a) (vBoundary b)) ||
                   (&& (hInterior a) (hBoundary b))
bi_intersect a b = (&& (vBoundary a) (vInterior b)) ||
                   (&& (hBoundary a) (hInterior b))
bb_intersect a b = (&& (nBoundary a) (nBoundary b))
```

Here the arguments a and b stand for two sub-arrays, one from each raster image, with the same index. That means that with this compact code at any raster position (i, j) the four intersection sets between two region interiors and boundaries are determined:

```
fourIntersectionIJ a b i j = [ii, ib, bi, bb] where
    ii = ii_intersect (get22Mat a i j) (get22Mat b i j)
    ib = ib_intersect (get22Mat a i j) (get22Mat b i j)
    bi = bi_intersect (get22Mat a i j) (get22Mat b i j)
    bb = bb_intersect (get22Mat a i j) (get22Mat b i j)
```

The remaining task is to move the 2-by-2 sub-arrays over both rasters in parallel. So the determination of the four-intersection is reduced to a convolution:

```
fourIntersection a b = ((map or).transpose)
    [ fourIntersectionIJ a b i j |
            i<-[begRow .. endRow], j<-[begCol .. endCol] ]
```

In the code a test is added to guarantee the identical image sizes. Also the patterns begRow and endCol are defined in the code, with exploitation of the outer band of 'Empty' in both images.

Let us consider the last function in more detail. Convolution yields a list of four-intersections for each raster position (right hand of the equation), which are realized as lists of four Booleans. Transposing this list of lists yields a list of four lists each containing all Booleans regarding one intersection set for the whole overlaid images. The map operation applies the argument – the or function – to all elements of the lists: we derive four Booleans for the global four intersection sets.

Extension of the procedure to the nine-intersection is straight forward.

3.4 Computational Improvements

In functional languages, the optimization is easily performed – but it is not even necessary. Languages like Gofer are 'lazy'; they evaluate functions only when needed, and only to a degree that is needed. While lazy evaluation optimizes program execution of the Gofer interpreter, the effects must be made explicit for translation to standard programming languages.

Partly the given Gofer code is already optimized: consider the limitation of evaluating intersection sets with hybrid elements of specific dimensions only. For example, ii_intersect evaluates only cells – no edges or nodes. That is sufficient because if the interior-interior-intersection set is not empty it must contain two-dimensional elements. – Open for optimization is the last function fourIntersect. The or, mapped to a list of Booleans, is true if at least one element is true. In principle it is sufficient to stop evaluation of each intersection set when the first true result is found.

Once optimization is done (and tested), the code can be translated into standard programming languages, like Pascal or C++.

4 Examples

Because Gofer is an executable (interpreted) language, one can run the code with some test cases. To generate such examples, first a constructor is called to deliver a raster image, initialized as 'Empty':

```
imgEmpty = binArray (-1) (-1) 2 3 False
```

Note that the bounds yield a 4-by-5 array, where the usable indices 0...1 or 2 guarantee the outer band of 'Empty' resels. – With the same constructor now two rectangular regions are created. Each region is combined with the empty image, creating the two raster images `imgA` and `imgB` (Figure 6):

```
boxA = binArray 0 0 1 0 True
boxB = binArray 0 1 0 2 True
imgA = imgEmpty // assocs boxA
imgB = imgEmpty // assocs boxB
```

More complex regions could be generated iteratively. Now we can formulate the query:

```
? fourIntersection imgA imgB
```

The result is: `[False, False, False, True]`. That means the only intersection set of Equation 1 (here in linear order) that is not empty is the intersection between the two (one-dimensional!) boundaries. The topological relationship between region A and B must be MEET therefore.

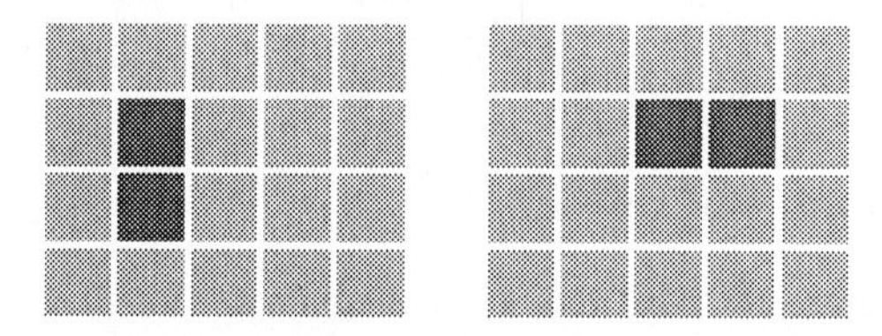

Fig. 6. The regions A (in the left raster image) and B (in the right raster image) meet along an implicit one-dimensional common boundary.

5 Conclusions

The systematic and conform extension of the topological relations, as defined by Egenhofer, from the vector representation to the raster representation can be achieved using the conceptual transformation of the raster representation into the hybrid raster, as a complete topological model. This seems not practical, but a careful examination shows that no representation for the hybrid raster representation must be constructed, and the necessary parts can be computed on the fly from a regular raster representation.

The approach to specify in a functional language yields a semantically clear piece of code that can be run with test application to demonstrate the correctness in the investigated test cases. The systematic development and the application of standard methods of program simplification and optimization leads from a conceptually simple and correct formalization to efficient operations, which can be coded in various languages. For example, a translation into C++ took only few hours including testing. Differences between Gofer specification and C++ implementation concern the conceptual change to an algorithmic language, and some adaptions to specific efficiency properties fo the target language. It is interesting to compare the codes.

In the paper effects are not investigated that originate in resolution of vector-raster conversion. We do not claim that an operation on a pair of vector regions results in the same topological relation than applied on the rasterized regions. Instead we claim in this paper that the behavior of vector and raster representation can be assimilated, by extending the raster with its skeleton. So far, the paper contributes to the merging of raster and vector operations. With the use the same conceptual operations in both representations, the difference between both can be hidden in one more instance.

It is to expect that in principle the ideas are applicable to quad-trees, too. But one has to take care of neighboring quad-tree leaves of different size. The construction of that skeleton is open to formalization. Furthermore, translation of the given specifications into standard programming languages is open for further elaboration. Only then evidence can be given for time consumption of the algorithms. We expect that the requirements are not bad, because the complexity of the problem is $O(n * m)$ with the number $n * m$ of raster elements.

References

1. Bird, R., and P. Wadler, *Introduction to Functional Programming*, Series in Computer Science, Prentice Hall International, New York, 1988.
2. Bittner, T., and A. U. Frank, *On representing geometries of geographic space*, in 8th International Symposium on Spatial Data Handling, T. K. Poiker and N. Chrisman, eds., Vancouver, 1998, International Geographical Union, pp. 111–122. 2774.
3. Dorenbeck, C., and M. J. Egenhofer, *Algebraic optimization of combined overlay operations*, in Auto-Carto 10, ACSM-ASPRS, D. M. Mark and D. White, eds., Baltimore, 1991, ACSM-ASPRS, pp. 296–312.
4. Egenhofer, M. J., *Reasoning about binary topological relations*, in Advances in Spatial Databases (SSD '91), O. Günther and H.-J. Schek, eds., Springer, 1991, pp. 143–160.
5. Egenhofer, M. J., E. Clementini, and P. di Felice, *Topological relations between regions with holes*, International Journal of Geographical Information Systems, 8 (1994), pp. 129–142.
6. Egenhofer, M. J., A. U. Frank, and J. P. Jackson, *A topological data model for spatial databases*, in Design and Implementation of Large Spatial Databases,

A. Buchmann, O. Günther, T. R. Smith, and Y.-F. Wang, eds., vol. 409 of Lecture Notes in Computer Science, Springer, New York, 1989, pp. 271–286.

7. Egenhofer, M. J., and R. D. Franzosa, *Point-set topological spatial relations*, International Journal of Geographical Information Systems, 5 (1991), pp. 161–174.
8. Egenhofer, M. J., and J. R. Herring, *A mathematical framework for the definition of topological relationships*, in 4th International Symposium on Spatial Data Handling, Zürich, 1990, International Geographical Union, pp. 803–813.
9. Egenhofer, M. J., and J. Sharma, *Topological relations between regions in* $\mathbb{R}^2$ *and* $\mathbb{Z}^2$, in Advances in Spatial Databases, D. Abel and B. Ooi, eds., vol. 692 of Lecture Notes of Computer Science, Springer, New York, 1993, pp. 316–336.
10. Frank, A. U., and W. Kuhn, *A specification language for interoperable GIS*, in Interoperating Geographic Information Systems, M. F. Goodchild, M. Egenhofer, R. Fegeas, and C. Kottman, eds., Kluwer, Norwell, MA, to appear.
11. Frank, A. U., W. Kuhn, W. Hölbling, H. Schachinger, and P. Haunold, eds., *Gofer as used at GeoInfo/TU Vienna*, vol. 12 of GeoInfo Series, Dept. of Geoinformation, TU Vienna, Vienna, Austria, 1997.
12. Hernández, D., *Qualitative Representation of Spatial Knowledge*, Springer, Berlin, 1994.
13. Kong, T. Y., and A. Rosenfeld, *Digital topology: Introduction and survey*, CVGIP 48, (1989), pp. 357–393.
14. Kovalevsky, V. A., *Finite topology as applied to image analysis*, Computer Vision, Graphics, and Image Processing, 46 (1989), pp. 141–161.
15. Samet, H., *The Design and Analysis of Spatial Data Structures*, Addison-Wesley, Reading, MA, 1990.
16. Winter, S., *Topological relations between discrete regions*, in Advances in Spatial Databases, M. J. Egenhofer and J. R. Herring, eds., vol. 951 of Lecture Notes in Computer Science, Springer, 1995, pp. 310–327.
17. Winter, S., *Location-based similarity measures for regions*, in ISPRS Commission IV Symposium, Stuttgart, Germany, 1998.

3D Synthetic Environment Representation Using the "Non-Manifold 3D Winged-Edge" Data Structure *

Roy Ladner[1], Kevin Shaw[1], and Mahdi Abdelguerfi[2]

[1] Naval Research Laboratory, Stennis Space Center, Mississippi
{rladner,shaw}@nrlssc.navy.mil
[2] University of New Orleans, Computer Science Department, New Orleans, LA
mahdi@cs.uno.edu

Abstract. A Non-Manifold data structure for the modeling of 3D synthetic environments is proposed. The data structure uses a boundary representation (B-rep) method. B-rep models 3D objects by describing them in terms of their bounding entities and by topologically orienting them in a manner that enables the distinction between the object's interior and exterior. Consistent with B-rep, the representational scheme of the proposed data structure includes both topologic and geometric information. The topologic information encompasses the adjacencies involved in 3D manifold and non-manifold objects, and is described using a new, extended Winged-Edge data structure. This data structure is referred to as "Non-Manifold 3D Winged-Edge Topology". The time complexity of the newly introduced data structure is investigated. Additionally, the Non-Manifold 3D Winged-Edge Topology is being prototyped in a Web-Based virtual reality application. The prototype data consists of Military Operation in Urban Terrain (MOUT) data for Camp LeJeune, North Carolina. The application is expected to be ideal for training and simulation exercises as well as actual field operations requiring on-site assistance in urban areas.

1 Introduction

This paper describes the research into the extension of the National Imagery and Mapping Agency's (NIMA's) current Vector Product Format (VPF)[1] by the Naval Research Lab's Digital Mapping, Charting, and Geodesy Analysis Program (DMAP). This work has been carried on with the support from the Defense Modeling Simulation Office (DMS0) and NIMA's Terrain Modeling Project Office (TMPO).

* This work was sponsored by the Defense Modeling Simulation Office (DMSO) and the National Imagery and Mapping Agency's (NIMA) Terrain Modeling Project Office (TMPO), under Program Element 0603832D, with Jerry Lenczowski and Ron Magee as program managers. The views and conclusions contained in this paper are those of the authors and should not be considered as representing those of DMSO and NIMA.

VPF's Winged-Edge topology, in its current form, is documented as not being capable of modeling a wide range of three dimensional objects that may be encountered in an integrated three-dimensional synthetic environment. This class of objects includes non-manifold objects and objects which may be transmitted and received through the Synthetic Environment Data Representation and Interchange Specification (SEDRIS)[2]. DMAP therefore proposes VPF+, an extension to VPF that provides for georelational modeling in 3D (including non-manifold objects) and which would benefit the Modeling and Simulation (M&S) community.

2 Non-Manifold 3D Winged-Edge Topology

The data structure relationships of the Non-Manifold 3D Winged-Edge topology are summarized in the object model of Fig. 1. Figure 1 uses a modified version of Raumbaugh notation (shown in Fig. 5). References to geometry are omitted.

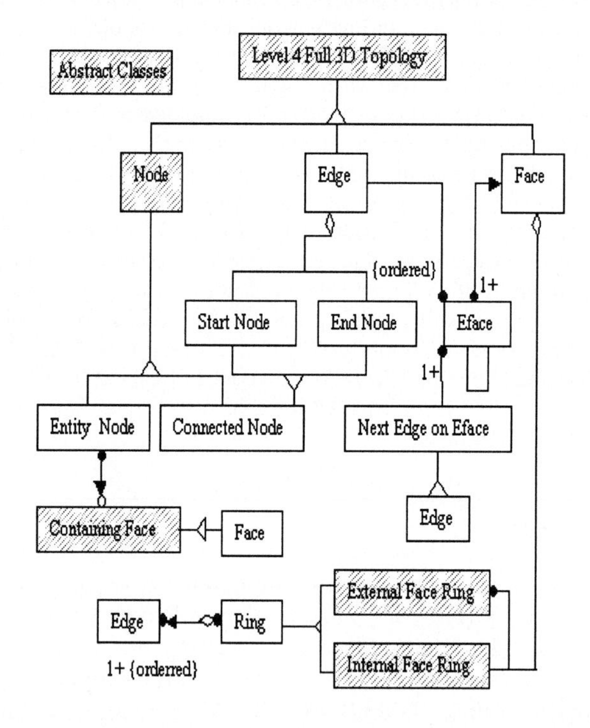

Fig. 1. Non-Manifold 3D Winged-Edge Topology Object Model.

2.1 Primitives

The main VPF+ primitives are:

– Entity node - used to represent isolated features.
– Connected node - used as endpoints to define edges.
– Edge - an arc used to represent linear features or borders of faces.
– Face - a two-dimensional primitive used to represent area features such as a lake.

Inside the primitive directory, a mandatory Minimum Bounding Box (MBB) table (not shown in Fig. 1 for clarity) will be associated with each edge and face primitive. Because of its simple shape, an MBB is easier to handle than its corresponding primitive. The primitives shown above have an optional spatial index. The spatial index is based on an adaptive grid-based 3D binary tree, which reduces searching for a primitive down to binary search. Due to its variable length records, the connected node table has a mandatory associated variable length index.

The ring table identifies the ring forming the outer boundary and all internal rings of a face primitive. This table (along with the face table) allows the extraction of all of the edges that form both the outer boundary and the internal rings of a face primitive.

The entity node and the external rings are not essential to the understanding of the 3D non-manifold data structure and will not be discussed further. For more information, the interested reader is referred to [1, 6].

The object model of Fig. 1 introduces a new structure called EFaces to resolve the ambiguities resulting from the absence of a fixed number of Faces adjacent to an Edge. The EFaces structure describes a use of a Face by an Edge and allows maintenance of the adjacency relationships between an Edge and zero, one, two or more Faces incident to an Edge. Each Connected_Node is related to one Edge in each manifold object to which the Node is attached and to each dangling Edge connected to the Node. As shown in Fig. 1, each Edge is related to its start and end Nodes and to its first and last EFaces. Each EFace is related to its Face, the Next_EFace in the ordered circular linked list of EFaces that the EFace is a member of, and to the Next_Edge_on_EFace. The Face in turn is linked to a Ring, which is related to its starting Edge.

2.2 Features

Traditional VPF defines five categories of cartographic features: Point, Line, Area, Complex and Text. Point, Line and Area features are classified as Simple Features, composed of only one type of primitive. Each Simple Feature is of differing dimensionality: zero, one and two for Point, Line and Area Features respectively. Unlike Simple Features, Complex Features can be of mixed dimensionality, and are obtained by combining Features of similar or differing dimension.

The object model of the feature level of VPF+ is shown in Fig. 2. VPF+ adds a new simple feature class of dimension three. The newly introduced feature, referred to as 3D Object Feature, is composed solely of Face primitives. This new feature class is aimed at capturing a wide range of 3D objects. The EFace Table is also added to the structural scheme. While the Ring Table provides a relationship between a Face and all the Edges that compose the Face's Rings, the EFace Table provides a relationship between an Edge and all the Faces that meet at that Edge.

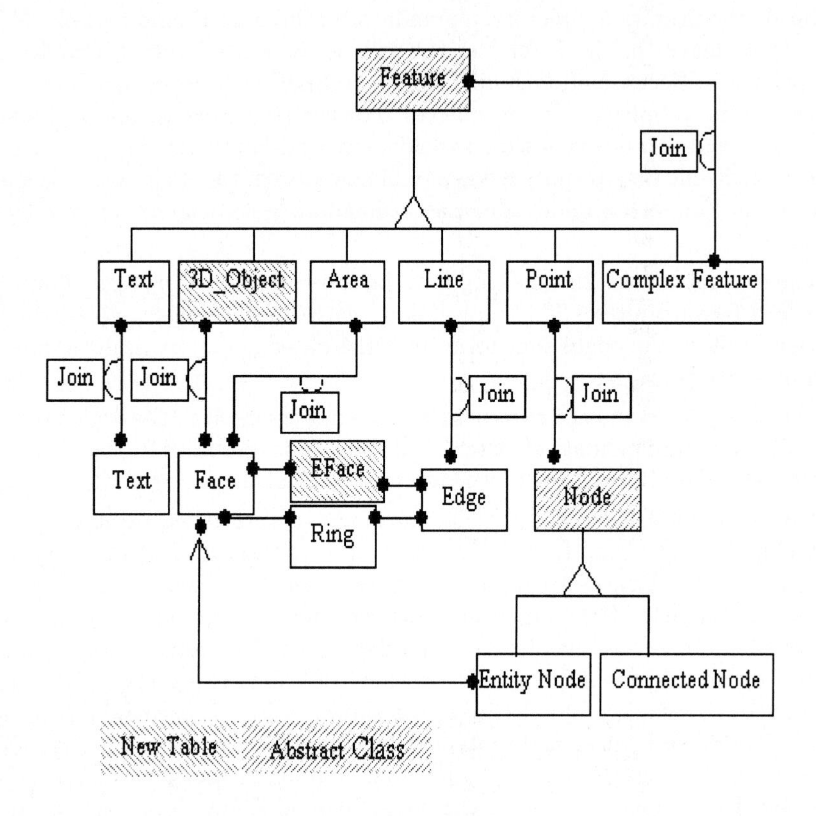

Fig. 2. VPF+ Feature Class Structural Schema

Although 3D Objects are restricted to primitives of one dimension, 3D Objects of mixed dimensionality can be modeled through Complex Features using Simple Features of similar or mixed dimensionality as building blocks.

3 Performance Analysis

A time performance analysis of the non-manifold winged-edge topology is performed in this section. The performance evaluation involves investigating the time complexity to implement nine access primitives. These access primitives, listed in Table 1, describe the retrieval of all topological adjacencies for each of the Face, Node and Edge primitives. A data structure that stores only a subset of all possible adjacency relations between primitives yet can satisfy queries to all nine access primitives is said to be topologically sufficient. Our Non-Manifold Winged-Edge data structure satisfies queries to all nine access primitives [6].

Table 1. The Nine Basic Access Primitives

Access Primitives (AP)	Description
AP1	Given face i find all n_i nodes around it
AP2	Given face i find all e_i edges around it
AP3	Given face i find f_i faces around it
AP4	Given node i find f_i faces around it
AP5	Given node i find all n_i nodes connected to it
AP6	Given node i find all e_i edges connected it
AP7	Given edge i find its two extreme vertices
AP8	Given edge i find all e_i edges connected to it
AP9	Given edge i find f_i faces intersecting it

The following notation will be used for the time complexity analysis:

$|APi|$ = Average time needed to implement access primitive APi.
K = Average time needed to access a row of an existing table using its primary key.
E_f = Average number of edges around a face.
E_n = Average number of edges around a node.
F_e = Average number of faces adjacent to an edge.
α = Average number of distinct objects connected to a node.

The performance study follows the methodology outlined in [3, 4 and 5]. A lower bound on the implementation of any access primitive is K. This is achieved when an item is retrieved using the table's primary key. For instance, given an edge id, the retrieval of its two associated entity nodes (access primitive AP7) requires an amount of time equal to K (referred to herein as constant access time). This is because each row in the edge table stores an edge id (primary key) and pointers to its two connected nodes. Therefore, using the edge table's primary key (edge id), the ids of the two associated connected nodes can be retrieved in an amount of time equal to K. The worst-case (upper bound) performance in the implementation of an access primitive occurs when a table is searched using a foreign key, which may require a linear scan of the whole table. Such a degradation of performance is never experienced by our non-manifold winged-edge data structure (see Table 3).

The average time complexity of implementing each access primitive is given in Table2. Table 3 (column 2) shows more explicit expressions of the time complexities.

Table 2. Average Time Complexity

Access Primitives (AP)	Description
AP1	$\|AP2\| + E_f K$
AP2	$F_e\ E_f\ K/2 + 3K$
AP3	$\|AP2\| + E_f\ \|AP9\|$
AP4	$\|AP6\| + E_n\ \|AP9\|$
AP5	$\|AP6\| + K\ E_n$
AP6	$K + \alpha\ (\ \|AP9\| + F_e\ \|AP2\|) + \alpha(F_e\ E_f\ \text{-}1)K$
AP7	K
AP8	$K + 2\ \|AP6\|$
AP9	$(F_e + 1)\ K$

Using the above values for α, F_e, E_f and E_n, average time complexities for the manifold case are derived in Table 3 (column 3). These expressions show that, in the manifold case, all nine access primitives can be performed in constant-time, on average.

For the manifold case, α=1 and F_e=2 (assuming the existence of a universal face). Additionally, in the manifold case, E_f and E_n have been shown in [4] to be such that:

$$E_f = E_n \leq 6(1 - 2(1 - G)/V). \qquad (1)$$

V and G represent the number of connected nodes and the number of holes respectively. The above expression implies that, in the manifold case, both E_f and E_n are approximately 6 on the average.

Table 3. Average Time Complexity: Manifold and Non-Manifold Cases

Access Primitives	**Average Time Complexity (Non-Manifold Case)**	**Average Time Complexity (Manifold Case)**
AP1	$(F_e\ E_f/2 + E_f + 3)K$	15K
AP2	$(F_e\ E_f/2 + 3)K$	9K
AP3	$(3F_e\ E_f/2 + E_f + 3)K$	27K
AP4	$K(\ 1 + \alpha(4F_e + F_e^2 E_f/2 + F_e E_f) + E_n(F_e + 1))$	51K
AP5	$K(1 + \alpha(4F_e + F_e^2\ E_f/2 + F_e\ E_f) + E_n)$	39K
AP6	$K + \alpha K(F_e^2\ E_f/2 + F_e\ E_f + 4\ F_e)$	33K
AP7	K	K
AP8	$3K + 2\alpha K(F_e^2\ E_f/2 + F_e\ E_f + 4\ F_e)$	67K
AP9	$(F_e + 1)\ K$	3K

4 The VPF+ Prototype

VPF+ is being prototyped in a Web-based virtual reality application for the Naval Research Laboratory at Stennis Space Center. In this application, a browser is used to select a user-defined extent of terrain and known features existing within that extent. The VPF+ database is queried and a 3D virtual world is generated using Virtual Reality Modeling Language (VRML). Functionality includes the ability to walk or fly through the terrain, move around objects, enter buildings, display floor plans, etc. The application should be ideal for training and simulation exercises as well as actual field operations requiring on-site assistance in urban areas.

The data flow in the creation of the VPF+ database is shown in Fig. 3. The prototyped data consists of Military Operations in Urban Terrain (MOUT) Data for Camp LeJeune, North Carolina. The elevation data is DTED Level 5, providing one meter elevation post spacing. The feature data consists of the coordinates (latitude and longitude) for the footprints of various buildings, the coordinates of the centerline of each of various roadways, and the coordinates of point features, all with descriptive attributes.

An interim elevation model was obtained using ArcInfo to produce a Triangulated Irregular Network (TIN) of the elevation data, using the lines forming the footprints of the feature buildings as constraints. The original terrain elevation data contained over 90,000 elevation points for an area of only approximately 600 meters square. TINning reduced the total elevation points to approximately 400, greatly improving performance. Since this geographic area is known to be relatively flat, the remaining elevation points were considered adequate to approximate the terrain. Using the buildings' footprints as constraints guaranteed the existence of nodes conforming to the coordinates of each of the footprints and also guaranteed a uniformly flat terrain under each of the buildings.

As a final preprocessing step, ArcInfo was used to convert the TIN into an ArcInfo Net file containing primitive data for all nodes, edges and faces in the terrain. Additionally, nodes, edges and faces were added for each building in the feature data, since only primitives for the buildings' footprints existed. Roads, previously existing only as line features defined by centerlines, were widened to their appropriate width, and nodes, edges and faces were added.

Library models were used to represent designated point features, such as lampposts and park benches. VPF+ data structures were then populated with all primitives and VPF+ topology using software tools specifically written for that purpose. Additional software tools were developed in Java to extract data from VPF+ tables for rendering into a 3D synthetic environment with VRML.

Although development of the prototype is not yet complete, sample virtual worlds have been tested. Fig. 4 below shows one such world including the terrain area and some of the buildings, roads and point features in the data set, but without application of textures. Each figure shows the same world, but from a different viewpoint. Included is the successful generation of buildings' interiors and the ability to "walk" into the building.

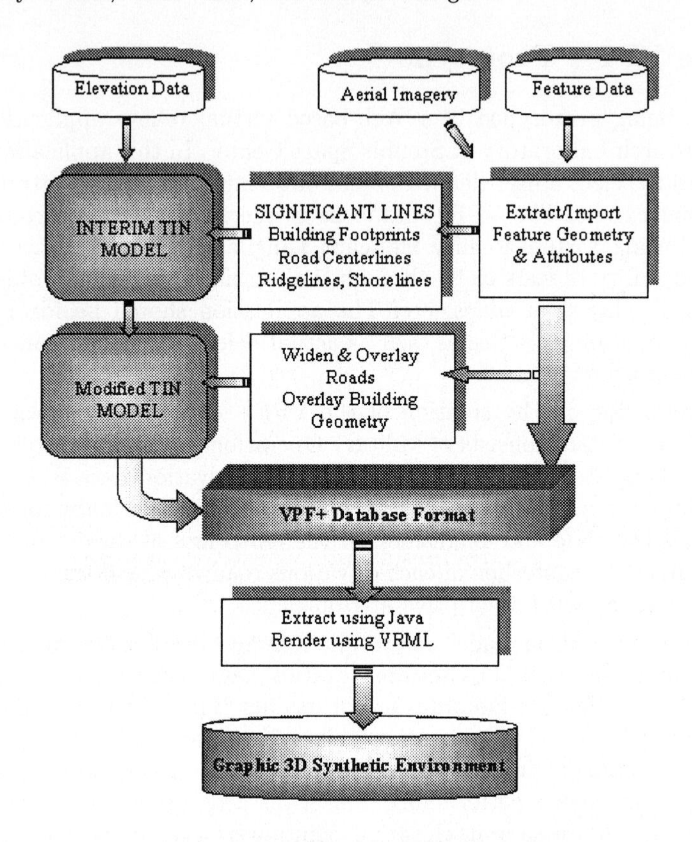

Fig. 3. Overview of Data Conversion and Rendering

References

1. Department of Defense, "Interface Standard for Vector Product Format", MIL-STD 2407, 28 June 1996.
2. SEDRIS Object Model, Release 1.02d.
3. Abdelguerfi, M., Cooper, E., Wynne, C., & Shaw, K., " An Extended Vector Product Format (EVPF) Suitable for the Representation of Three-Dimensional Elevation in Terrain Databases", International Journal of Geographical Information Science, Vol. 11, No.7, 1997, pp. 649-676.
4. Abdelguerfi, M., Cooper, E., Wynne, C., & Shaw, K., "A Terrain Database Representation Using an Extended Vector Product Format (EVPF)", Fifth International Conference on Information and Knowledge Management of Data", November 1996, Maryland, pp. 27-33.
5. Woo, T.C., "A Combinational Analysis of Boundary Data Structure Schemata", IEEE Computer Graphics and Applications, Vol.5, No.3, 1985, pp.19-27.
6. Abdelguerfi, M., Ladner, R., Shaw, K., Chung, M.J., Wilson, R., VPF+: A Vector Product Format Extension Suitable for Three-Dimensional Modeling and Simulation, Naval research Laboratory, Stennis Space center, MS, Technical Report: NRL/FR/7441-98-9683, 1998

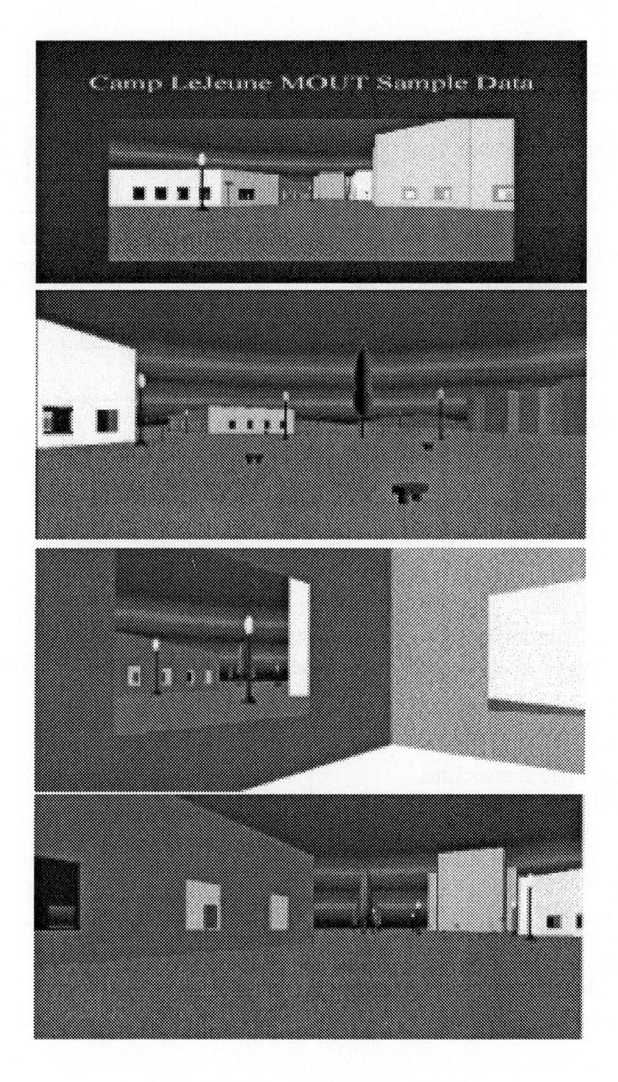

Fig. 4. Virtual World Samples

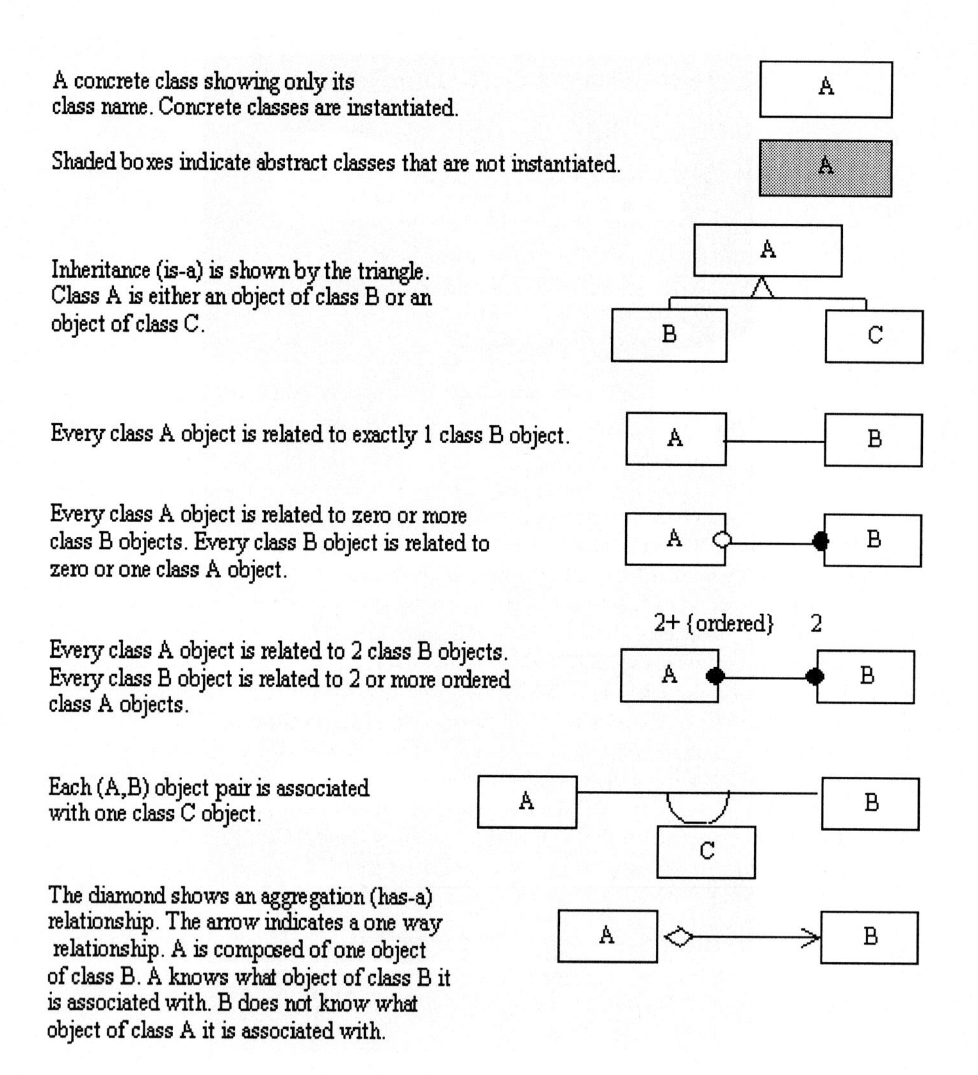

A concrete class showing only its
class name. Concrete classes are instantiated.
A
Shaded boxes indicate abstract classes that are not instantiated.
A
Inheritance (is-a) is shown by the triangle.
Class A is either an object of class B or an
object of class C.
A
B
C
Every class A object is related to exactly 1 class B object.
A
B
Every class A object is related to zero or more
class B objects. Every class B object is related to
zero or one class A object.
A
B
2+ {ordered}
2
Every class A object is related to 2 class B objects.
Every class B object is related to 2 or more ordered
class A objects.
A
B
Each (A,B) object pair is associated
with one class C object.
A
B
C
The diamond shows an aggregation (has-a)
relationship. The arrow indicates a one way
relationship. A is composed of one object
of class B. A knows what object of class B it
is associated with. B does not know what
object of class A it is associated with.
A
B

Fig. 5. Appendix A. Modified Rambaugh Notation

Pluggable Terrain Module – Moving Digital Terrain Modelling to a Distributed Geoprocessing Environment

Daria Martinoni, Bernhard Schneider

Department of Geography
University of Zuerich
Winterthurerstrasse 190
CH-8057 Zuerich
Phone +41 1 635 52 56
Fax +41 1 635 68 48
{dariam, benni}@geo.unizh.ch
http://www.geo.unizh.ch/{~dariam, ~benni}

Abstract. Reliable digital terrain modelling necessarily needs to be based upon a set of rules which first prescribe how to discretise the continuous terrain surface, and second define the assumptions determining the subsequent interpolation. Unresolved problems in automated DTM analysis are mostly due to a failure of today's approach to fully appreciate the implications of these rules. The functional limitations imposed by monolithic GISs are identified as a major reason for this failure. The Pluggable Terrain Module (PTM) presented in this paper is proposed to overcome these drawbacks by using a modular approach. The basic idea is to move the processing of terrain information from GISs to the PTM; communication with other software components is specified by a set of interfaces, shifting terrain modelling towards distributed and interoperable geoprocessing. Based on the notions of the OpenGIS Geodata Model (OGM), the PTM design is proposed and its implications and benefits to terrain modelling are discussed.

1 Motivation – Digital Terrain Modelling until Now

Because the terrain surface is of great influence for most environmental processes and many human activities, digital terrain models (DTMs) provide an important basis for many types of analysis within GIS. Likewise, DTM analysis and characterisation is one of the major tasks of GIS applications.

Despite the improvements in automated DTM analysis made over the past years, there are a number of, as yet, unresolved problems. These are based upon both, conceptual problems which result when considering DTMs as models of a continuous surface, and drawbacks of today's approach to modelling within GIS.

1.1 Conceptual Problems when Using DTMs as Basis for Spatial Modelling

First, despite being models of continuous surface form, DTMs are mostly created from discrete elevation values. This causes conceptual difficulties because proper and logically consistent DTM analysis requires continuous surface representation.[1]

A second conceptual problem is the question what each sample data element represents within the model. This problem focuses on semantic uncertainties arising from questions such as:

- What is the area represented by each sample data element, e. g. does a given gridded elevation matrix represent a point grid or a cell grid, or are there any forms of implicit interpolation?
- Are any additional assumptions made about the spatial relationships between recorded sample values, e.g. are irregularly distributed points related to each other only based on proximity?
- Which is the information content of each sample data element, e. g. does it carry only height information or also further information such as representing a peak, a pass, or part of a slope break line?

Third, DTMs implicitly model at a certain scale. Hence, the information derived is relevant to the scale implied by the model. Since this scale is often arbitrary and not necessarily related to the scale of analysis, derived results may not always be appropriate.[6]

Finally, although the importance of quality information to spatial modelling and decision making is theoretically well known, most DTM applications neglect the management of error and uncertainties inherent in the model.

1.2 Drawbacks of Today's Approach to Modelling Within GIS

Currently, most of the available GISs provide certain data models and functionality to support the implementation of a spatial modelling task. Since GISs can not be infinite in complexity, they offer a limited selection of functionality that is considered to be a flexible and comprehensive tool set. Thus, the translation of a spatial model into a respective digital representation is limited by the information system's capabilities. It is not surprising that modelling mostly occurs with the help of few but common functions. Consequently, the phenomena to be modelled have a much lower influence on the choice of the functions than the available GIS and its limitations. Since the limitations of monolithic GISs hinder fully considering the conceptual problems mentioned above, appropriate environmental modelling with available GISs is a difficult task.

[1] Such difficulties may be interpreted as communication problems between different information communities (data producer and data user communities) who describe terrain – or, to be more general, geographic information – in different ways.[1]

1.3 Modular Approach for Digital Terrain Representation and Processing

Since monolithic GISs can not provide all the functionality needed to perform sound applications of DTMs, it is logical to process the digital terrain representation outside the GIS. This calls for a modular approach. For such an approach, the so-called Pluggable Computing Model as proposed by the OpenGIS Guide [1] forms an excellent conceptual basis. In this concept, Tool Services, implemented as so-called Pluggable Tools, provide sets of functionalities to fulfil specific tasks. Well-defined and commonly known and accepted interfaces allow exchanging of messages and data objects with other Tool Services or other software components.

In this paper, a modular approach to process terrain information is proposed which overcomes the conceptual problems mentioned above. Being based on the concept of Pluggable Tools, our approach follows the notions of distributed and interoperable geoprocessing. First, an essential model for digital terrain models is proposed, followed by a discussion of its implications in the context of environmental modelling. After these preliminary considerations, the Pluggable Terrain Module (PTM) is presented. It is shown how its concept derives from the Virtual Data Set (VDS, [4]). The internal design of the module is explained, its benefits are highlighted.

2 Essential Model

The purpose of a DTM is to provide *a sound digital description of a portion of the earth's surface.* Based on this understanding of digital terrain representations, we suggest that the essential model[2] of a DTM consist of the triple (D, $\{rules\}$, V) [5] where

- D denotes the spatial extent of the area of interest (i.e. the *portion*),
- $\{rules\}$ is a set of rules which define the information content of the sample data set and the assumptions made about the true surface (i.e. *sound description*),
- V is the range of the terrain representation.

D is a usually rectangular area of the earth's – or another planet's – surface, given in some geographic reference system. Except for the task of joining two or more different terrain models, D does not cause conceptual problems.

The same holds true for the value type of V, because as long as the modelling of errors and uncertainties is neglected elevation can be represented as a simple scalar. However, the value type of V becomes much more complex (e.g. intervals,

[2] According to the object-oriented modelling approach proposed by [2], the term 'essential model' denotes model design at the highest level of abstraction; the essential model provides a description of some real or imaginary situation by purposefully considering entities and phenomena situated into a context, rather than by trying to describe all of some supposedly objective reality.

tuples, or probability functions) when quality issues are introduced. Since this paper focuses on the integration of quality information in terrain representation and analysis, and not on the assessment and representation of quality itself, a broad discussion of V is beyond the scope of this paper.

The set of *rules* determines how the data that support the terrain representation have to be sampled and arranged and how elevation values and other topographic information is to be derived, i.e. how the actual terrain surface representation is to be built upon the data. These tasks are often considered to be mere questions of surveying techniques and instruments, sampling density, and algorithmic and computational efficiency. The conceptual problems mentioned above, though being inherent to the *rules*, are less frequently discussed and very rarely taken into account in DTM applications. Especially the effects from the absence of appropriate assumptions are barely perceived. Therefore, the *rules* need further discussion.

As terrain is a continuous surface form, its appropriate representation by digital means necessarily involves two steps:

- *Discretisation*, when sampling the data,
- simulation of continuity by *interpolation*.

In the context of digital terrain modelling, each of these two steps raises a question that seems simple, but that is of substantial importance:

- *How should the terrain surface be discretized?*
  This question asks about the information content of the sample data set. Should the sample consist only of elevation measurements, for example, or should it include further terrain-specific information, such as rivers, breaks in slope lines, etc.? And second, what should its modelling scale be?
  The discretisation has serious implications for the features represented by the DTM and their levels of detail, because information which is not explicitly indicated by the sample data can not be modelled (nor extracted) reliably (Fig. 1).
  Though measurement is a crucial source of uncertainty, a review of the actual data collection (in the literal sense of measurement of the sample data) is beyond the scope of this paper (for a detailed discussion of this task see [5]).
- *How should the surface be interpolated?*
  As the real terrain surface is not a mathematical surface, we are forced to make *assumptions* about the true surface. Reasonable assumptions could be that the surface must be G1-continuous[3] except where break lines occur or that the surface must not contain sinks where not explicitly indicated.
  A direct consequence of interpolating based on assumptions is that the resulting DTM bears distinct properties. In the above example, one property of the resulting surfaces would be that a derivative exists for every point on the surface (expect at break lines), or, in the latter case, that the resulting model is hydrologically sound.

[3] 'Geometric continuity': for every point of the surface one and only one tangent plane exists.

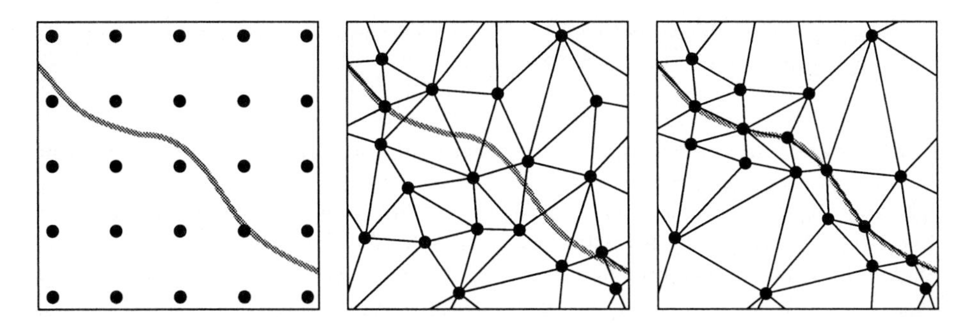

Fig. 1. Reliable modelling of a topographic feature. The rigid structure of the regular point raster is not able to model the stream as reliably as irregularly distributed and triangulated points. However, the triangulated irregular network (TIN) does not guarantee an appropriate modelling; smart placement of the data points as well as a triangulation that follows geomorphologic constraints are essential.

As assumptions are, by their nature, uncertain to some degree, further uncertainty is introduced in this step. This uncertainty is (mostly) independent from the sample data and may be denoted as 'semantic uncertainty'.

3 Implications on Modelling Terrain Surfaces

3.1 Information Content of Sample Data Set

The base data (consisting of geometric, semantic, and quality information) implies the information content as well as the scale of the model (where 'scale' refers to level of detail):

- The *information content* has a direct influence on the reliability of the model. First, the more information the modelling is based upon, the less vague the result of the interpolation step, namely the shape of the digital terrain surface, will be (where 'more information' does not refer to an increased number of samples, but to more meaningful data samples). Second, as mentioned above, reliable feature extraction means that there must be some evidence in the base data for the results. Only topographic information that is contained by the base data can be made derivable with the help of interpolation and extraction methods.
- Extraction of *scale dependent* topographic information requires that the 'scale of analysis' agrees with the 'scale implied by the model'. If for instance a spatial model calls for sinks, the term 'sink' will always be (explicitly or implicitly) defined at a certain scale, i.e. a sink must have a certain minimal size to be considered. Consequently, the terrain model must contain all sinks of that minimal size, and the extraction methods must be aware that

smaller sinks – if represented by the terrain model – are to be ignored, i.e. generalised.[4]

3.2 Interpolation Scheme – Properties of the Terrain Representation

The digital terrain surface generated by the interpolation has distinct properties. Digital terrain models are abstractions of the true terrain – not the true terrain itself. Therefore, to avoid semantic ambiguities, the interpolation schemes together with the underlying assumptions must be explicitly stated.

It is argued that such *semantic ambiguities* may lead to choosing inappropriate terrain analysis methods. Take hydrologic modelling as an example: Current algorithms for automated drainage network extraction implicitly assume an idealised water discharge of zero volume. This assumption will work well if the DTM used is hydrologically sound. If this is not the case, for instance if small sinks occur, the idealised approach is inappropriate. The extraction will yield interrupted and disconnected watercourses unless the approach is adjusted, that is by making different assumptions about the rate of stream flow. Hence, an explicitly stated interpolation scheme allows the comparison of the spatial modelling approaches with the properties of the DTM to ensure that the terrain representation supports the selected approaches.

3.3 Quality (in the sense of 'fitness for use')

Quality does not only express metric accuracy. It determines the usability of the abstraction underlying a digital terrain model by specifying for instance:

- completeness: Does the model represent all required topographic information?
- currency: Are base data and model current (though generally constant, terrain may change quickly in specific environments, for instance glaciers and dunes)?
- attribute accuracy: Do the base data represent what they are supposed to?

When terrain models are applied in spatial models, a major focus of quality reports is to express uncertainty. Uncertainty is inherent to all the data (persistent and virtual). Additionally, uncertainty is introduced through inappropriate usage of data, methods, and models. To minimise uncertainties introduced within the scope of a spatial modelling project, the following requirements must be fulfilled:

- The terrain model has to be suitable for the intended modelling (i.e. all the required topographic information has to be appropriately represented by the DTM).

[4] From the latter point, it follows that a denser sampling of base data does not necessarily make the results of the analysis more reliable.

- Information about properties and quality of the terrain model have to be available independently from the actual topographic data.
- The user of the DTM has to understand the terrain model, i.e. to know the abstraction as well as the related properties of the model.

4 Pluggable Terrain Module (PTM)

The above discussion leads to the following list of requirements to terrain modelling:

- The real terrain must be appropriately represented, i.e. as a continuous surface,
- quality information must be available,
- metadata to avoid semantic ambiguities must be available,
- appropriate modelling tools to derive topographic information must be available.

As already stated, these requirements can not be fully fulfilled when modelling terrain within a monolithic GIS. We therefore propose to perform terrain modelling and analysis within a module outside the GIS. This module holds the actual terrain data as well as comprehensive information about quality and semantics; it simulates the continuous surface with the help of sound modelling functionality, and makes topographic and meta-information available through a well-defined interface. In this way, the DTM becomes an autonomous module that is able to communicate with GISs.

The proposed module simulates, on the one hand, a complete continuous surface with the help of base data and appropriate modelling methods, both hidden inside the module. On the other hand, it makes all derivable information available via an interface. Hence, it is in accordance with the notion of the Virtual Data Set[5] (VDS) as presented by [4]. The design of the module, described in detail in the following sections, builds upon the OpenGIS Data Model (OGM, [1]). Its function in the context of interoperable geodata processing coincides with the Pluggable Tool in the Pluggable Computing Model environment [1] – thus the name *Pluggable Terrain Module* (PTM).

[5] The basic idea is to extend sample data with methods to provide any derivable or predictable information. Instead of transforming original data to a standard format and storing them, the original data are enhanced with persistent (i. e. explicit) methods that only will be executed upon request.[4] In other words, the data exchange is not specified by a standardised data structure (e. g., a physical file format) but by a set of interfaces. These interfaces provide data access methods to retrieve the actual data values contained in a data set or a query result. An application that uses a VDS does therefore not 'read' the data from a physical file (or query database), but will call a set of corresponding methods defined in the VDS which return the data requested.[5] As its name indicates, a virtual data set contains virtual data. Virtual data is information which is not physically present, that is, which is not persistent. This data is computed upon request at run time.[4]

4.1 Internal Design

The PTM consists of 3 components (see Fig. 2):

- the Property Set,
- the Specification Schema,
- and the Identification Schema.

Property Set. The terrain module is basically a collection of identified properties. *Geospatial properties* consist of geometry and semantic, the latter expressed through attributes. To each geospatial property one or more *quality properties* are attached.

- *Geospatial Properties*:
  Geospatial properties represent the (geospatial) phenomena and entities modelled by the module. Geospatial properties may be persistently stored data – in which case we speak of *supporting geospatial properties* – or persistently stored methods[6], i.e. virtual data in the sense that it is derived upon request – therefore *virtual geospatial properties.*
- *Quality Properties*:
  Quality property tuples describe quality aspects of the terrain module. They may be related to single geospatial properties, to groups of geospatial properties, or to the whole terrain model. Quality properties may be persistently stored (if related to persistent data), or virtual, i.e. persistently stored methods for assessing the respective quality property (if related to virtual data).

Specification Schema. The *specification schema* exposes the exact specifications of all data elements and methods (i.e. of all properties) constituting the module. The schema itself has three components:

- A *Geometry Schema* which exactly specifies all the geometric structures used, as well as the spatial reference system(s) referred to.
- An *Attribute Schema* which exactly specifies all properties.
  Each geospatial property is specified with its exact property name – in the case of virtual data together with the name of the corresponding method –, its value type – which in case of virtual data is at the same time the return value type of the respective method –, and its quality properties.
  The quality properties are specified with the exact property names and value types. Specification must also be provided for the method to be used for quality assessment and the reference system to be referred to.
- A *Method Schema* which contains specifications of all the methods used, including all the methods involved in the generation of virtual data and methods for quality assessment.

[6] These persistent methods basically correspond to what is denoted by 'stored functions' from [1].

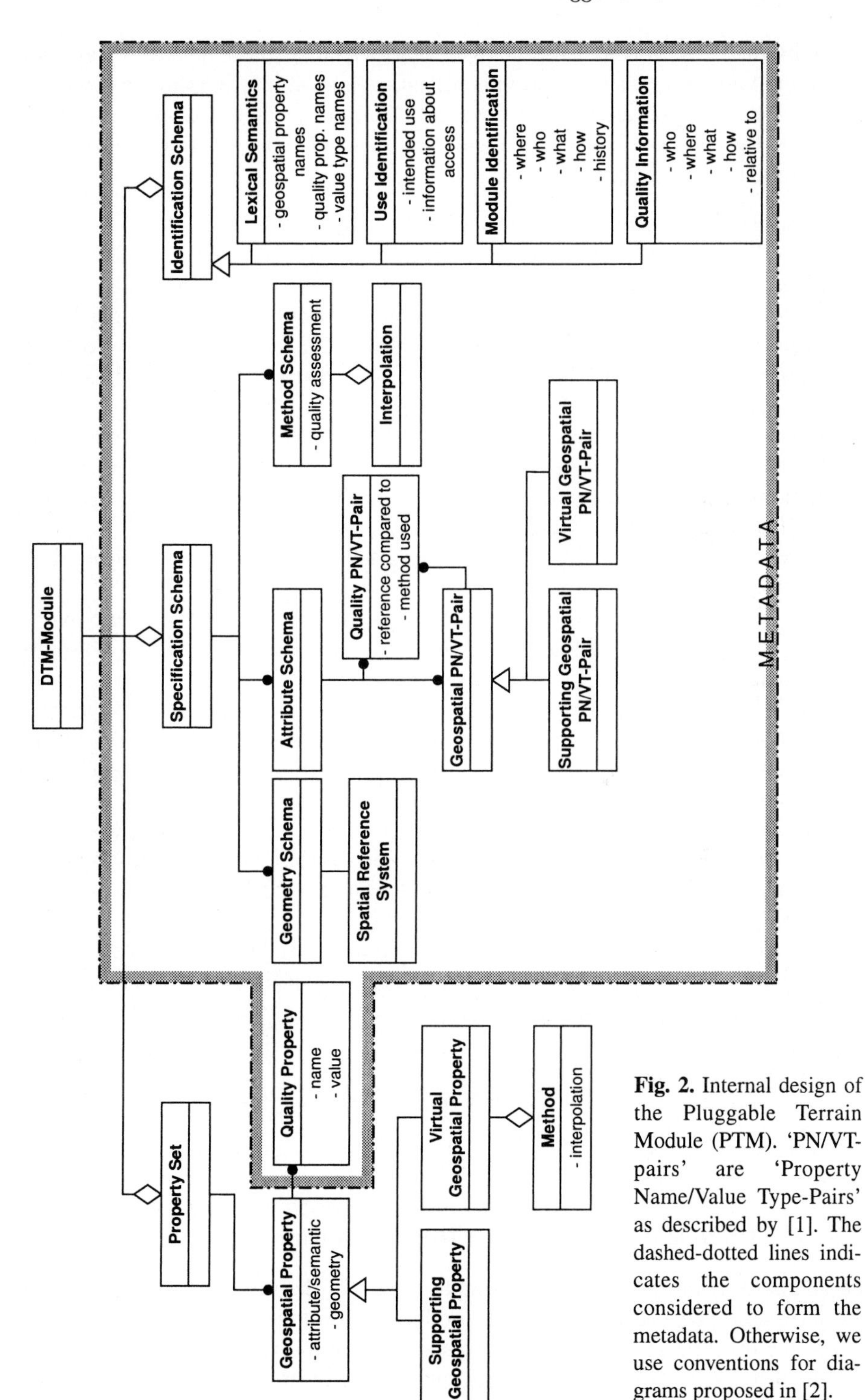

Fig. 2. Internal design of the Pluggable Terrain Module (PTM). 'PN/VT-pairs' are 'Property Name/Value Type-Pairs' as described by [1]. The dashed-dotted lines indicates the components considered to form the metadata. Otherwise, we use conventions for diagrams proposed in [2].

Identification Schema. The *Identification Schema* is basically an explanation of the module. It serves to avoid misunderstandings and to help understand the intended meaning of the terrain representation, i.e. what topographic information does the module contain, and which it does not. The identification schema itself consists of four parts:

- *Lexical Identification* ('Lexical Semantics', [1])
  Contains the dictionary of terms used in the specification schema. The lexical identification must provide sufficiently rich definitions and examples so that there will be no ambiguity concerning the meaning of the terms used. Put in more concrete terms, the lexical identification must explain all property names (of both geospatial and quality properties) and all value types.
- *Use Identification* ('Use Semantics', [1])
  Provides details on how to exploit the PTM. Included are the intended uses for the information as the terrain model was generated (if any) as well as options and limitations for obtaining the module.
- *Module Identification* ('Project Semantics', [1])
  Describes how the DTM was conceptualised and how it was generated. It must contain information concerning:
  - *Where*: What is the region of the earth covered (i.e. physical extent)?
  - *Who* was responsible for data collection and DTM generation, who distributes the data.
  - *What* phenomena are modelled in the terrain module, and what is their approximate scale. The underlying assumptions and conceptualisations must be included. This item explains the geospatial properties of the terrain model.
  - *How* were the data acquired, and how was the terrain model generated?
  - *History*: When did data collection and terrain model generation take place? What transformations occurred, and when?
- *Quality Identification*
  Detailed description of the quality information available about the terrain model. This serves to catalogue the quality information available and helps to avoid misleading interpretations. Included must be the following information:
  - *Who* was responsible for the quality assessment.
  - *When* was quality investigated?
  - *What* are the quality property names and value types?
  - *How* was quality assessed?
  - What is the reference system *referred to*?

4.2 Continuous Terrain Surface Representation

The actual terrain representation is realised within the property set based upon the geospatial properties. One way to model a continuous surface is, for instance, to cluster or arrange the supporting geospatial properties (that is, the persistently stored sample data) into a pattern of non-overlapping geometries. This step is usually denoted as 'tesselation'. For each tile of the tesselation, elevation is then modelled by means of an explicit and persistent interpolation scheme. While the sample data define elevations only for the supporting geospatial properties, the domain of the model is extended to the entire tesselation tiles by interpolation.

Being part of the necessarily explicit interpolation scheme, the tesselation is likely to be stored persistently. It is, however, also possible to compute the tesselation only upon request, that is, that the tesselation shows virtual behaviour. The decision, whether the tesselation shall be persistent or not is left to the implementation (and, in the end, is likely to result as a matter of efficiency).

Interpolation does not necessarily need to be preceded by tesselation, as may seem suggested by the above example. However, the point is, that the proposed approach allows the terrain surface to be specified directly from the sample data, thereby avoiding lossy transformations of the original data.

4.3 PTM Query

Being based on the notion of Virtual Data Sets [4], the main idea underlying the approach taken here is that the query and exchange of derivable terrain data is not specified by standardised data formats but by a set of interfaces. These interfaces provide the methods necessary to retrieve the actual information represented by the terrain model. For this purpose, a PTM needs to compute – if the requested information is not persistent – and to expose its geospatial property values at locations specified by applications. This requires the specification of persistent methods which, upon request, compute the virtual geospatial properties. The proposed PTM design fulfils this requirement by means of the method schema which is part of the specification schema. (Consequently, the method schema contains the specifications of the query interface).

Therefore, PTMs lend themselves to implementation within distributed computing environments as objects or services which can be queried by applications requesting terrain information

5 Conclusions

5.1 Benefits of the Proposed Approach

- The PTM presents the abstraction of the true terrain as a continuous surface.
- With the terrain surface being specified directly from the sample data, transformations of the original data set are avoided, thereby preventing that information is lost and uncertainty introduced by (unnecessary) transformations.
- The assumptions the terrain representation is implicitly based upon are made explicit by means of the persistent methods. Likewise, metadata is provided to detail the structures and terms used. Together, this information essentially contributes to prevent ambiguities in data interpretation.
- The major uses of metadata are to help the assessment of fitness for use and to provide the information needed to acquire, process and properly interpret data. The proposed PTM design allows the metadata to be separately

available. However, this requires that the persistent methods are designed to provide quality information for each derived geospatial property.
- The PTM basically includes the topics of metadata identified in the 'Content Standard for Digital Geospatial Metadata'.[3]
- Unlike DTMs in a conventional sense, a PTM comprehends more than the terrain surface description itself. Based on persistent methods, it also represents derivable terrain properties. By embedding the functionality needed for terrain analysis, a PTM provides an application with proper tools[7] and, thereby, prevents it from choosing inappropriate approaches to handle the terrain information.
- Being themselves geospatial properties – virtual geospatial properties – the methods provided for terrain analysis, or strictly speaking their results, are necessarily accompanied by quality properties. Therefore, at least on a design level, no terrain information should be derivable without corresponding quality descriptions.
- A PTM may provide several different terrain representations for a single sample data set. That is, different abstractions of the true terrain surface can be simultaniously realised (multiple terrain representations).
- The notion underlying a PTM implies a shift of responsability from data users to data producers.[5] It is argued that the expert knowledge forming a prerequisite to terrain representation and analysis is available more in the data producer domain than in the data user communities. It therefore seems obvious to leave the specification of such tasks to terrain modelling experts (which, following the above argumentation, are likely to be part of the data producer domain). Data users on the other hand get the terrain information they need by PTM query through well-defined interfaces. Thus, they can concentrate on their application and need not be aware of low-level details of the actual implementation. However, they can query a PTM for such information.
- Of course, it makes no sense to have data producers write the methods constituting a PTM from scratch for every implementation. Instead, predefined methods, e.g. from a library, could be used and specialised.[8] In that sense, a PTM offers an efficient implementation approach based on reusable software components.

5.2 Drawbacks of the Approach

- As quality properties are added to the geospatial properties, the data volume tends to be multiplied.
- Computational and communication costs are substantial and may heavily impact runtime efficiency.

---

[7] That is, the actual analysis happens inside the PTM. The derived results can be quieried by the GIS trough well-defined interfaces.

[8] For persistent methods to be shared, they need to be encapsulated (or, in terms of OpenGIS, to become well known, and their parameters to become Well-Known Structures (WKS, [1])).

6 Outlook

The modular approach where information is computed on request from inside the module offers the usage of *symbolic computation*. Symbolic computation would, for example, allow the reduction of the numeric error introduced during computation. However, as pointed out by [5], symbolic computation is still available only within specialized software packages such as Maple or Mathematica.

With regard to full interoperability, technical questions need further discussion. Application of the module presented in an interoperable environment makes sense only if the PTM can be transferred to the client and executed in the client's context.[5]

Future research will concentrate on the implementation of a prototype PTM. Hereto, many unresolved problems, mainly concerning the representation and analysis of continuous terrain surfaces as well as the assessment and management of quality information, must be addressed. Research efforts in these directions, especially in the field of quality management, are currently underway at the Geographic Institute at the University of Zurich.

7 Acknowledgments

This research project has partly been funded by the Swiss National Science Foundation under contract No. 20-47153.96. We would also like to thank Frank Brazile for editing the English.

References

1. Buehler, K., McKee, L.: The OpenGIS Guide. OpenGIS Consortium, Inc., Wayland, Massachussetts, OGIS TC Document 96-001 (1996)
   Available at: `http://www.opengis.org/techno/guide.html`
2. Cook, S., Daniels, J.: Designing Object Systems – Object-oriented Modelling with Syntropy. Prentice Hall, New York, London (1994)
3. Federal Geographic Data Committee: Content Standard for Digital Geospatial Metadata (revised June 1998). Federal Geographic Data Committee, Washington, D. C. (1998)
4. Stephan, E.-M., Vckovski, A., Bucher, F.: Virtual Data Set – An Approach for the Integration of Incompatible Data. Proceedings of the AUTOCARTO 11 Conference (1993) 93–102
5. Vckovski, A.: Interoperable and Distributed Geoprocessing. Taylor and Francis, London, New York (1998)
6. Wood, J.: Scale-Based Characterisation of Digital Elevation Models. In: Parker, D. (eds.): Innovations in GIS 3, Taylor and Francis, London (1996) 163–175
7. Wood, J.: Activating Scale-Based Uncertainty in Digital Elevation Models. Paper presented at the First Cassini Workshop on Data Quality in Geographic Information, April 21st – 23rd 1997, Paris (1997)
   Available at: `http://www.geog.le.ac.uk/jwo/research/conferences/Paris97/index.html`

Author Index

Lecture Notes in Computer Science

For information about Vols. 1–1488
please contact your bookseller or Springer-Verlag

Vol. 1489: J. Dix, L. Fariñas del Cerro, U. Furbach (Eds.), Logics in Artificial Intelligence. Proceedings, 1998. X, 391 pages. 1998. (Subseries LNAI).

Vol. 1490: C. Palamidessi, H. Glaser, K. Meinke (Eds.), Principles of Declarative Programming. Proceedings, 1998. XI, 497 pages. 1998.

Vol. 1491: W. Reisig, G. Rozenberg (Eds.), Lectures on Petri Nets I: Basic Models. XII, 683 pages. 1998.

Vol. 1492: W. Reisig, G. Rozenberg (Eds.), Lectures on Petri Nets II: Applications. XII, 479 pages. 1998.

Vol. 1493: J.P. Bowen, A. Fett, M.G. Hinchey (Eds.), ZUM '98: The Z Formal Specification Notation. Proceedings, 1998. XV, 417 pages. 1998.

Vol. 1494: G. Rozenberg, F. Vaandrager (Eds.), Lectures on Embedded Systems. Proceedings, 1996. VIII, 423 pages. 1998.

Vol. 1495: T. Andreasen, H. Christiansen, H.L. Larsen (Eds.), Flexible Query Answering Systems. IX, 393 pages. 1998. (Subseries LNAI).

Vol. 1496: W.M. Wells, A. Colchester, S. Delp (Eds.), Medical Image Computing and Computer-Assisted Intervention – MICCAI'98. Proceedings, 1998. XXII, 1256 pages. 1998.

Vol. 1497: V. Alexandrov, J. Dongarra (Eds.), Recent Advances in Parallel Virtual Machine and Message Passing Interface. Proceedings, 1998. XII, 412 pages. 1998.

Vol. 1498: A.E. Eiben, T. Bäck, M. Schoenauer, H.-P. Schwefel (Eds.), Parallel Problem Solving from Nature – PPSN V. Proceedings, 1998. XXIII, 1041 pages. 1998.

Vol. 1499: S. Kutten (Ed.), Distributed Computing. Proceedings, 1998. XII, 419 pages. 1998.

Vol. 1500: J.-C. Derniame, B.A. Kaba, D. Wastell (Eds.), Software Process: Principles, Methodology, and Technology. XIII, 307 pages. 1999.

Vol. 1501: M.M. Richter, C.H. Smith, R. Wiehagen, T. Zeugmann (Eds.), Algorithmic Learning Theory. Proceedings, 1998. XI, 439 pages. 1998. (Subseries LNAI).

Vol. 1502: G. Antoniou, J. Slaney (Eds.), Advanced Topics in Artificial Intelligence. Proceedings, 1998. XI, 333 pages. 1998. (Subseries LNAI).

Vol. 1503: G. Levi (Ed.), Static Analysis. Proceedings, 1998. IX, 383 pages. 1998.

Vol. 1504: O. Herzog, A. Günter (Eds.), KI-98: Advances in Artificial Intelligence. Proceedings, 1998. XI, 355 pages. 1998. (Subseries LNAI).

Vol. 1505: D. Caromel, R.R. Oldehoeft, M. Tholburn (Eds.), Computing in Object-Oriented Parallel Environments. Proceedings, 1998. XI, 243 pages. 1998.

Vol. 1506: R. Koch, L. Van Gool (Eds.), 3D Structure from Multiple Images of Large-Scale Environments. Proceedings, 1998. VIII, 347 pages. 1998.

Vol. 1507: T.W. Ling, S. Ram, M.L. Lee (Eds.), Conceptual Modeling – ER '98. Proceedings, 1998. XVI, 482 pages. 1998.

Vol. 1508: S. Jajodia, M.T. Özsu, A. Dogac (Eds.), Advances in Multimedia Information Systems. Proceedings, 1998. VIII, 207 pages. 1998.

Vol. 1510: J.M. Zytkow, M. Quafafou (Eds.), Principles of Data Mining and Knowledge Discovery. Proceedings, 1998. XI, 482 pages. 1998. (Subseries LNAI).

Vol. 1511: D. O'Hallaron (Ed.), Languages, Compilers, and Run-Time Systems for Scalable Computers. Proceedings, 1998. IX, 412 pages. 1998.

Vol. 1512: E. Giménez, C. Paulin-Mohring (Eds.), Types for Proofs and Programs. Proceedings, 1996. VIII, 373 pages. 1998.

Vol. 1513: C. Nikolaou, C. Stephanidis (Eds.), Research and Advanced Technology for Digital Libraries. Proceedings, 1998. XV, 912 pages. 1998.

Vol. 1514: K. Ohta, D. Pei (Eds.), Advances in Cryptology – ASIACRYPT'98. Proceedings, 1998. XII, 436 pages. 1998.

Vol. 1515: F. Moreira de Oliveira (Ed.), Advances in Artificial Intelligence. Proceedings, 1998. X, 259 pages. 1998. (Subseries LNAI).

Vol. 1516: W. Ehrenberger (Ed.), Computer Safety, Reliability and Security. Proceedings, 1998. XVI, 392 pages. 1998.

Vol. 1517: J. Hromkovič, O. Sýkora (Eds.), Graph-Theoretic Concepts in Computer Science. Proceedings, 1998. X, 385 pages. 1998.

Vol. 1518: M. Luby, J. Rolim, M. Serna (Eds.), Randomization and Approximation Techniques in Computer Science. Proceedings, 1998. IX, 385 pages. 1998.

1519: T. Ishida (Ed.), Community Computing and Support Systems. VIII, 393 pages. 1998.

Vol. 1520: M. Maher, J.-F. Puget (Eds.), Principles and Practice of Constraint Programming - CP98. Proceedings, 1998. XI, 482 pages. 1998.

Vol. 1521: B. Rovan (Ed.), SOFSEM'98: Theory and Practice of Informatics. Proceedings, 1998. XI, 453 pages. 1998.

Vol. 1522: G. Gopalakrishnan, P. Windley (Eds.), Formal Methods in Computer-Aided Design. Proceedings, 1998. IX, 529 pages. 1998.

Vol. 1524: G.B. Orr, K.-R. Müller (Eds.), Neural Networks: Tricks of the Trade. VI, 432 pages. 1998.

Vol. 1525: D. Aucsmith (Ed.), Information Hiding. Proceedings, 1998. IX, 369 pages. 1998.

Vol. 1526: M. Broy, B. Rumpe (Eds.), Requirements Targeting Software and Systems Engineering. Proceedings, 1997. VIII, 357 pages. 1998.

Vol. 1527: P. Baumgartner, Theory Reasoning in Connection Calculi. IX, 283. 1999. (Subseries LNAI).

Vol. 1528: B. Preneel, V. Rijmen (Eds.), State of the Art in Applied Cryptography. Revised Lectures, 1997. VIII, 395 pages. 1998.

Vol. 1529: D. Farwell, L. Gerber, E. Hovy (Eds.), Machine Translation and the Information Soup. Proceedings, 1998. XIX, 532 pages. 1998. (Subseries LNAI).

Vol. 1530: V. Arvind, R. Ramanujam (Eds.), Foundations of Software Technology and Theoretical Computer Science. XII, 369 pages. 1998.

Vol. 1531: H.-Y. Lee, H. Motoda (Eds.), PRICAI'98: Topics in Artificial Intelligence. XIX, 646 pages. 1998. (Subseries LNAI).

Vol. 1096: T. Schael, Workflow Management Systems for Process Organisations. Second Edition. XII, 229 pages. 1998.

Vol. 1532: S. Arikawa, H. Motoda (Eds.), Discovery Science. Proceedings, 1998. XI, 456 pages. 1998. (Subseries LNAI).

Vol. 1533: K.-Y. Chwa, O.H. Ibarra (Eds.), Algorithms and Computation. Proceedings, 1998. XIII, 478 pages. 1998.

Vol. 1534: J.S. Sichman, R. Conte, N. Gilbert (Eds.), Multi-Agent Systems and Agent-Based Simulation. Proceedings, 1998. VIII, 237 pages. 1998. (Subseries LNAI).

Vol. 1535: S. Ossowski, Co-ordination in Artificial Agent Societies. XV; 221 pages. 1999. (Subseries LNAI).

Vol. 1536: W.-P. de Roever, H. Langmaack, A. Pnueli (Eds.), Compositionality: The Significant Difference. Proceedings, 1997. VIII, 647 pages. 1998.

Vol. 1537: N. Magnenat-Thalmann, D. Thalmann (Eds.), Modelling and Motion Capture Techniques for Virtual Environments. Proceedings, 1998. IX, 273 pages. 1998. (Subseries LNAI).

Vol. 1538: J. Hsiang, A. Ohori (Eds.), Advances in Computing Science – ASIAN'98. Proceedings, 1998. X, 305 pages. 1998.

Vol. 1539: O. Rüthing, Interacting Code Motion Transformations: Their Impact and Their Complexity. XXI,225 pages. 1998.

Vol. 1540: C. Beeri, P. Buneman (Eds.), Database Theory – ICDT'99. Proceedings, 1999. XI, 489 pages. 1999.

Vol. 1541: B. Kågström, J. Dongarra, E. Elmroth, J. Waśniewski (Eds.), Applied Parallel Computing. Proceedings, 1998. XIV, 586 pages. 1998.

Vol. 1542: H.I. Christensen (Ed.), Computer Vision Systems. Proceedings, 1999. XI, 554 pages. 1999.

Vol. 1543: S. Demeyer, J. Bosch (Eds.), Object-Oriented Technology ECOOP'98 Workshop Reader. 1998. XXII, 573 pages. 1998.

Vol. 1544: C. Zhang, D. Lukose (Eds.), Multi-Agent Systems. Proceedings, 1998. VII, 195 pages. 1998. (Subseries LNAI).

Vol. 1545: A. Birk, J. Demiris (Eds.), Learning Robots. Proceedings, 1996. IX, 188 pages. 1998. (Subseries LNAI).

Vol. 1546: B. Möller, J.V. Tucker (Eds.), Prospects for Hardware Foundations. Survey Chapters, 1998. X, 468 pages. 1998.

Vol. 1547: S.H. Whitesides (Ed.), Graph Drawing. Proceedings 1998. XII, 468 pages. 1998.

Vol. 1548: A.M. Haeberer (Ed.), Algebraic Methodology and Software Technology. Proceedings, 1999. XI, 531 pages. 1999.

Vol. 1550: B. Christianson, B. Crispo, W.S. Harbison, M. Roe (Eds.), Security Protocols. Proceedings, 1998. VIII, 241 pages. 1999.

Vol. 1551: G. Gupta (Ed.), Practical Aspects of Declarative Languages. Proceedings, 1999. VIII, 367 pgages. 1999.

Vol. 1552: Y. Kambayashi, D.L. Lee, E.-P. Lim, M.K. Mohania, Y. Masunaga (Eds.), Advances in Database Technologies. Proceedings, 1998. XIX, 592 pages. 1999.

Vol. 1553: S.F. Andler, J. Hansson (Eds.), Active, Real-Time, and Temporal Database Systems. Proceedings, 1997. VIII, 245 pages. 1998.

Vol. 1555: J.P. Mueller, A. Rao, M.P. Singh (Eds.), Intelligent Agents V. Proceedings, 1998. XXIV, 455 pages. 1999. (Subseries LNAI).

Vol. 1557: P. Zinterhof, M. Vajteršic, A. Uhl (Eds.), Parallel Computation. Proceedings, 1999. XV, 604 pages. 1999.

Vol. 1558: J.v.d. Herik, H. Iida (Eds.), Computers and Games. Proceedings, 1998. XVIII, 337 pages. 1999.

Vol. 1559: P. Flener (Ed.), Logic-Based Program Synthesis and Transformation. Proceedings, 1998. X, 331 pages. 1999.

Vol. 1560: K. Imai, Y. Zheng (Eds.), Public Key Cryptography. Proceedings, 1999. IX, 327 pages. 1999.

Vol. 1561: I. Damgård (Ed.), Lectures on Data Security.VII, 250 pages. 1999.

Vol. 1563: Ch. Meinel, S. Tison (Eds.), STACS 99. Proceedings, 1999. XIV, 582 pages. 1999.

Vol. 1567: P. Antsaklis, W. Kohn, M. Lemmon, A. Nerode, S. Sastry (Eds.), Hybrid Systems V. X, 445 pages. 1999.

Vol. 1569: F.W. Vaandrager, J.H. van Schuppen (Eds.), Hybrid Systems: Computation and Control. Proceedings, 1999. X, 271 pages. 1999.

Vol. 1570: F. Puppe (Ed.), XPS-99: Knowledge-Based Systems. VIII, 227 pages. 1999. (Subseries LNAI).

Vol. 1572: P. Fischer, H.U. Simon (Eds.), Computational Learning Theory. Proceedings, 1999. X, 301 pages. 1999. (Subseries LNAI).

Vol. 1575: S. Jähnichen (Ed.), Compiler Construction. Proceedings, 1999. X, 301 pages. 1999.

Vol. 1577: J.-P. Finance (Ed.), Fundamental Approaches to Software Engineering. Proceedings, 1999. X, 245 pages. 1999.

Vol. 1578: W. Thomas (Ed.), Foundations of Software Science and Computation Structures. Proceedings, 1999. X, 323 pages. 1999.

Vol. 1580: A. Včkovski, K.E. Brassel, H.-J. Schek (Eds.), Interoperating Geographic Information Systems. Proceedings, 1999. XI, 329 pages. 1999.